FORMAL METHODS FOR DISTRIBUTED SYSTEM DEVELOPMENT

IFIP - The International Federation for Information Processing

IFIP was founded in 1960 under the auspices of UNESCO, following the First World Computer Congress held in Paris the previous year. An umbrella organization for societies working in information processing, IFIP's aim is two-fold: to support information processing within its member countries and to encourage technology transfer to developing nations. As its mission statement clearly states,

IFIP's mission is to be the leading, truly international, apolitical organization which encourages and assists in the development, exploitation and application of information technology for the benefit of all people.

IFIP is a non-profitmaking organization, run almost solely by 2500 volunteers. It operates through a number of technical committees, which organize events and publications. IFIP's events range from an international congress to local seminars, but the most important are:

- The IFIP World Computer Congress, held every second year;
- open conferences;
- working conferences.

The flagship event is the IFIP World Computer Congress, at which both invited and contributed papers are presented. Contributed papers are rigorously refereed and the rejection rate is high.

As with the Congress, participation in the open conferences is open to all and papers may be invited or submitted. Again, submitted papers are stringently refereed.

The working conferences are structured differently. They are usually run by a working group and attendance is small and by invitation only. Their purpose is to create an atmosphere conducive to innovation and development. Refereeing is less rigorous and papers are subjected to extensive group discussion.

Publications arising from IFIP events vary. The papers presented at the IFIP World Computer Congress and at open conferences are published as conference proceedings, while the results of the working conferences are often published as collections of selected and edited papers.

Any national society whose primary activity is in information may apply to become a full member of IFIP, although full membership is restricted to one society per country. Full members are entitled to vote at the annual General Assembly, National societies preferring a less committed involvement may apply for associate or corresponding membership. Associate members enjoy the same benefits as full members, but without voting rights. Corresponding members are not represented in IFIP bodies. Affiliated membership is open to non-national societies, and individual and honorary membership schemes are also offered.

FORMAL METHODS FOR DISTRIBUTED SYSTEM DEVELOPMENT

FORTE / PSTV 2000
IFIP TC6 WG6.1 Joint International Conference on Formal Description Techniques for Distributed Systems and Communication Protocols (FORTE XIII) and Protocol Specification, Testing and Verification (PSTV XX)
October 10–13, 2000, Pisa, Italy

Edited by

Tommaso Bolognesi
Consiglio Nazionale delle Ricerche (CNR)
Istituto di Elaborazione dell'Informazione (IEI)
Italy

Diego Latella
Consiglio Nazionale delle Ricerche (CNR)
Istituto CNUCE
Italy

SPRINGER SCIENCE+BUSINESS MEDIA, LLC

Library of Congress Cataloging-in-Publication Data

IFIP TC6 WG6.1 Joint International Conference on Formal Description Techniques for Distributed Systems and Communication Protocols and Protocol Specification, Testing and Verification (2000 : Pisa, Italy)

Formal methods for distributed system development : October 10-13, 2000, Pisa, Italy / FORTE/PSTV 2000, IFIP TC6 WG6.1 Joint International Conference on Formal Description Techniques for Distributed Systems and Communication Protocols and Protocol Specification, Testing and Verification ; edited by Tommaso Bolognesi, Diego Latella.

p. cm. — (IFIP ; 55)

Includes bibliographical references.

DOI 10.1007/978-0-387-35533-7

1. Electronic data processing—Distributed processing—Congresses. 2. Formal methods (Computer science)—Congresses. I. Bolognesi, Tommaso. II. Latella, Diego.

QA76.9.D5 I33845 2000
004'.36—dc21 00-061057

Printed on acid-free paper.
www.springer.com/mycopy

Contents

Part Three
MODEL CHECKING - THEORY

Part Four
MODEL CHECKING - APPLICATIONS

Part Five
MULTICAST PROTOCOL ANALYSIS AND SIMULATION

Part Six
EXHAUSTIVE AND PROBABILISTIC TESTING

Contributors

K. Ajami
M. Benerecetti
M. Bernardo
E. Borcoci
M. Boreale
N. Busi
A. Cerpa
G. Delzanno
R. De Nicola
L. Durante
D. Estrin
K. Etessami
F. Giunchiglia
N. Griffeth
S. Gupta
S. Haddad
R. Hao
J. He
A. Helmy
T. Higashino
J.-M. Ilié
H. Katagiri
M. Kirimura
N. Klarlund
D. Lee
M. Majster-Cederbaum
N. Martí-Oliet
M. Panti
D. Peled
A. Petrenko
H. Pfeifer
R. Pugliese
B. Rayan
M. Ribaudo
G. Roessler
J. Rushby
F. Salger
A. Santone
R. K. Shyamasundar
R. K. Sinha
R. Sisto
M. A. Smith
M. Sorea
L. Spalazzi
T. Steinert
K. Taniguchi
J. Templemore-Finlayson
K. J. Turner
U. Ultes-Nitsche
A. Valenzano
A. Verdejo
D. Vlah
M. von der Beeck
M. Yannakakis
K. Yasumoto
N. Yevtushenko
Y. Yu

Preface

The 20th anniversary of the IFIP WG6.1 Joint International Conference on Formal Methods for Distributed Systems and Communication Protocols (FORTE XIII / PSTV XX) was celebrated by the year 2000 edition of the Conference, which was held for the first time in Italy, at Pisa, October 10-13, 2000. In devising the subtitle for this special edition -- 'Formal Methods – Implementation Under Test' -- we wanted to convey two main concepts that, in our opinion, are reflected in the contents of this book. First, the early, pioneering phases in the development of Formal Methods (FM's), with their conflicts between evangelistic and agnostic attitudes, with their over-optimistic applications to toy examples and over-skeptical views about scalability to industrial cases, with their misconceptions and myths..., all this is essentially over. Many FM's have successfully reached their maturity, having been 'implemented' into concrete development practice: a number of papers in this book report about successful experiences in specifying and verifying real distributed systems and protocols. Second, one of the several myths about FM's – the fact that their adoption would eventually eliminate the need for testing – is still quite far from becoming a reality, and, again, this book indicates that testing theory and applications are still remarkably healthy.

A total of 63 papers have been submitted to FORTE/PSTV 2000, out of which the Programme Committee has selected 22 for presentation at the Conference and inclusion in the Proceedings. The central themes of this book are the theory and practice of distributed system Verification (with Model Checking playing a predominant role), and Testing. Several papers deal with the analysis of communication protocols, and some of them address, in particular, multicast protocols. Other papers deal with the specification, implementation and testing of hardware systems. The issue of formal semantics is also covered, but to a lesser extent than in past editions of the Conference, as a further indication of the shift of emphasis from the definition to the application of FM's. Three internationally recognized speakers, namely Rocco De Nicola, Fausto Giunchiglia, and John Rushby, have provided invited contributions on Verification and on Security protocols: these are collected in the last part of the book.

The Conference has been complemented by two tutorial/advanced seminar tracks including a total of six presentations:

- E. Allen Emerson: Model Checking.
- Jan Tretmans: Specification based testing with formal methods -- from theory via tools to applications.
- Marta Kwiatkowska: Modelling and verification of probabilistic real time systems using probabilistic timed automata.
- Hartmut Koenig, Peter Langendoerfer: Automated Derivation of Efficient Implementations from Formal Protocol Specifications.
- R. Gotzhein, E. Börger, A. Prinz: Abstract State Machines and their Applications.
- C. Petitpierre: Bridging the Gap Between Formal Methods and the Implementation Process.

Finally, a multidisciplinary Satellite Workshop -- 'Formal Methods Elsewhere' – was held on October 10, devoted to applications of FM's to areas other than communication protocols and distributed systems, such as physics, chemistry, biology, social sciences, arts and humanities, music. After two decades, FM's are perhaps ready to spread out of their native territory and, at the turn of the millenium, invade new exciting areas of research, for a wider exploitation of the huge intellectual investment behind their definition.

Tommaso Bolognesi
Diego Latella

Programme Committee and Referees

Programme Committee Co-Chairmen

Tommaso Bolognesi (CNR – IEI, Italy)
Diego Latella (CNR – CNUCE, Italy)

Programme Committee

G. v. Bochmann (Univ. of Ottawa, Canada), H. Bowman (Univ. of Kent at Canterbury, UK), E. Brinksma (Univ. of Twente, The Netherlands), S. Budkowski (INT, France), A. Cavalli (INT, France), S. T. Chanson (Hong Kong Univ. of Sci. & Tech., China), J.-P. Courtiat (LAAS-CNRS, France), D. De Frutos Escrig (UCM, Spain), P. Dembinski (IPI-PAN, Poland), C. J. Fidge (Univ. of Queensland, Australia), S. Fischer (International University, Bruchsal, Germany), F. Giunchiglia (IRST and Trento Univ., Italy), S. Gnesi (CNR – IEI, Italy), R. Gorrieri (Univ. of Bologna, Italy), R. Gotzhein (Univ. of Kaiserslautern, Germany), R. Groz (France Telecom CNET, France), T. Higashino (Univ. of Osaka, Japan), D. Hogrefe (Univ. of Luebeck, Germany), G. J. Holzmann (Bell Labs, USA), M. C. Kim (Information and Communications University, Korea), H. Koenig (BTU Cottbus, Germany), G. Leduc (Univ. of Liege, Belgium), D. Lee (Bell Labs Research, China), L. Logrippo (Univ. of Ottawa, Canada), E. Najm (ENST, France), A. Petrenko (CRIM, Canada), J. Quemada (UPM, Spain), G. Reggio (Univ. of Genova, Italy), S. Smolka (SUNY at Stony Brook, USA), K. Suzuki (KDD, Japan), K. Tarnay (Nokia, Hungary), R. Tenney (Univ. of Massachusetts, USA), K. J. Turner (Univ. of Stirling, UK), A.Valenzano (CNR-IRITI, Italy), S. T. Vuong (Univ. of British Columbia, Canada), J. Wu (Tsinghua Univ., China).

Referees

A. Aldini
A. Cavalli
A. Cerone
A. Bailly
C. Bergerot
C. Bernardeschi
G. v. Bochmann
M. Boreale
S. Boroday
H. Bowman
R. Bruni
J.W. Bryans
S. Budkowski
C. Carrez
R. Castanet
O. Catrina
S. Chanson
R. G. Clark
R. Cleaveland
P. D'Argenio
D. de Frutos Escrig
P. Dembinski
J. Drissi
L. Durante
K. El-Fakih
A. Fantechi
A. Felty
C. Fidge
S. Fischer
R. Focardi
F. Giunchiglia
S. Gnesi
R. Gotzhein
J. Grabowski
P. Gradit
R. Grosu
R. Groz
A. G. Hamilton
R. Hao
T. Hasegawa
H. Hermanns
S. Heymer
T. Higashino
D. Hogrefe
G. Holzmann
D. Howe
A. Idoue
M. Ionescu
A. Janowska
P. Janowski
S. Kang
J.-P. Katoen
K. Khordoc
M. Kim
B. Koch
H. Koenig
O. Koné
R. Langerak
G. Leduc
D. Lee
S. Leue
Y. Levené
L. Llana Diaz
L. Logrippo
S. Maag
S. Maharaj
O. Marroquin Alonso
F. Mazzanti
W. Monin
C. Munoz
E. Najm
A. Nakata
M. Nuñez
W. Penczek
A. Petrenko
Y.-M. Quemener
G. Reggio
C. Rinderknecht
M. Schmitt
R. K. Sinha
R. Sisto
S. Smolka
P. Strooper
K. Suzuki
K. Tarnay
J.Templemore-Finlayson
R. L. Tenney
K. J. Turner
A. Valenzano
V. Valero Ruiz
F. Vernadat
S. T. Vuong
J. Wu
K. Yasumoto
W. Zuberek

Acknowledgements

The Editors wish to express their gratitude to the Programme Committee members of FORTE/PSTV 2000, the referees, the invited speakers, the tutorial and advanced seminar speakers, the tutorials chairman Alessandro Fantechi, and the paper authors for their contribution to the Conference and to this book.

The Conference is sponsored by IFIP WG6.1 (Architectures and Protocols for Distributed Systems), which is chaired by Guy Leduc; we express special gratitude to Guy for his continuous advice during a period of over 12 months before the event.

The Conference has been kindly supported by the European Commission (DGXII, Human Potential Programme, High-Level Scientific Conferences, contract HPCF-2000-00084), by ERCIM (European Research Consortium for Informatics and Mathematics), by CNR (Italian National Research Council) and CNR Institutes I.E.I. and CNUCE, and by the Informatics Department of the University of Pisa. Generous contributions have been provided by Marconi Communications, Bell Labs Research China (Lucent Technologies), Ansaldo Segnalamento Ferroviario, and KDD R&D Laboratories Inc.

Many people have contributed to the local organisation of the Conference; we are especially grateful to Gabriele Lenzini, Maurizio 'Icio' Caneve, Giuseppe Lami, Cristina Lazzerini, Mieke Massink, Grazia Carrai, Ettore Ricciardi, Maria Vittoria Masserotti, Sergio Barsocchi.

Yana Lambert, at Kluwer, deserves special thanks for her patience and assistance in preparing the manuscript.

Part One

VERIFICATION AND THEOREM PROVING

FORMAL VERIFICATION OF THE TTP GROUP MEMBERSHIP ALGORITHM

Holger Pfeifer
Universität Ulm
Fakultät für Informatik
D-89069 Ulm, Germany
pfeifer@informatik.uni-ulm.de

Abstract This paper describes the formal verification of a fault-tolerant group membership algorithm that constitutes one of the central services of the Time-Triggered Protocol (TTP). The group membership algorithm is formally specified and verified using a diagrammatic representation of the algorithm. We describe the stepwise development of the diagram and outline the main part of the correctness proof. The verification has been mechanically checked with the PVS theorem prover.

Keywords: Deductive verification, Time-Triggered Architecture, fault-tolerant distributed algorithms, safety-critical control.

1. INTRODUCTION

The Time-Triggered Architecture (TTA) developed by the University of Vienna and TTTech provides an integrated set of services for dependable distributed real-time systems [5, 6]. TTA is intended for devices controlling safety-critical electronic systems without mechanical backup, so-called "by-wire" systems such as those for automotive steering, braking, and suspension control [15]. It has been argued that the kind of reliability required in such situations cannot be achieved without a careful formal analysis of the mechanisms and algorithms involved [3, 12].

The Time-Triggered Protocol (TTP) [7] is the core of the communication level of TTA. Group membership is central service of TTP as it provides to all non-faulty processors a consistent view of which nodes are operational and which are not at any given moment. Distributed fault-tolerant algorithms are inherently difficult to reason about. In order to make formal verification feasible it is essential that the various aspects of an algorithm are specified and verified at appropriate levels of abstraction that capture the essence of the property under study and abstract from irrelevant details. The group membership algo-

Tommaso Bolognesi and Diego Latella (Eds.), *Formal Methods for Distributed System Development.*

rithm [1] is modeled as a synchronous system and it abstracts from the clock synchronization service that justifies the synchrony assumption. Its verification is significantly more difficult than other fault-tolerant algorithms because information about the failure of processors is not available immediately but only with a certain delay. Therefore one has to be very careful when reasoning about possibly failed components. Verification of safety properties, like the requirement that all (non-faulty) processors of a system should have the same opinion about the current membership status of other processors, is typically accomplished by an induction proof. In order to establish the induction step, however, one generally has to strengthen the invariant because often enough the property of interest is not inductive. Usually repeated strengthening is necessary before an inductive invariant is found and although some of the strengthenings can be generated automatically this becomes the main task when performing a mechanized verification. Experience with a membership algorithm similar to that of TTP [4] showed that this verification strategy is infeasible for our purpose.

Therefore, we take a different approach recently proposed by John Rushby: instead of expressing the correctness property as one large conjunction, we use a set of disjunctively connected formulas that can be seen as the description of an abstract state machine [14]. Each disjunct contains the desired property and represents a particular configuration the membership algorithm can reach. To establish the correctness of the algorithm one has to show that at every point in time the system is in one of these configurations. For the TTP group membership algorithm, we have formally proved both an agreement property, i.e., that every non-faulty node considers the same set of processors to be part of the membership, and a liveness property that states that a faulty processor will eventually remove itself from the membership. All definitions and proofs have been developed and mechanically checked with the PVS specification and verification system [10]. We present fragments of the PVS specification in this paper. To increase readability, however, the syntax has liberally been modified by replacing some ASCII codings with a more familiar mathematical notation. The full specification together with all proofs can be obtained via WWW [16].

The next section explains the group membership algorithm of TTP and provides a formal specification. Section 3 illustrates the approach we take to formally verify the algorithm, and Section 4 describes the main parts of the proof. The last section contains concluding remarks and directions for future work.

2. GROUP MEMBERSHIP IN TTP

In TTP, access to the broadcast bus is implemented by a *time division multiple access (TDMA)* schema: each processor is assigned a certain time interval, a so-called *slot*, in which it is allowed to send a message on the bus while the other processors listen. Slots are numbered and can be seen as an abstraction of a global time base that is provided by a separate clock synchronization algorithm [8, 11]. In our model we assume a set `proc` of n processors, labeled $0, 1, \ldots, n-1$, that are arranged in a logical ring. The period of n successive slots is called a *TDMA round.*

Every processor p maintains a set $\mathtt{mem}_p^t$—the membership set of processor p—that contains all processors that p considers operational at time t. In slot t the processor with label $t \bmod n$ is the broadcaster, denoted $\mathit{broadcaster}(t)$. In addition to the message data, the broadcaster sends those parts of its internal state that are critical for the protocol to work properly. More precisely, a CRC checksum that is calculated over the message data and the critical state information—which includes the membership set—is appended to the message. For the analysis of the group membership algorithm it is sufficient to assume that a message contains the broadcaster's local view $\mathtt{mem}_b^t$ on the membership.

As the order of messages is statically defined there is no need for special membership messages. Instead, a successfully received message is interpreted as a life-sign of the sender and a receiver will maintain the broadcaster in its local membership set if it agrees with the broadcaster's critical state information and hence with its membership set.[1] Conversely, if a processor does not receive an expected message or does not agree with the broadcaster's view on the membership, the broadcaster will be considered faulty and the receiver removes it from its membership set.

The group membership algorithm is designed to operate in the presence of faults. A processor can be *send-faulty*, in which case it will fail to broadcast in its next slot, while a *receive-faulty* processor will not succeed in receiving the message of the next non-faulty processor. This restricted fault model is appropriate since other protocol services of TTP ensure that other fault modes manifest themselves as either send or receive faults by enforcing a faulty processor to fail silently. For example, the bus guardian, a special hardware element of the TTP controller, prevents a processor that has lost synchrony of its clock from accessing the broadcast bus outside its designated slots. We use $\mathcal{NF}^t$ to denote the set of non-faulty processors at time t, and $p \notin \mathcal{NF}^t$ indicates that p is either send-faulty or receive-faulty at time t. Furthermore, sends_b^t describes that the current broadcaster b sends a message on the bus, while $\mathit{arrives}_p^t$ means that the message sent by the broadcaster arrives at the receiver p.

A non-faulty broadcaster b will only send a message if b is contained in its own membership set; if b has removed itself from the membership set (due to diagnosing a fault) it will stop sending message in its broadcast slot. The

[1] Technically, the receiver calculates a CRC checksum over the received message data and its own critical state information and compares the result with the checksum sent by the broadcaster.

following specification [1] shows the axiomatization of $sends_b^t$ as defined in PVS:[2]

```
1
NF^t       : set[proc]
sends_b^t  : bool
arrives_p^t : bool

sending : AXIOM
  LET b = broadcaster(t) IN
    b ∈ NF^t ∧ b ∈ mem_b^t ⇒ sends_b^t

fail_silence : AXIOM
  LET b = broadcaster(t) IN
    b ∉ mem_b^t ⇒ ¬sends_b^t
```

A message sent by the current broadcaster b will arrive at a non-faulty processor p. Of course, there is no generation of spontaneous messages and hence messages arrive only if they have been sent. These axioms also imply that broadcasts are consistent: a message arrives either at all non-faulty processors or, if the broadcaster is send-faulty, at none of them. The PVS specification is given in [1].

```
2
arrival : AXIOM
  LET b = broadcaster(t) IN
    sends_b^t ∧ p ∈ NF^t ⇒ arrives_p^t

nonarrival : AXIOM
  LET b = broadcaster(t) IN
    ¬sends_b^t ⇒ ¬arrives_p^t
```

In our model we consider faults to occur only if they affect the system in the next slot; e. g. a send fault of a processor p is not considered to occur until immediately before p's broadcasting slot. If the broadcaster at time $t+1$ becomes faulty in its sending slot it will fail to successfully send a message. Similarly, a newly faulty processor will fail to receive a message. Note that the behaviour of faulty processors is unspecified for the slots after the fault occurred: processors may or may not succeed in sending or receiving subsequent messages.

```
3
faulty_broadcaster : AXIOM
  LET b = broadcaster(t+1) IN
    b ∈ NF^t ∧ b ∉ NF^(t+1) ⇒ ¬sends_b^(t+1)

faulty_receiver : AXIOM
  LET b = broadcaster(t+1) IN
    p ∈ NF^t ∧ p ∉ NF^(t+1) ∧ p≠b ⇒ b ∈ NF^(t+1) ∧ ¬arrives_p^(t+1)
```

[2] PVS allows variables such as t to occur free in formulas; these are implicitly universally quantified. Moreover, the type of variables can be dropped if it has been introduced to PVS by a corresponding variable declaration. Throughout this paper, we use b, p, q, x, y, and z to denote processors (of type `proc`), and t and s are of type `time`.

Finally, once a processor becomes faulty it will be considered faulty forever.

The task of a group membership algorithm is to diagnose the failure of a faulty processor and to inform all non-faulty processors about it. In order to cause a broadcaster to realize that it is send-faulty the TTP group membership algorithm uses an (implicit) acknowledgment mechanism. A processor p that is the broadcaster in slot t checks whether the next non-faulty broadcaster, say q, that sends in the next[3] slot has the same membership set as q and in particular contains p in its membership set. If so, p can conclude that its broadcast was successful. Otherwise, either p failed to broadcast or q is receive-faulty. To resolve this ambiguity p waits for the next non-faulty broadcaster following q, say r. If r contains p in its membership set but not q while having the same view considering other processors, the original message of p was sent correctly and q failed. If p is not in r's membership set, but q is (and the rest of the membership sets of p and r are the same), then q and r agree that p failed to send. In this case, p will remove itself from its own membership set and fail silently.

A similar mechanism could be used for diagnosing receive faults: if a processor p does not receive an expected message it could check whether the next non-faulty broadcaster maintained the original sender in its membership set in which case p must realize that it has suffered from a receive fault. However, TTP employs a slightly different mechanism that is also used to avoid the formation of disjoint cliques at the same time. A clique is a group of processors where agreement on the current state is reached only within the group. Each processor p maintains two counters, $\mathtt{acc}_p^t$ and $\mathtt{rej}_p^t$, which keep track of how many messages p has *accepted* (successfully received) and *rejected*, respectively. A processor p will increment the counter $\mathtt{rej}_p^t$ if p does not agree with the broadcaster's view on the membership. In p's next broadcast slot it checks whether it has accepted more messages in the last round than it has rejected. If so, p resets the counters and broadcasts; the other case indicates that p suffered from a receive fault and therefore p removes itself from the membership and by not broadcasting its message p can inform the other processors about its failure.

Formally, the group membership algorithm is described by a set of guarded commands. In every slot t, every processor executes exactly one of these commands. The guards are evaluated in a top-down order. The formal description involves two additional boolean state variables, $\mathtt{prev}_p^t$ and

[3] More generally, if there are already faulty processors that were scheduled to send between p and q, the latter is the broadcaster in slot $t+i$ for some $i > 1$.

$\mathtt{doubt}_p^t$. If a processor p was the previous broadcaster and now waits for being acknowledged, $\mathtt{prev}_p$ is set to true, while $\mathtt{doubt}_p$ is true if p did not get acknowledged by its successor and waits for the second successor to resolve the conflict. In this case, the variable $\mathtt{succ}_p^t$ holds p's first successor which refused to acknowledge p. In the following definition state components that are not mentioned explicitly do not change.

Broadcaster: Let b be the current broadcaster, i. e. $b = t \bmod n$.

(1) $\mathtt{acc}_b^t > \mathtt{rej}_b^t \rightarrow \mathrm{mem}_b^{t+1} := \mathrm{mem}_b^t,$
$\mathtt{prev}_b^{t+1} := \mathsf{T},\ \mathtt{acc}_b^{t+1} := 1,\ \mathtt{rej}_b^{t+1} := 0$

(2) $\mathtt{acc}_b^t \leq \mathtt{rej}_b^t \rightarrow \mathrm{mem}_b^{t+1} := \mathrm{mem}_b^t \setminus \{b\}$

Receiver: Every processor p different from the current broadcaster executes the first of the following commands whose guard evaluates to true:

(3) $p \notin \mathrm{mem}_p^t \rightarrow$ no change

(4) $\mathtt{prev}_p^t \wedge \mathit{arrives}_p^t \wedge \mathrm{mem}_p^t = \mathrm{mem}_b^t \cup \{p\} \rightarrow \mathrm{mem}_p^{t+1} := \mathrm{mem}_p^t,$
$\mathtt{prev}_p^{t+1} := \mathsf{F},\ \mathtt{acc}_p^{t+1} := \mathtt{acc}_p^t + 1$

(5) $\mathtt{prev}_p^t \wedge \mathit{arrives}_p^t \wedge \mathrm{mem}_p^t = \mathrm{mem}_b^t \setminus \{p\} \rightarrow \mathrm{mem}_p^{t+1} := \mathrm{mem}_p^t \setminus \{b\},$
$\mathtt{prev}_p^{t+1} := \mathsf{F},\ \mathtt{doubt}_p^{t+1} := \mathsf{T},$
$\mathtt{rej}_p^{t+1} := \mathtt{rej}_p^t + 1,\ \mathtt{succ}_p^{t+1} := b$

(6) $\mathtt{prev}_p^t \wedge \mathit{sends}_b^t \rightarrow \mathrm{mem}_p^{t+1} := \mathrm{mem}_p^t \setminus \{b\},$
$\mathtt{prev}_p^{t+1} := \mathsf{T},\ \mathtt{rej}_p^{t+1} := \mathtt{rej}_p^t + 1$

(7) $\mathtt{prev}_p^t \rightarrow \mathrm{mem}_p^{t+1} := \mathrm{mem}_p^t \setminus \{b\},$
$\mathtt{prev}_p^{t+1} := \mathsf{T}$

(8) $\mathtt{doubt}_p^t \wedge \mathit{arrives}_p^t \wedge \mathrm{mem}_p^t = \mathrm{mem}_b^t \cup \{p\} \setminus \{\mathtt{succ}_p^t\} \rightarrow \mathrm{mem}_p^{t+1} := \mathrm{mem}_p^t,$
$\mathtt{doubt}_p^{t+1} := \mathsf{F},\ \mathtt{acc}_p^{t+1} := \mathtt{acc}_p^t + 1$

(9) $\mathtt{doubt}_p^t \wedge \mathit{arrives}_p^t \wedge \mathrm{mem}_p^t = \mathrm{mem}_b^t \cup \{\mathtt{succ}_p^t\} \setminus \{p\} \rightarrow \mathrm{mem}_p^{t+1} := \mathrm{mem}_p^t \setminus \{p\},$
$\mathtt{doubt}_p^{t+1} := \mathsf{F},\ \mathtt{acc}_p^{t+1} := \mathtt{acc}_p^t + 1$

(10) $\mathtt{doubt}_p^t \wedge \mathit{sends}_b^t \rightarrow \mathrm{mem}_p^{t+1} := \mathrm{mem}_p^t \setminus \{b\},$
$\mathtt{doubt}_p^{t+1} := \mathsf{T},\ \mathtt{rej}_p^{t+1} := \mathtt{rej}_p^t + 1$

(11) $\mathtt{doubt}_p^t \rightarrow \mathrm{mem}_p^{t+1} := \mathrm{mem}_p^t \setminus \{b\},$
$\mathtt{doubt}_p^{t+1} := \mathsf{T}$

(12) $\mathit{arrives}_p^t \wedge \mathrm{mem}_p^t = \mathrm{mem}_q^t \rightarrow \mathrm{mem}_p^{t+1} := \mathrm{mem}_p^t,$
$\mathtt{acc}_p^{t+1} := \mathtt{acc}_p^t + 1$

(13) $\neg \mathit{sends}_b^t \rightarrow \mathrm{mem}_p^{t+1} := \mathrm{mem}_p^t \setminus \{b\}$

(14) otherwise $\rightarrow \mathrm{mem}_p^{t+1} := \mathrm{mem}_p^t \setminus \{b\},$
$\mathtt{rej}_p^{t+1} := \mathtt{rej}_p^t + 1$

The group membership algorithm has to fulfill three major correctness requirements:

- **Validity:** At all times, non-faulty processors should have all and only the non-faulty processors in their membership sets, while faulty processors should have removed themselves from their sets. This requirement is, however, impossible to satisfy as it may take some time to diagnose the faultiness of a processor. We therefore must allow a single faulty processor to be included in the membership sets of non-faulty processors, while faulty processors may have (a subset of) the non-faulty processors plus themselves in their sets.

5
```
validity : THEOREM
```
$$p \in \mathcal{NF}^t \Rightarrow \text{mem}_p^t = \mathcal{NF}^t \lor \exists x:\ x \notin \mathcal{NF}^t \land \text{mem}_p^t = \mathcal{NF}^t \cup \{x\}$$
$$\land\ p \notin \mathcal{NF}^t \Rightarrow p \notin \text{mem}_p^t \lor \text{mem}_p^t \subseteq \mathcal{NF}^t \cup \{p\}$$

- **Agreement:** All non-faulty processors should have the same membership sets.

6
```
agreement : THEOREM
```
$$p \in \mathcal{NF}^t \land q \in \mathcal{NF}^t \Rightarrow \text{mem}_p^t = \text{mem}_q^t$$

- **Self-diagnosis:** A faulty processor should eventually diagnose its fault and remove itself from its own membership set.

7
```
self_diagnosis : THEOREM
```
$$x \in \mathcal{NF}^t \land x \notin \mathcal{NF}^{t+1} \Rightarrow \exists s:\ s>0 \land x \notin \text{mem}_x^{t+s}$$

These properties are subject to two additional assumptions that constitute the *fault hypothesis*. First, as processors will be able to diagnose a fault only if no new fault occurs during that process, the specification of the Time-Triggered Protocol requires the membership algorithm to work properly only if two successive failures occur at least two TDMA rounds apart [1]. More frequent fault arrivals are dealt with by other protocol mechanisms of TTP. In our formal verification the fault arrival assumption is expressed in a different way that is adapted to the verification approach we take. We assume that a fault only occurs if the system is in a certain configuration called *stable*. If a fault occurs the system leaves this configuration and the fault arrival assumption states that no new fault occurs until the system reaches the *stable* configuration again.

8
```
fault_may_occur_when_stable : AXIOM
```
$$\text{stable}(t,y,z) \Rightarrow \mathcal{NF}^{t+1} = \mathcal{NF}^t \lor \exists x:\ x \in \mathcal{NF}^t \land \mathcal{NF}^{t+1} = \mathcal{NF}^t \setminus \{x\}$$

```
no_faults_when_not_stable : AXIOM
```
$$\neg\text{stable}(t,y,z) \Rightarrow \mathcal{NF}^{t+1} = \mathcal{NF}^t$$

It is necessary to rule out the trivial solution to the *validity* and *agreement* requirements where the system never returns to *stable*. Otherwise only one fault could occur which would not even be tolerated. We are therefore forced to prove a *liveness* property saying that after a fault occured the *stable* configuration will eventually be reached again.

```
liveness : THEOREM                                                        9
  stable(t,y,z)  ⇒  ∃s,y,z: s>0  ∧  stable(t+s,y,z)
```

We believe that a stronger property could also be proved, where s is constrained to be less or equal $2n$. Then the system would return to *stable* no later than two TDMA rounds after a fault occured which would yield the original fault arrival assumption of TTP. Moreover, the *self-diagnosis* requirement listed above is actually a simple corollary of *liveness*.

The second part of the fault hypothesis refers to the number of processors that are required for the algorithm to work correctly. A set of three processors would be sufficient to tolerate a single fault, however, in order to prevent a non-faulty processor from being fooled by a faulty broadcaster at least 4 processors are required to be present in the system with at least three being non-faulty at all times. Otherwise a non-faulty processor might mistakenly diagnose a fault of its own and remove itself from its membership set (thereby violating both the validity and the agreement property).

```
n    : { m:ℕ | m≥4 }                    % -- number of processors          10
proc : TYPE+ = { m:ℕ | 0≤m<n }          % -- the set of processors (labels)

three_non_faulty_exist : AXIOM
  ∃x,y,z: x≠y ∧ x≠z ∧ y≠z ∧ x∈NF' ∧ y∈NF' ∧ z∈NF'
```

3. APPROACH TO VERIFYING THE ALGORITHM

For the verification of the TTP group membership algorithm we apply a method recently proposed by John Rushby [14]. The requirements *validity* and *agreement* express properties that should hold for all reachable states of the system. Such invariants, or safety-properties, are usually verified by some form of induction proof; one demonstrates that the property holds in the initial state(s) and that all state transitions preserve the property. The problem is, however, that the property of concern is not inductive in general and hence, in order to establish the proof of the induction step, one has to strengthen it by conjoining additional properties to it which themselves have to be invariants. Usually, this process has to be repeated several times before the induction proof can be accomplished. Rushby reports in [14] that the number and complexity of additional invariants that had to be discovered during the proof defeated several determined attempts to mechanically verify a group membership algorithm [4] similar to that of TTP. Because of its peculiaritity of managing counters for accepted and rejected messages and the way how acknowledgment of messages is done, the group membership algorithm of TTP is considered quite tricky

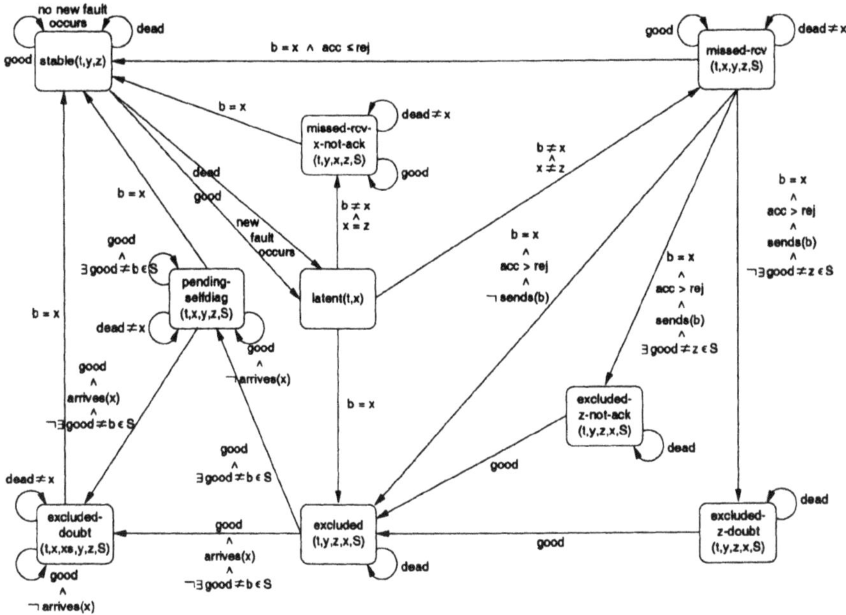

Figure 1 Configuration diagram for the TTP group membership algorithm.

and a mechanical proof using the traditional approach is very likely to be even more complex.

Instead of expressing the property to be verified as a conjunction of predicates Rushby's method exploits a disjunctively connected set of formulas—which in turn may be conjunctions of predicates—to prove the property of concern. Each disjunct can be seen as the description of a certain *configuration*, a property of the current global state of the whole system. The configurations are defined such that every single configuration implies the desired property, and to verify that property one has to show that at all times the system is in one of these configurations. Thus, the main part of the proof can be represented as a *configuration diagram*. The diagram for the group membership algorithm is shown in Figure 1. The nodes of the diagram represent the *configurations*, and arrows denote transitions from one configuration to others and are labeled with transition conditions. Configurations are parameterized by the time t and describe the global state the system is in. Configurations can have additional parameters such as processors $(x, y, \ldots)$ that behave differently from the rest of system, or additional entities necessary to describe the system state. The labels of transitions express the preconditions for the system to move from one configuration to another. For example, the label $b = x$ from the transistion from *latent* to *excluded* means that the system takes this transistion if x is the current broadcaster, while a transistion with the label $dead \neq x$ is taken whenever the current broadcaster is already faulty but different from x. The transition conditions leading from one configuration need not necessarily be disjoint, but one has to show that they are complete in the sense that their disjunction is true.

The diagram can be developed step-by-step. One usually starts by defining some initial configuration or the one in which the system stays under normal circumstances, i. e. as long as no fault occurs. For TTP, this central configuration is the one labeled *stable*. By symbolically evaluating the algorithm in the current configuration and by splitting on possible cases we generate some new configurations, and the transitions from the original configuration are labeled with the appropriate conditions. By repeatedly applying this construction on each transition and each new configuration one aims to develop a closed diagram. To prove safety properties like *validity* or *agreement* one then has to demonstrate that every configuration implies the desired property and that the disjunction of the transition conditions leading from any one configuration evaluates to true; this ensures that there is no other configuration the system can possibly get into. In order to prove liveness properties like *self-diagnosis* one has to establish that the system can not loop forever on a configuration other than *stable*.

There are several benefits to this approach: first, the diagram can be developed incrementally and in a totally systematic way by symbolically executing one step of the algorithm in every configuration. Second, the completed diagram is a suitable means of analysing the difficult special cases of the algorithm and to explain how and why it works (or doesn't). Last, it seems that the creative steps in developing the proof can be accomplished easier than by using the traditional way of repeated invariant strengthening. The configurations as presented here generally are not invariants and are therefore identified more easily.

The next section describes how the configuration diagram for the TTP group membership algorithm is gradually developed and outlines the verification of the three correctness requirements for the TTP group membership algorithm.

4. DEVELOPING THE CONFIGURATION DIAGRAM

The system is said to be in a *stable* configuration if the membership set of all non-faulty processors p is equal to $\mathcal{NF}^t$, the set of all non-faulty processors at time t, and all faulty processors have already diagnosed their fault and thus removed themselves from their own membership set. For *stable* the two safety properties *validity* and *agreement* follow immediately from these definitions. Moreover, *stable* is the initial configuration of the system.

```
                                                                    11
stable(t,y,z) : bool =
        recent(t,y,z)
  ∧ ∀p:    p ∉ NF^t  ⇒  p ∉ mem_p^t
       ∧ p ∈ NF^t  ⇒     mem_p^t = NF^t
                       ∧ p=z  ⇔  acc_p^t = rej_p^t + 1
                       ∧ p≠z  ⇒  acc_p^t > rej_p^t + 1
                       ∧ prev_p^t = ⊤  ⇔  p=z
                       ∧ doubt_p^t = ⊥
initial : AXIOM
  ∃y,z: stable(0,y,z)
```

In the configuration *stable*(t, y, z) the counters of non-faulty processors are set such that $\mathtt{acc}_p^t > \mathtt{rej}_p^t + 1$. This is to allow for a non-faulty processor p to cope with a send fault of one of the other broadcaster in the next round, in which case the counter $\mathtt{rej}_p^t$ will be increased; this should not lead to p removing itself from its own membership set in its next sending slot, for which $\mathtt{acc}_p^t > \mathtt{rej}_p^t$ must hold. However, the most recent non-faulty broadcaster, say z, cannot satisfy this condition as in its sending slot z sets the counters to $\mathtt{acc}_z^t = 1$ and $\mathtt{rej}_z^t = 0$. Suppose z was the broadcaster at time $t - 1$ and the broadcaster at time t committed a send fault. Now z increases $\mathtt{rej}_z^t$ and has $\mathtt{acc}_z^t = \mathtt{rej}_z^t$. If z was scheduled to be the next broadcaster at time $t + 1$ the condition $\mathtt{acc}_p^t > \mathtt{rej}_p^t$ would not hold and thus z would execute command (2), thereby wrongly removing itself from its membership set. In order to avoid this case, it is crucial that z can be the next broadcaster. Therefore we must ensure that there is another non-faulty processor, say y, that will send before z does. For similar, yet more involved reasons there has to be another non-faulty processor broadcasting before both z and y. As a result, this requires at least 4 processors to be present in the system in order for it to correctly tolerate a single fault. Finally, in *stable*, z is the only processor that has its `prev` flag set, while the `doubt` flag is cleared for all processors.

The characterization of the two most recent non-faulty broadcasters is covered by the following predicate, which also expresses that neither y nor z can be the broadcaster at time t.

```
                                                                    12
recent(t,y,z) : bool =
        y ∈ NF^t  ∧  z ∈ NF^t
  ∧ before(t,y,z)
  ∧ ∃p: p≠y  ∧  p ∈ NF^t  ∧  before(t,p,z)
  ∧ ∀p: before(t,zp)  ⇒  p ∉ NF^t
  ∧ ∃p: p ∈ NF^t  ∧  before(t,p,y)
  ∧ ∀p: before(t,y,p)  ∧  before(t,p,z)  ⇒  p ∉ NF^t
```

The expression *before*(t, y, z) denotes that at time t, processor y will broadcast again before z does.

```
                                                                    13
before(t,y,z) : bool =
  ∀(i:time): z=broadcaster(t+i)  ⇒  ∃(s:time): s<i  ∧  y=broadcaster(t+s)
```

In order to determine the transition conditions from the *stable* configuration, we consider whether or not a new fault occurs. If no processor becomes faulty, that is $\mathcal{NF}^{t+1} = \mathcal{NF}^t$ holds, the system remains in *stable*, because neither the broadcaster nor the receivers change their membership set: the broadcaster will execute command (1), while the receivers will execute command (12), except for z which will execute command (4). However, depending on whether the current broadcaster b is non-faulty or an already faulty one, the values of the parameters of *stable* might change. In the first case, z and b now become the two most recent non-faulty broadcaster and the new configuration is *stable*$(t+1, z, b)$, while in the latter case nothing changes, i. e. the new configuration is *stable*$(t+1, y, z)$.

```
                                                                        14
stable_to_stable_non_faulty : LEMMA
  LET b = broadcaster(t) IN
    stable(t,y,z) ∧ b ∈ NF^t ∧ NF^{t+1} = NF^t ⇒ stable(t+1,z,b)

stable_to_stable_faulty : LEMMA
  LET b = broadcaster(t) IN
    stable(t,y,z) ∧ b ∉ NF^t NF^{t+1} = NF^t ⇒ stable(t+1,y,z)
```

If a processor, say x, which was non-faulty at time t becomes faulty at time $t+1$, the same commands as above will be executed, but the system will change into a new configuration, that we call *latent*.

```
                                                                        15
latent(t,x,y,z) : bool =
     x ∈ NF^{t-1} ∧ NF^t = NF^{t-1} ∖ {x} ∧ recent'(t,x,y,z)
  ∧ ∀p:    (p ∉ NF^t ∧ p≠x) ⇒ p ∉ mem_p^t
        ∧ (p ∈ NF^t ∨ p=x) ⇒    mem_p^t = NF^t ∪ {x}
                              ∧ p=z ⇔ acc_p^t = rej_p^t + 1
                              ∧ p≠z ⇒ acc_p^t > rej_p^t + 1
                              ∧ prev_p^t = ⊤ ⇔ p=z
                              ∧ doubt_p^t = ⊥
```

This configuration is very similar to *stable* with the exception that both the non-faulty processors and x now do not only contain non-faulty processors in their membership sets, but also the newly faulty processor x. It is the task of the group membership algorithm to ensure that eventually all processors, including x, become aware of x's faultiness and remove it from their membership sets and thus return to the *stable* configuration. The fault hypothesis states that no new fault will occur during that time until the system will be *stable* again. The predicate *recent'*(t, x, y, z) differs from *recent*(t, y, z) in that we must allow y or z be actually identical to x and hence the first two conjuncts are changed to $y \neq x \Rightarrow y \in \mathcal{NF}^t$ and $z \neq x \Rightarrow z \in \mathcal{NF}^t$, respectively.

Again, it is a simple matter to establish the properties *validity* and *agreement* for the configuration *latent*. All non-faulty processors have the same membership sets which contain only one faulty processor, and the faulty processors have already removed themselves from their membership sets, apart from x, which still has itself in its membership. An additional verification condition, which is proved equally easily, is that *latent* is different from *stable* (and therefore faults are assumed not to occur).

As for *stable* there are two transition conditions for the system to transit to *latent*: either the broadcaster b at time $t+1$—when x becomes faulty—is an already faulty processor, or it is non-faulty. In the first case, the new configuration is $latent(t+1, x, z, b)$, while in the latter case the system will be in $latent(t+1, x, y, z)$.

16

```
stable_to_latent_non_faulty : LEMMA
  LET b = broadcaster(t) IN
    stable(t,y,z) ∧ x ∈ NF^t ∧ b ∈ NF^t ∧ NF^{t+1} = NF^t ∖ {x}
      ⇒ latent(t+1,x,z,b)

stable_to_latent_faulty : LEMMA
  LET b = broadcaster(t) IN
    stable(t,y,z) ∧ x ∈ NF^t ∧ b ∉ NF^t ∧ NF^{t+1} = NF^t ∖ {x}
      ⇒ latent(t+1,x,y,z)
```

From *latent* two cases must be distinguished: one where the faulty processor x is the next broadcaster, and one where it is not. In the former case, x executes command (1), however, it will not be able to successfully transmit its message and hence the receivers will execute their command (13), thereby removing x from their membership sets. This leads to the configuration *excluded.* In the latter case, x is a (faulty) receiver and hence will not succeed to receive the message sent by the broadcaster. The other receivers will execute command (12), or (4) in the case of the previous broadcaster. On the other hand, x executes either command (6) or (14), depending on whether or not x was the previous broadcaster (z, that is) and the system moves into the configurations *missed-rcv* or *missed-rcv-x-not-ack*, respectively.

By systematically analyzing the possible cases for a given configuration one proceeds to develop the configuration diagram. Every transition either leads to a new configuration or to an already existing one. In some cases it may be necessary to generalize an existing configuration in order to establish the proof of a transition. The ultimate goal in this process is to end up with a configuration diagram which is closed. Due to lack of space we have to omit the detailed description of the remaining states and transitions of the diagram in this paper. The complete specification of the algorithm and all proofs are available via WWW [16].

Once the configuration diagram is closed the most difficult and complex part of the proof of the two safety properties *validity* and *agreement* is done. To formally established these properties one has to show basically two things: first, every configuration implies the property under consideration. This is quite easy to see as all the non-faulty processors always have the same membership sets and there is always at most one faulty processor, namely x, that has not yet removed itself from its membership set. Second, there is no other configuration the system could possibly turn into. In other words, the transition conditions from any configuration must cover all the possible cases. Once this part of the proof is done, which is mainly a task of simple case analyses, these two facts can be combined to finally yield the desired safety properties.

The third correctness property, *self-diagnosis*, is a liveness property. Intuitively, one has to show that once the system has left *stable* it can not be trapped in one of the loops of the other configurations. Every of these configurations with a loop can be left in either of two cases: first, if the next broadcaster is non-faulty or, second, if it is x's slot to broadcast. The latter case is easy to demonstrate as every n slots a given processor will be the current broadcaster. The former case can be established using the fault model, see PVS box 10, that states that there exist at least three non-faulty processors at all times. Hence, eventually one of these will become the next broadcaster.

5. CONCLUSIONS

The group membership protocol presented here is one of a whole suite of algorithms for safety-critical real-time control implemented in the Time-Triggered Protocol. Industrial needs to minimize cost enforces the group membership algorithm to be heavily intertwined with other protocol services of TTP. We have isolated the core of the group membership algorithm and subjected it to thorough formal analysis. The proofs of the main correctness properties of the algorithm have been developed and mechanically checked with the assistance of the PVS specification and verification system. The complete verification comprises more than 160 proved lemmas and theorems and it takes almost one and a half hours to re-run all proofs even on a fast Sparc Ultra-II. These numbers might give an impression of the immense complexity of fault-tolerant group membership algorithms in general and of the TTP instance in particular. One source of this complexity is the requirement of fault-tolerance itself as one has to be very careful when reasoning about possibly failed components. Another reason why the TTP algorithm is significantly more difficult than other similar group membership protocols is that for optimization reasons there are still other services present as parts of the TTP group membership algorithm; for example, the way how self-diagnosis is accomplished in TTP by using the two special state variables $\texttt{acc}_p$ and $\texttt{rej}_p$ is in fact used for doing clique avoidance at the same time. This results in a self-stabilizing group membership algorithm, a property not considered in this paper.

The verification of the TTP group membership protocol applies a method recently proposed by John Rushby [14] who has used it to mechanically verify a similar algorithm [13]. Several attempts to formally verify that algorithm, which was flawed in its original publication [4], failed because of the number and complexity of additional invariants needed to establish the proof. This proof method, which is quite closely related to the verification diagrams of Manna and Pnueli [2,9], made it both possible and easy to verify the (corrected) algorithm. This promising experience with an apparently difficult problem made us decide to apply Rushby's method rather than to attempt a traditional invariance proof of the correctness of the TTP algorithm, which is considered even more complicated. It turned out that one of its main advantages over the usual method of repeated invariant strengthening is that the proof can be developed incrementally and in a very systematic way. Thus, one can break down the overall proof into small manageable steps.

Further research is concerned with formally specifying other protocol services of the Time-Triggered Architecture, such as initialization or re-integration of nodes, and analyzing their interrelationships. For example, the group membership algorithm presented here is specified at the level of a synchronous system; for this model to be adequate, we have to assume fault-tolerant clock synchronization [11]. Conversely, the clock synchronization mechanism of TTP also relies on the group membership service being able to avoid the formation of cliques of processors. Finding ways to clearly identify the relationships and interfaces of the various protocol services in order to avoid these circular dependencies remain a challenging and interesting problem.

Acknowledgments

I would like to thank John Rushby for introducing me to his method and for sharing both experience and PVS theories, which served as a starting point for the verification described here. I am also grateful for the constructive comments of the anonymous referees and of my colleagues at Ulm that substantially improved the presentation. Last, but not least, Michael Paulitsch and Günter Bauer from Vienna were of great assistance in understanding all the subtle details of TTP group membership.

References

[1] G. Bauer and M. Paulitsch. An Investigation of Membership and Clique Avoidance in TTP/C. In *Proc. of 19th IEEE Symposium on Reliable Distributed Systems*. IEEE, Oct. 2000. To appear.

[2] I. A. Browne, Z. Manna, and H. Sipma. Generalized Temporal Verification Diagrams. In *15th Conference on the Foundations of Software Technology and Theoretical Computer Science*, volume 1026 of *Lecture Notes in Computer Science*, pages 484–498. Springer-Verlag, 1995.

[3] R. W. Butler and G. B. Finelli. The Infeasibility of Quantifying the Reliability of Life-Critical Real-Time Software. *IEEE Trans. on Software Engineering*, 19(1):3–12, Jan. 1993.

[4] S. Katz, P. Lincoln, and J. Rushby. Low-Overhead Time-Triggered Group Membership. In M. Mavronicolas and P. Tsigas, editors, *11th International Workshop on Distributed Algorithms: WDAG'97*, volume 1320 of *Lecture Notes in Computer Science*, pages 155–169, 1997.

[5] H. Kopetz. The Time-Triggered Approach to Real-Time System Design. In B. Randell, J.-C. Laprie, H. Kopetz, and B. Littlewood, editors, *Predictably Dependable Computing Systems*. Springer-Verlag, 1995.

[6] H. Kopetz. *Real-Time Systems: Design Principles for Distributed Embedded Applications*. Engineering and Computer Science. Kluwer, 1997.

[7] H. Kopetz and G. Grünsteidl. TTP – A Time Triggered Protocol for Fault-Tolerant Real-Time Systems. *IEEE Computer*, 27(1):14–23, 1994.

[8] H. Kopetz and W. Ochsenreiter. Clock Synchronization in Distributed Real-Time Systems. *IEEE Trans. Computers*, 36(8):933–940, 1987.

[9] Z. Manna and A. Pnueli. Temporal Verification Diagrams. In M. Hagiya and J. C. Mitchell, editors, *International Symposium on Theoretical Aspects of Computer Software: TACS'94*, volume 789 of *Lecture Notes in Computer Science*, pages 726–765. Springer-Verlag, 1994.

[10] S. Owre, J. Rushby, N. Shankar, and F. von Henke. Formal Verification for Fault-Tolerant Architectures: Prolegomena to the Design of PVS. *IEEE Trans. on Software Engineering*, 21(2):107–125, February 1995.

[11] H. Pfeifer, D. Schwier, and F. W. von Henke. Formal Verification for Time-Triggered Clock Synchronization. In C. B. Weinstock and J. Rushby, editors, *Dependable Computing for Critical Applications 7*, volume 12 of *Dependable Computing and Fault-Tolerant Systems*, pages 207–226. IEEE Computer Society, January 1999.

[12] J. Rushby. Formal Methods and their Role in the Certification of Critical Systems. In R. Shaw, editor, *Safety and Reliability of Software Based Systems (Twelfth Annual CSR Workshop)*. Springer-Verlag, 1995.

[13] J. Rushby. Formal Verification of a Low-Overhead Group Membership Algorithm, 2000. In Preparation.

[14] J. Rushby. Verification Diagrams Revisited: Disjunctive Invariants for Easy Verification. In *Computer Aided Verification (CAV 2000)*, Chicago, IL, July 2000. To appear.

[15] C. Scheidler, G. Heiner, R. Sasse, E. Fuchs, H. Kopetz, and C. Temple. Time-Triggered Architecture. In J.-Y. Roger, B. Stanford-Smith, and P. T. Kidd, editors, *Advances in Information Technologies: The Business Challenge. Proceedings of EMMSEC'97*. IOS Press, 1997.

[16] `http://www.informatik.uni-ulm.de/ki/PVS/membership.html`.

VERIFICATION OF A SLIDING WINDOW PROTOCOL USING IOA AND MONA

Mark A. Smith*
IRISA
35042 Rennes Cedex, France
Mark.Smith@irisa.fr

Nils Klarlund
AT&T Labs Research
Florham Park, NJ 07932, U.S.A.
klarlund@research.att.com

Abstract We show how to use a decision procedure for WS1S (the MONA tool) to give automated correctness proofs of a sliding window protocol under assumptions of unbounded window sizes, buffer sizes, and channel capacities. We also verify a version of the protocol where the window size is fixed. Since our mechanized target logic is WS1S, not the finite structures of traditional model checking, our method employs only two easy reductions outside the decidable framework. Additionally, we formulate invariants that describe the reachable global states, but the bulk of the detailed reasoning is left to the decision procedure. Because the notation of WS1S is too low-level to describe complicated protocols at a reasonable level of abstraction, we use a higher level language for the protocol description, and then build a tool that automatically translates this language to the MONA syntax. The higher level language we use is IOA.

Keywords: Automated Verification, Formal Methods, Sliding Window Protocol, MONA, I/O Automata

1. INTRODUCTION

Network protocols are complicated. Protocol developers often determine the validity of their protocols by running simulations. While simu-

*Most of this work was done while the author was at AT&T Labs Research, NJ, U.S.A..

Tommaso Bolognesi and Diego Latella (Eds.), *Formal Methods for Distributed System Development.*

lations can be useful, simulations generally cannot show that a protocol is correct for all possible cases. Mechanical formal verification is another way to check the validity of protocols. The traditional benchmark used for testing mechanical verification of protocols is the alternating bit protocol [2]. For this protocol, many papers have shown that mechanical verification is viable (see for example [4, 9, 15, 1]).

In this article we demonstrate that an abstract algorithm description language can be effectively linked to the MONA tool [12, 10] for verification purposes. In particular, with only a couple of simple abstraction steps, the verification of invariants can be fully automated for network protocols that are substantially more complicated than the alternating bit protocol. We also argue that the state machine description IOA language [6] allows us to express the protocol at a convincing level of abstraction, and we discuss a translator from IOA to MONA.

We use these tools to verify a sliding window protocol. The protocol allows a dynamic and unbounded window size; and it assumes unbounded, lossy channels, and unbounded buffers. Our automated tools also allow us to verify the protocol with a fixed window size, and we verify the protocol with a window size as large as 256. The proof depends on two simple, meta-logical simplifications in the spirit of abstract interpretation. The proof itself consists of an invariant whose correctness is established automatically by MONA.

The main contribution of our work is to effectively demonstrate that the MONA tool can be used to verify non-trivial protocols with unbounded data structures as well as bounded data structures. Unbounded data structures cannot be handled by traditional Binary Decision Diagrams (BDDs) [3] based tools. On the other hand, theorem provers (for example [8, 16]) can also handle unbounded data structures. However, given the complexity of the protocol we verify in this work, we believe our verification process requires a lot less user interaction than is typical with theorem provers. By linking the MONA tool with IOA via the translation tool, we also allow users to describe protocols in a reasonably intuitive manner.

1.1. THE MONA TOOL

The MONA tool is an efficient implementation of the decision procedures for *WS1S (Weak Second-order Logic of One Successor)* and *WS2S (Weak Second-order Logic of Two Successor)*. WS1S is a variation of first-order logic with quantifiers and boolean connectives. Its interpretation is tied to a somewhat weakened version of arithmetic. The WS1S first order variables denote natural numbers, which can be compared and subjected to addition by constants. The logic also allows second-order variables which can be interpreted as a finite set of numbers. WS2S is a generalization of WS1S interpreted over the infinite binary tree.

A MONA program consists of a number of declarations and formulas. While the WS1S/WS2S logics have a simple and natural notation, the notation is too low-level to define protocols in a readable manner. Thus, we decided to use a high level language for describing the protocols. The protocol written in the high level language would then be translated to an equivalent MONA program.

1.2. THE IOA LANGUAGE

We decided to use the IOA language of Garland and Lynch [6] as the high level language. The IOA language is a precise language for describing Input/Output (I/O) Automata and stating their properties. I/O Automata [14], models components in asynchronous concurrent systems as labeled transitions systems. The model provides a mathematical framework for describing and reasoning about system components that interact with each other in an asynchronous manner. An I/O automaton consists of five components, a set of states, a nonempty set of start states, a set of actions, a set of steps, and a optional set of tasks. The set of actions can be partitioned into three disjoint sets, input, output, and internal. The set of tasks is a partition of the output and internal actions of the automaton.

1.3. RELATED WORK

The *sliding window* or *go back n* protocol is used in various standard data link control (DLC) protocols such as HDLC (high level DCL) and SDLC (synchronous DCL). It also forms the basis of reliable transport layer protocols like TCP. The sliding window protocol has been studied extensively and there are several works that present manual verification of the protocol. For example see [13, 19].

There has been a lot less work on mechanical verification of the sliding window protocol. In [18] Richier *et al.* use temporal logic and model checking to verify safety properties of the sliding window protocol. Because of the state explosion problem, verification is only possible for a fairly restricted set of values of parameters. Another paper in which a mechanical verification of the sliding window protocol is carried out is [11] by Kaivola. In this paper both safety and liveness proofs are performed. In order to alleviate the state explosion problem, Kaivola uses compositional property preserving equivalences and preorders which allows one to replace components of a system with smaller ones. The verification can then be carried out on the smaller systems. Using *non-divergent failures divergences* (NDFD) preorders, safety and liveness properties of the sliding window protocol for arbitrary channel lengths and "realistic" parameter values are verified semi-automatically. The largest parameter value presented in Kaivola's paper is a window size of seven, which is used with two other smaller fixed values that represent buffer sizes. Godefroid and Long [7] verify a full duplex sliding window protocol with the queue size as large as 11. Queue BDDs which they introduce and use in their verification of the protocol are very similar to BDD-represented automata used by the MONA tool.

One way in which our work here differs from the other papers on mechanical verification of the sliding window protocol is that our model of the protocol is different from the others. First, we assume that the sender and receiver have unbounded buffers and that the sender uses an unbounded set of sequence numbers. Because we assume unbounded sequence numbers, we do not require that the channels in our model be FIFO. While the assumption of unbounded buffers and sequence numbers is not a realistic one, in the situation where the sliding window protocol is used in the transport layer of the Internet (a la TCP [17]), the assumption of FIFO channels is not realistic either. TCP does not assume FIFO channels, instead it uses timeout mechanisms to simulate unbounded sequence numbers. This timeout mechanism is quite complicated, so we decide to just assume unbounded sequence numbers. Nevertheless, we believe our model more closely resembles the actual workings of TCP that the models used in the related work discussed above. Another way in which our work differs is that we do not have to assume a fixed window size. Our model allows the window size to be set nondeterministically. However, if we assume a fixed window size, we have verified the protocol with a window size of 256.

1.4. ORGANIZATION OF THE PAPER

In Section 2 we briefly describe the program that translates IOA to the MONA syntax. Section 3 and 4 contain descriptions of two versions of the sliding window protocol with unbounded window sizes, and also describes the abstractions we use. The proof of safety is discussed in Section 5. In Section 6 we discuss the case where we fix the window size, and we also present some experimental results. Section 7 contains some concluding remarks.

2. THE TRANSLATOR PROGRAM

The IOA language is Turing complete. Therefore, we cannot translate every IOA program into a valid equivalent WS1S/WS2S formula. However, we are able to translate a subset of the IOA language that is sufficiently expressive to allow the formulation of reasonable protocol descriptions.

We treat the simple built in types of the IOA language in a straightforward manner. IOA types boolean and (non-negative) integer translate to boolean and integer in WS1S. However, we cannot translate reals, and we do not translate IOA types character or string.

The IOA language also has several built in "compound" types. These types are, `Array[I, E]`, `Map[I, E]`, `Seq[E]`, `Set[E]` and `Mset[E]`. For these types, `I` is the index type, and `E` is the type of the elements. Each compound type has a set of built in operators. We are able to translate these compound types when the size of `E` has a known bound. However, for type `Set[E]`, where `E` is the set of integers, we translate directly to the MONA representation of sets of integers. The other types we represent in MONA as a group of sets where one set contains the valid indices, when the data type has indices, and the other set(s) are used to represent the elements. The IOA language also has an enumeration type. We represent this type in MONA using boolean valued variables, where the bit value of the variables taken together as binary number corresponds to the enumerated values.

We illustrate the translation schema with a small example. If we have an IOA type `Animals` which is an enumeration of `dog, cat, rat`, and `bat`, it is translated to MONA boolean (zero'th-order) variables, say `S0` and `S1`, where the values `S0 = false` and `S1 = false` represent `dog`, `S0 = false` and `S1 = true` represent `cat`, `S0 = true` and `S1 = false` represent `rat`, and `S0 = true` and `S1 = true` represent `bat`. A compound IOA variable `animalBuf` that has type `Seq[Animals]` is translated into three second order MONA variables, say `S2, S3`, and `S4`. The first set `S2` holds the set of valid indices for the sequence. The other two sets encode the values in these positions in the following manner: for each valid index in `S2` the presence or absence of that index in sets `S3` and `S4` encodes the values in the position. Thus, if `dog` is the value of position 0 in the sequence, then sets `S3` and `S4` do not contain the value 0. However, if `cat` is the value in position 0, then set `S3` does not contain 0 while set `S4` does. Thus, `animalBuf = [cat, rat, dog, bat, rat]` where `cat` is at index position 0 is represented by the MONA sets `S2` $= \{0, 1, 2, 3, 4\}$, `S3` $= \{1, 3, 4\}$, and `S4` $= \{0, 3\}$. IOA operators on sequences are translated accordingly. For example, the MONA translation of the `head` operator checks if the value 0 is present in the first set representing the sequence; this is a test for emptiness, and if the sequence is empty `head` is undefined. If the sequence is not empty, the two boolean values returned by the MONA operations `0 in S3` and `0 in S4` gives the value of the head of the sequence. In this example the values returned would be `false` and `true`, which is our MONA representation for `cat`.

IOA transitions are translated to MONA predicates. We used unprimed and primed MONA variables to represent the pre and post states of the IOA variables. IOA transitions are atomic, but the assignments in a transition happen sequentially. Thus, in our MONA translation we add temporary variables to hold intermediate values. The translator also generates a MONA predicate that represents the assignment of initial values to the IOA variables.

To facilitate the verification process we add a predicate, **preds**, section to the IOA language constructs. The **preds** section of an IOA program contains named predicates on the state of the program variables. We translate the predicates to unprimed and primed MONA predicates of the same name. The unprimed state represents the values before a transition and the primed states represent the values after the transition. Two MONA predicates `Inv` and `Inv'` that respectively are the conjunction of the unprimed and primed predicates in the **preds** section are automatically generated. `Inv` is the Invariant we wish to verify holds for the IOA protocol. The translator also creates a separate file for each transition in the IOA program. Each file contains a MONA predicate of the form $\texttt{Inv} \wedge \textit{step} \Rightarrow \texttt{Inv}'$, where step is the name of the IOA transition. Additionally, a MONA file which contains the MONA formula to test that `Inv` holds in the initial state is generated. These are the necessary files for the MONA tool to automatically verify the invariant. To verify that the invariant holds, we run MONA on each of the files. If MONA tells us that all the formulas are valid, then the invariant holds.

3. THE PROTOCOL DESCRIPTION

In this section we describe our general IOA model of a sliding window protocol. The sliding window protocol is a mechanism for reliable message delivery in a distributed setting where a sender and a receiver communicate over lossy channels. The basic structure of these protocols is shown in Figure 1. In addition to the four components we model, there are two external *users*. The user on the sender side inputs the data to be sent by the sender, and the user on the receiver side gets the data that the receiver delivers.

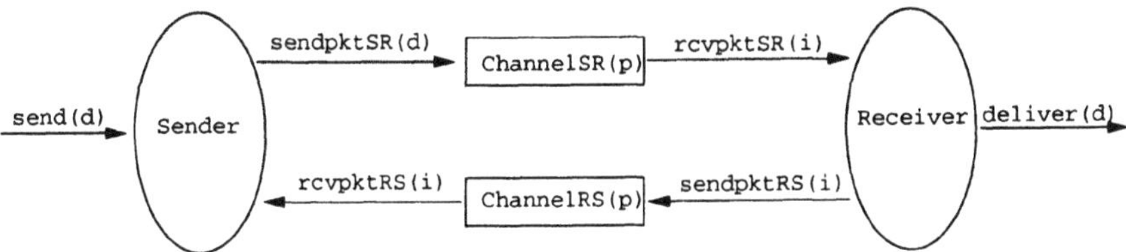

Figure 1 The basic structure and components of a sliding window protocol.

The basic idea of the sliding protocol is that a *window* of size $n \geq 0$ determines how many successive packets of data can be sent in the absence of a new acknowledgment. The window size may be fixed or may vary depending on the conditions of the different components of the protocol. In our model we will assume that n varies nondeterministically. Each packet of data is sequentially numbered, so the sender is not allowed to send packet $i+n$ before packet i has been acknowledged. Thus, if i is the sequence number of the packet most recently acknowledged by the receiver, there is a window of data numbered $i+1$ to $i+n$ which the sender can transmit. As successively higher-numbered acknowledgments are received, the window slides forward. The acknowledgment mechanism is cumulative in that if the receiver acknowledges packet k, where $k \geq i+1$, it means it has successfully received all packets up to k. Packet k is acknowledged by sending a request for packet $k+1$. Typically, transmitted data is kept on a retransmission buffer until it has been acknowledged. Thus, when k is acknowledged, packets with sequence number less than or equal to k are removed from the retransmission buffer. Packets that are not acknowledged are eventually retransmitted. For our modeling of the sliding window protocol we assume that sequence numbers are unbounded, we do not assume that the channels are FIFO.

Below we present two IOA versions of the sliding window protocol. The first version is a general formulation, where the data domain is unspecified. The second version of the protocol, is a parameterized transition system expressible in WS1S; it relies on two additional assumptions: that the data domain consists of two values and that the channels are approximated (in the sense to be explained). This version of the protocol is in the full paper [20].

3.1. A GENERAL PROTOCOL MODEL

In this section we define a sliding window protocol as an I/O Automaton. The I/O automaton is defined using the IOA language. The general IOA model is shown in Figure 2.

```
type D  % data types
type A =  enumeration of ack
automaton SlidingWindow(D:  Type)
signature
 input send(d:  D)
 output deliver(d:  D)
 internal sendpktSR(d:  D), prepareNewSeg(d:  D),prepareRetranSeg(d:  D),
          rcvpktRS(i:  Int), rcvpktSR(i:  Int), sendpktRS(i:  Int)
states
 SendBuf:  Seq[D] := {},  hSendBuf:  Int := 1,
 W: Int :=  choose n  where (n > 0),
 RetranBuf:  Seq[D] := {}, hRetranBuf:  Int := 1,
 readyToSend:  Bool := false, segment:  D, seqNum:  Int := 0,
 RcvBuf:  Seq[D] := {}, hRcvB: Int := 1, sendAck:  Bool := false,
 temp:  D, transitSR: Map[Int, Mset[D]] := empty,
 transitRS: Map[Int, Mset[A]] := empty
transitions
input send(d)
eff SendBuf := SendBuf |- d
internal prepareNewSeg(d)
pre readyToSend = false  ∧ hSendBuf <= len(SendBuf) ∧ d = SendBuf[hSendBuf]
   ∧ len(RetranBuf) < W
eff RetranBuf := RetranBuf |- d;
   seqNum := hSendBuf;
   hSendBuf := hSendBuf + 1;
   readyToSend := true;
   segment := d
internal sendpktSR(d)
pre readyToSend = true ∧ d = segment
eff readyToSend := false;
   transitSR := update(transitSR, seqNum, insert(d, transitSR[seqNum]))
internal rcvpktRS(i)
pre defined(transitRS, i) ∧ transitRS[i]  = {}
eff transitRS :=  update(transitRS, i, delete(ack, transitRS[i]));
    if hRetranBuf < i  then
       hRetranBuf := i;
       W :=  choose n  where (n >= W ∧ n >= hRetranBuf)  fi
internal prepareRetranSeg(d)
pre readyToSend = false ∧ d = RetranBuf[hRetranBuf]
   ∧ hRetranBuf <= len(RetranBuf)
eff readyToSend := true;
   seqNum := hRetranBuf;
   segment := d
```

Figure 2 The general IOA model of the sliding window protocol.

3.1.1 Types. The type `D` represents a data domain whose internal structure is unspecified. The type `A` is an acknowledgment type, and has only one possible value, `ack`.

3.1.2 Signature and States. The **signature** is a partition of the actions of the automaton into input, output and internal actions. The actions `send(d)`, `sendpktSR(d)`, `prepareNewSeg(d)`, `prepareRetranSeg(d)` and `rcvpktRS(i)` all happen on the sender side. The actions `rcvpktSR(i)`, `sendpktRS(i)`, and `deliver(d)` happen on the receiver side, and the actions `dropSR(i)` and `dropRS(i)` happen in the channel from the sender to the receiver and the channel from the receiver to the sender respectively. We explain the functionality of each action when we describe the transitions below.

The **states** part of the IOA program represents the set of states and the set of initial values. In the listing, variables `SendBuf` through `hRetranBuf` are variables of the sender, variables `RcvBuf` through `temp` are receiver variables, variable `transitSR` is the state variable of the channel from the sender to the receiver, and variable `transitRS` is the state variable of the channel from the receiver to the sender. The variable `SendBuf` is a sequence that holds elements of type `D` and is initially empty. This buffer holds the input data, and the position of the element in the sequence corresponds to the sequence number of that piece of data. The buffer has an associated pointer called `hSendBuf`; the section from `hSendBuf` to `len(SendBuf)` is called the *unsent* part of `SendBuf`. The variables `RetranBuf` and `hRetranBuf` represent the retransmission buffer in the same way the send buffer is represented. The variable `W` represents the current edge of the window, that is, the highest index allocated for use by the retransmission buffer. The non-deterministic setting of the initial value allows us to have an unbounded window size. Variable `readyToSend` is a boolean flag. It is set whenever a datum in the `segment` variable is ready to be transmitted; `seqNum` is then the sequence number of that segment.

On the receiver side, the variable `RcvBuf` holds received data before it is passed to the external user. Variable `hRcvBuf` is the index of the first element of the *undelivered* part of this buffer. The variable `sendAck` is a boolean flag that enables the sending of an acknowledgment packet, and `temp` is a data variable that is used to hold a value received from the channel before it is added to `RcvBuf`.

The variables `transitSR` and `transitRS` are map types that represent packets in transit from the sender to the receiver and from the receiver to the sender respectively. Both maps are indexed by integers. The `transitSR` map holds elements that are multisets with elements of type

D. These elements are multisets because there may be duplicates of data packets on the channel. The index of the defined elements in **transitSR** represents the sequence number of the packet(s). For **transitRS** the element type is a multiset of type A. The index of an element of **transitRS** represents the acknowledgment number of the packet.

3.1.3 Transitions. The send(d) input causes the data item d from the external user to be appended to the end of SendBuf. The internal action prepareNewSeg(d) prepares the data item d to be sent. This is data that has not been sent before. The precondition hSendBuf $\leq$ len(SendBuf) means the active part of the send buffer, SendBuf, is not empty. The data item d is the element index by hSendBuf. There must also be space available in the allowable window; that is, hSendBuf $<$ W. The effect of the prepareNewSeg(d) action is to append d to the retransmission buffer, RetranBuf. The sending of this data item in a packet is also enabled by the other assignments. The internal action sendpktSR(d) sends the data item d if readyToSend is true and d = segment. This action assigns readyToSend to false so that no more data is sent until some preparation occurs. An element is placed on the channel by inserting it into the multiset in the position in the transitSR map indexed by seqNum. The internal action rcvpktRS(i) passes an acknowledgment packet from the receiver to the sender. The position of the element in the transitRS map represents the acknowledgment number. Valid acknowledgments (hRetranBuf $<$ i) cause hRetranBuf to be assigned the value i, which effectively removes the acknowledged packets from the active part of RetranBuf. When packets are acknowledged, we also non-deterministically update the value of W, the edge of the window. This new value of W must of course be no less than the previous value, and it must also be no less than the head of the retransmission buffer. The prepareRetranSeg(d) is similar to prepareNewSeg(d) action except that the data item comes from RetranBuf and not SendBuf.

The rcvpktSR(i), deliver(d), sendpktRS(i) actions happen on the receiver side. These are the actions where the receiver receives a packet from the channel, delivers data to the user, and send an acknowledgment respectively. Actions dropSR(i) and dropRS(i) are the actions by the respective channels for dropping a copy of a packet. These actions are not shown here but are in the full paper [20], where they are also described further.

4. A WS1S SLIDING WINDOW MODEL

We desire to prove that what comes out is what goes in; that is, for all reachable states:

Predicate omega *For all i if i* $\leq$ `len(RcvBuf)` *then* `SendBuf[i] = rcvBuf[i]`.

We establish this safety property by exhibiting an invariant that implies **omega**. However, we are not able to carry out any automated proof of invariance for the general model. Instead, we will bend the general model into an *abstract model* whose transitions can be represented in WS1S.

4.1. FIRST SIMPLIFYING ASSUMPTION

The general model is parameterized with the D type. Our first simplifying assumption is based on idea of data-independence:

Proposition 1 *[21] The* **omega** *property holds for all reachable states of all abstract programs if and only if it holds for* D $=$ {`white`, `red`}.

Proof: The program itself contains no comparisons of data. Thus, if there is some D and some reachable state that violates the **omega** property by virtue of `SendBuf[i]` $= d'$ and `rcvBuf[i]` $= d''$ with $d' \neq d''$, then the same program on the same domain D but with all `send` (d) actions replaced by `send`$(\hat{d})$, where $\hat{d} = d$ if $d \in \{d', d''\}$ and, arbitrarily, $\hat{d} = d'$ if $d \notin \{d', d''\}$, will result in a reachable state violating the **omega** property. Thus, a two-value data domain suffices to establish **omega** for all domains. (This proof is due to Wolper [21].) ■

The *concrete model* is the general model, where D $=$ {`white`, `red`}.

4.2. SECOND SIMPLIFYING ASSUMPTION

Our second simplifying assumption is that the channel structure of the concrete model can be simplified so that each sequence number is associated not with a multiset, but with an element in a fixed, finite set. The method is that of abstract interpretation [5], where the computation is modeled in an abstract domain, which approximates the concrete domain.

4.2.1 The Abstract Domain.

The abstract domain **multiD** models a multiset over D by four different values: **D_empty** represents the empty multiset; **D_r** represents a multiset that contains only **red** occurrences; **D_w** represents a multiset that contain only **white** occurrences; and **D_rw** represents a multiset that have occurrences of both **red** or **white**. This correspondence can be expressed mathematically by an *abstraction function* $\alpha_{\texttt{multiD}}$: `Mset`$[\{red, white\}] \rightarrow$ `MultiD` that maps a multiset to an abstract value. Similarly, the correspondence can be expressed by a *concretization func-*

tion $\gamma_{\texttt{multiD}}$: MultiD $\rightarrow$ Set[Mset[{*red*, *white*}]] that maps an abstract value to the set of corresponding multisets. Note that if a is not **D_empty**, then $\gamma_{\texttt{multiD}}(a)$ is an infinite set of multisets.

Also for any multiset M, $M \in \gamma_{\texttt{multiD}} \circ \alpha_{\texttt{multiD}}(M)$. In this sense, the abstract domain approximates the concrete values of the multisets. The approximation is not exact, of course, since generally there will be many other multisets in $\gamma_{\texttt{multiD}} \circ \alpha_{\texttt{multiD}}(M)$ than M.

The multisets of the `transitRS` channel are modeled by the domain **multiA** whose values are **A_empty**, denoting the empty multiset, and **A_pres**, denoting a non-empty multiset.

In the second IOA model, we call a global state a an abstract state. It contains two maps modeling the channels, and each may contain an unbounded number of abstract values denoting multisets.

The concretization functions above define for an abstract state a a set $\gamma(a)$ of states c of the concrete IOA model.

All operations involving the multisets are then modified to work on the abstract values. For example, the dropping of a message as specified in `dropSR(i)`, where `i` is a sequence number, involves the deletion of an element `v` from the multiset `transitSR(i)`. In the WS1S version, the deletion is modeled by changing the `multiD` value non-deterministically. For example, `D_rw` is changed to either `D_r` (there is only one occurrence of a white message, which is dropped), `D_w` (there are only white occurrences after dropping the single red message), or `D_rw` (there are still occurrences of both red and white messages after a message has been dropped).

This treatment of the abstractions entails that for any computation in the concrete model $c_0, c_1, \ldots$, we can find a corresponding computation in the abstract domain $a_0, a_1, \ldots$, whose states include those of the concrete computation in the sense that $c_i \in \gamma(a_i)$. In particular, if c_0 is the initial state of the concrete model, then we can let a_0 be the initial state of the abstract model.

Proposition 2 *Let ϕ be a formula that does not refer to channels. If ϕ holds for all reachable states of the abstract model, then ϕ holds for all reachable states of the concrete model.*

Proof: (1) We omit the details which establish on a transition by transition basis that the abstract operations are such that any concrete computation corresponds to an abstract one. (2) By the restriction on ϕ, all variables ϕ mentions have the same value in c as in $\psi(c)$. Therefore, ϕ holds about c if and only if it holds about $\psi(c)$.

The statement of the proposition is implied by (1) and (2). ■

4.3. THE ABSTRACT MODEL

The abstract model is obtained from a transformation of the concrete model. The IOA version of the abstract model is in the full paper [20]. This model includes a **preds** section that contains predicates on the state of the automaton. We prove in Section 5, where these predicates are discussed further, that their conjunction is an invariant of the IOA `SlidingWindow`.

5. PROOF OF SAFETY

We use the standard inductive meta-argument for proving safety properties through invariants. We find a formula ϕ_I that implies the safety property, and we show that ϕ_I holds for the initial state, and that for every step (s, a, s') if ϕ_I holds in state s then it also holds in state s'. All such proofs are completely automated.

Our main task at hand is to find ϕ_I.

5.1. THE PREDICATES

Predicate omega asserts the safety property, but by itself this predicate is not invariant, so a stronger condition is needed. The other predicates are found by a combination of intuition and interaction with the MONA tool. When the MONA tool is run on a predicate that is not an invariant, the counter example provided by MONA gives some indication of what is needed to strengthen the predicate. Typically, the counter example involves some state that intuition tells one is not reachable from any start state. A predicate is then written to verify this intuition. The verification of this new predicate may in turn lead to the formulation of other predicates. The predicates are shown in Figure 3.

We claim the conjunction of all the predicates above is invariant for `SlidingWindow`. The claim is proved by using MONA to automatically verify the following:

$$\textbf{omega} \wedge \textbf{alpha} \wedge \textbf{beta} \wedge \textbf{gamma} \wedge \textbf{delta} \wedge \textbf{epsilon} \wedge \textbf{zeta}.$$

This invariant clearly implies the safety property. The verification of the protocol with unbounded window sizes took 68.26 seconds. The largest number of states generated by the MONA tool during the process was 18886, and the largest number of BDD nodes was 277859. We ran the tools on a SUNW,Ultra-5_10.

```
Preds
alpha1 hSendBuf <= (len(SendBuf) + 1);
alpha2 hSendBuf > seqNum;
alpha3 hSendBuf = len(RetranBuf) + 1;
alpha4 (hRetranBuf > 0) ∧ (readyToSend => seqNum > 0);
alpha5 ∀ i: Int (defined(transitSR,i) => (hSendBuf > i ));
alpha6 hSendBuf > len(RcvBuf)  ∧ len(SendBuf) >= len(RcvBuf);
beta  ∀ i:  Int ((i <= len(RetranBuf)) => (SendBuf[i] = RetranBuf[i]));
gamma ∀ d:  D ((segment = d ∧ readyToSend) => SendBuf[seqNum] = d);
delta1 ∀ i:  Int ((transitSR[i] = D_r) => SendBuf[i] = red);
delta2 ∀ i:  Int ((transitSR[i] = D_w) => SendBuf[i] = white);
epsilon1 ∀ i:  Int ((defined(transitSR, i) ∧ (RetranBuf[i] = red)) =>
(transitSR[i] = D_r ∨ transitSR[i] = D_empty));
epsilon2 ∀ i:  Int ((defined(transitSR, i) ∧ (RetranBuf[i] = white)) =>
(transitSR[i] = D_w ∨ transitSR[i] = D_empty));
zeta1 (defined(transitSR, seqNum) ∧ segment = red) =>
transitSR[seqNum]  = D_w;
zeta2 (defined(transitSR, seqNum) ∧ segment = white) =>
transitSR[seqNum]  = D_r;
zeta3  ∀ i:  Int (defined(transitSR, i) => transitSR[i]  = D_rw);
omega  ∀ i:  Int (i <= len(RcvBuf)  => (SendBuf[i] = RcvBuf[i]))
```

Figure 3 A series of predicates the conjunction of which is invariant on the reachable states of the protocol.

6. FIXED WINDOW SIZE

In this section we discuss our results for the protocol where the window size is fixed. We make two changes in the protocol to use a fixed window size. For the `prepareNewSed(d)` transition the test to check if there is available space in the allowable window is changed to **hSendBuf** $\leq$ (**hRetranBuf** + **256**), where 256 is the fixed window size. For the `rcvpktRS(i)` transition, the change is removing the update of the variable `W`.

Table 1 lists our experimental results. In the table, the states and BDD nodes columns holds the largest number of states and the largest number of BDD nodes generated by the MONA tool during the verification respectively. The results show that the processing time, the largest number of states, and the maximum number of BDD nodes generated by the MONA tool grows linearly as a function of the window size. The linear growth is to be expected: the amount of information that the automaton for the unbounded case carries is bounded (by the size of the state space). The automaton for the bounded case must, in addition, count up to the window size, and that makes the resulting automaton size linear in the window size. In our experiments we note that the size of the automaton and the running time in the bounded case surpasses that of the unbounded case for window sizes greater than 44.

Table 1 The results of our experiments verifying the protocol with different fixed window sizes.

window size	time (sec's)	states	BDD nodes
16	30.98	4838	147223
32	46.67	5983	258583
64	86.63	11103	481303
128	342.01	21343	926743
256	1286.45	41823	1817623

7. CONCLUSION AND FUTURE WORK

In this paper we have demonstrated that a decision procedure for WS1S can be used to verify safety properties of a non-trivial network protocol.

Although our work indicates that for network protocols a good deal of nitty-gritty reasoning about queues and indices can be carried out automatically, more research into the following problems is needed: handling of sequence numbers that wrap around modulo the length of the window; handling of a more concrete presentation of queues, such as those actually found in a real programming language; automated support for abstraction steps; and handling of liveness properties.

Acknowledgments

Thanks to Dennis Dams and Richard Trefler for helpful comments.

References

[1] P.A. Abdulla, A. Aniinichini, S. Bensalem, A. Bouajjani, P.Habermehl, and Y. Lakhnech. Verification of infinite-state systems by combining abstraction and reachability analysis. In *Computer Aided Verification. 11th International Conference, CAV '99*, volume 1633 of *LNCS*. Springer-Verlag, July 1999.

[2] K. A. Barlett, R. A. Scantlebury, and P. C. Wilkinson. A note on reliable transmission over half duplex links. *Communications of the ACM*, 12, 1969.

[3] R. E. Bryant. Symbolic boolean manipulation with ordered binary-decision diagrams. *ACM Computing Surveys*, 24(3):293–318, September 1992.

[4] E.M. Clarke and E.A. Emerson. Design and synthesis of synchronization skeletons using branching time temporal logic. In *Workshop on Logics of Programs*,

volume 131 of LNCS. Springer-Verlag, 1981.

[5] P. Cousot and R. Cousot. Abstract interpretation: A unified lattice model for static analysis of programs by construction or approximation of fixpoints. In *Proceedings of 4th ACM Symposium on Principles of Programming Languages*, pages 238–252, Los Angeles, California, 1977.

[6] Stephen J. Garland and Nancy A. Lynch. The IOA language and toolset: Support for designing, analyzing, and builiding distributed systems. Technical Report MIT/LCS/TR-762, M.I.T., August 1998.

[7] P. Godefroid and D.E. Long. Symbolic protocol verification with Queue BDDs. *Formal Methods in System Design*, 14(13):257–271, may 1999.

[8] M. J. C. Gordon and T. F. Melham, editors. *Introduction to HOL: A Theorem Proving Environment for Higher Order Logic*. Cambridge University Press, 1993.

[9] K. Havelund and N. Shankar. Experiments in theorem proving and model checking for protocol verification. In *Formal Methods Europe (FME)*, volume 1051 of *Lecture Notes in Computer Science*. Springer-Verlag, July 1996.

[10] J. G. Henriksen, J. Jensen, M. Jørgensen, N. Klarlund, R. Paige, T. Rauhe, and A. Sandholm. Mona: Monadic second-order login in practice. In *Tools and Algorithms for the Construction and Analysis of Systems, First International Workshop, TACAS '95 LCNS 1019*, 1995.

[11] Roope Kaivola. Using compositional preorders in the verification of sliding window protocol. In *Computer Aided Verification. 9th International Conference, CAV'97*, volume 1254 of *LNCS*, pages 48–59, Haifa, Israel, June 1997. Springer-Verlag.

[12] Nils Klarlund. Mona & Fido: The logic-automaton connection in practice. In *CSL '97 Proceedings*, volume 1414 of *LNCS*. Springer-Verlag, 1998.

[13] Donald Knuth. Verification of link-level protocols. *BIT*, 21:31–36, 1981.

[14] Nancy Lynch and Mark Tuttle. An introduction to input/output automata. *CWI Quarterly*, 3(2), September 1989.

[15] Panagiotis Manolios, Kedar Namjoshi, and Robert Sumners. Linking theorem proving and model-checking with well-founded bisimulations. In *Computer Aided Verification. 11th International Conference, CAV '99*, volume 1633 of *LNCS*. Springer-Verlag, July 1999.

[16] S. Owre, J. Rushby, N. Shankar, and F. von Henke. Formal verification for fault-tolerant architectures: Prolegomena to the design of PVS. *IEEE Transactions on Software Engineering*, 21(2):107–125, 1995.

[17] Jon Postel. Transmission Control Protocol - DARPA Internet Program Specification (Internet Standard STC-007). Internet RFC-793, September 1981.

[18] J. L. Richier, C. Rodriguez, J. Sifakis, and J. Voiron. Verification in Xesar of the sliding window protocol. In *Protocol Specification, Testing and Verification VII*, pages 235–248. North-Holland, 1987.

[19] A. Udaya Shankar and S. S. Lam. An HDLC protocol specification and verification using image protocols. *ACM Transactions on Computer Systems*, 1(4):331–368, 1983.

[20] Mark A. Smith and Nils Klarlund. Verification of a sliding window protocol using IOA and MONA. Technical report, IRISA. To appear.

[21] P. Wolper. Synthesis of communication processes from temporal logic specifications. In *Proceedings 13th ACM Symposium on POPL*, pages 184–193, January 1986.

A PRIORI VERIFICATION OF REACTIVE SYSTEMS

On Simultaneous Syntactic Action Refinement for TCSP and the Modal Mu-Calculus

Mila Majster-Cederbaum, Frank Salger and Maria Sorea

Universität Mannheim

Fakultät für Mathematik und Informatik,

D7, 27, 68131 Mannheim, Germany

{mcb, fsalger,msorea }@pi2.informatik.uni-mannheim.de

Abstract In this paper we present a refinement operator $\cdot[a \rightsquigarrow Q]$, defined both on *TCSP*-like process terms P and formulas φ of the *Modal Mu-Calculus*. We show that

> the system induced by a term P satisfies a specification φ if and only if the system induced by the refined term $P[\alpha \rightsquigarrow Q]$ satisfies the refined specification $\varphi[\alpha \rightsquigarrow Q]$

where Q is a process term from an appropriate sublanguage of *TCSP*. We explain how this result can be used to reason about reactive systems. In particular it supplies a method to verify systems a priori: Provided $P \models \varphi$ holds, the refinement of φ into $\varphi[a \rightsquigarrow Q]$ induces a transformation of P into $P[a \rightsquigarrow Q]$ such that $P[a \rightsquigarrow Q] \models \varphi[a \rightsquigarrow Q]$. The above result holds under the restriction that the set of actions that occur in the term Q has to be disjoint from the set of actions occurring in the term P and the formula φ. Though such restrictions on alphabet disjointness are commonly adopted in approaches to syntactic action refinement for process term languages they preclude the possibility to carry out certain refinement steps that might be necessary in the stepwise development of reactive systems. We show, that while dropping the above restriction, the validity of the two implications comprised in the above result can still be established independently for different interesting fragments of the Modal Mu-Calculus.

Keywords: Verification, Syntactic Action Refinement, Process Algebra, Reactive Systems, Temporal Logic.

Tommaso Bolognesi and Diego Latella (Eds.), *Formal Methods for Distributed System Development.*

1. INTRODUCTION

One possibility to model *reactive systems* is provided through various *process calculi* like e.g. CCS [16] or TCSP [2, 17]. On the other hand *temporal logics* [9] can be used to reason about attributes of such systems. A particular expressive temporal logic is the *Modal Mu-Calculus* [14] as it subsumes many logics that are commonly used to specify properties of reactive systems [10, 7]. The basic building blocks of process calculi and modal logics are *(atomic) actions* considered to be the conceptual entities at a chosen level of abstraction. The concept of action refinement (see e.g. [1, 11, 12]) is sometimes regarded as one of the fundamental methods to develop reactive systems. A particularly interesting concept of action refinement is *syntactic action refinement* (e.g. [1]) where an action a which occurs in a process term P is refined by a more complex term Q thereby yielding a more detailed process description $P[a \rightsquigarrow Q]$. Due to its definitional clarity, syntactic action refinement can easily be used by system developers without a thorough knowledge about reactive system semantics.

Until recently, the connection between syntactic action refinement in process languages and syntactic action refinement in logics has rarely been adressed in the literature (see the last section for related work). However, syntactic action refinement in logics has various interesting applications to the verification of reactive systems as already discussed in [15]. We show the validity of the assertion

$$P \models \varphi \Leftrightarrow P[a \rightsquigarrow Q] \models \varphi[a \rightsquigarrow Q] \quad (*)$$

where $\cdot[a \rightsquigarrow Q]$ denotes the refinement operator in the considered process languages and logics. P is a process term from a $TCSP$-like language, Q is a term from an appropriate sublanguage and φ is a Modal Mu-Calculus-formula that in addition might contain refinement operators.

The simultaneous application of a refinement operator to process terms and formulas (as in assertion (*)) will be called *simultaneous syntactic action refinement* (SSAR for short). Assertion (*) says, that the system induced by a term P[1] satisfies a specification φ if and only if the system induced by the refined term $P[\alpha \rightsquigarrow Q]$ satisfies the refined specification $\varphi[\alpha \rightsquigarrow Q]$. As an application of assertion (*) we consider the following 'assembly line' process in a car factory shown in Figure 1 which can be 'implemented' by a $TCSP$-like process expression P. The job of P is to adjust the motor and the gear of a car and to mount the

[1]In assertion (*), we identified the program (process term) P with its semantics. This convention will be continued throughout the paper if the context avoids ambiguity.

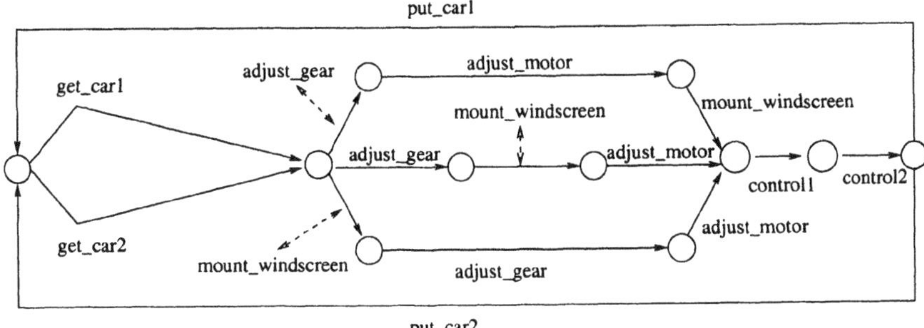

Figure 1 The Process P

windscreen. To reach a defined system status before the car is carried back to the conveyer band two control actions are executed by P. The process P has the temporal property that 'whenever a car is taken from the conveyer band (either get_car1 or get_car2 is executed), the control actions will eventually be executed' which can be denoted by a formula φ in the Modal Mu-Calculus. We propose that the verification task be facilitated by integrating verification into the procedure of program development via assertion (*). To this end we observe, that the process P arises from the process P_s (shown in Figure 2) by the application of four successive refinement steps i.e.

$$\begin{aligned} P_1 &= P_s[control \rightsquigarrow (control1; control2)] \\ P_2 &= P_1[put\_car \rightsquigarrow (put\_car1 + put\_car2)] \\ P_3 &= P_2[adjust \rightsquigarrow (adjust\_gear; adjust\_motor)] \\ P &= P_3[get\_car \rightsquigarrow (get\_car1 + get\_car2)] \end{aligned}$$

Now let us assume that we had already established $P_s \models \varphi_s$ where the Modal Mu-Calculus formula φ_s denotes the property 'whenever a car is taken from the conveyer band, the control action will eventually be executed'.

In this paper we show how to define a *logical refinement* operation $\cdot[a \rightsquigarrow Q]$ on the Modal Mu-Calculus where

$$\begin{aligned} \varphi_1 &= \varphi_s[control \rightsquigarrow (control1; control2)] \\ \varphi_2 &= \varphi_1[put\_car \rightsquigarrow (put\_car1 + put\_car2)] \\ \varphi_3 &= \varphi_2[adjust \rightsquigarrow (adjust\_gear; adjust\_motor)] \\ \varphi &= \varphi_3[get\_car \rightsquigarrow (get\_car1 + get\_car2)] \end{aligned}$$

such that $P_s \models \varphi_s$ iff $P \models \varphi$.

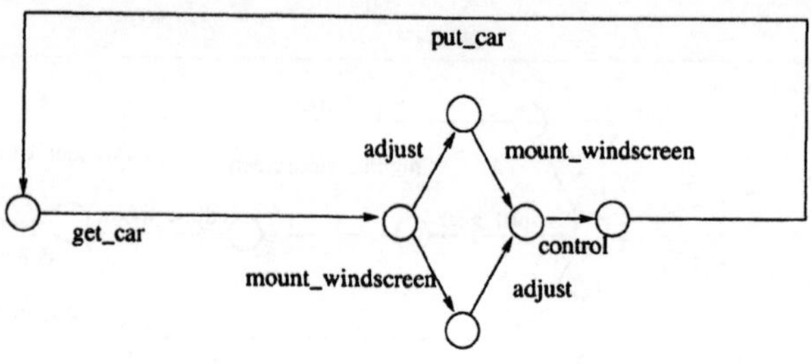

Figure 2 The Process P_s

As indicated by the above example, assertion (∗) can be used for the following:

- It offers the possibility of simplifying the verification task by repeatedly applying (∗) as was shown in the example. Instead of checking that

$$P = P_s[a_1 \rightsquigarrow Q_1] \dots [a_n \rightsquigarrow Q_n] \models \varphi_s[a_1 \rightsquigarrow Q_1] \dots [a_n \rightsquigarrow Q_n] = \varphi$$

  we only have to check that $P_s \models \varphi_s$.

- Let us assume that the specification of a system is developed incrementally and that φ_s is the actual specification which is now given more details by refining it to $\varphi = \varphi_s[a_1 \rightsquigarrow Q_1] \dots [a_n \rightsquigarrow Q_n]$. When we look for an implementation of φ, we only have to supply an implementation P_s for φ_s thereby automatically obtaining a process $P = P_s[a_1 \rightsquigarrow Q_1] \dots [a_n \rightsquigarrow Q_n]$ such that $P \models \varphi$. This can be seen as a concept of *a priori-verification.* For the example above, suppose the car factory was equipped with the assembly line P_s and $P_s \models \varphi_s$ has already been established. Due to new requirements on the productivity (speed) of P_s the conveyor band *get_car* needs to be replaced by two conveyor bands *get_car*1, *get_car*2. This change of resources can be specified by refining φ_s to the specification $\phi = \varphi_s[get\_car \rightsquigarrow (get\_car1 + get\_car2)]$. The system $\tilde{P} = P_s[get\_car \rightsquigarrow (get\_car1 + get\_car2)]$ which is given via assertion (1) is (a priori) correct with respect to ϕ, i.e. $\tilde{P} \models \phi$. The other refinement steps in the example can easily been interpreted accordingly.

- As a determination of the logical consequences of process term refinements: If $P \models \varphi$ then the process refinement $P[a \rightsquigarrow Q]$ automatically yields a refined specification $\varphi[a \rightsquigarrow Q]$ via assertion (∗) such that $P[a \rightsquigarrow Q] \models \varphi[a \rightsquigarrow Q]$.

Some restrictions have to be obeyed to guarantee that assertion $(*)$ holds, one of which is *alphabet disjointness*[2] of P from Q. Alphabet disjointness of P and Q for terms $P[a \rightsquigarrow Q]$ is commonly used in approaches to syntactic action refinement[3]. However alphabet disjointness rules out the possibility to conduct particular refinement steps that can become important in the development of reactive systems:
Suppose the system P can execute the atomic actions a, b. At the current level of abstraction, the action a (b) is considered to be the name of a procedure Q_a (Q_b resp.) which is not yet implemented. In an intermediate development step, Q_a and Q_b are implemented making use of a common subsystem S which we might assume has been provided by a system library. Hence, alphabet disjointness of Q_a and Q_b does not hold. Consequently we cannot use assertion $(*)$ to establish

$$P \models \varphi \Leftrightarrow P[\rho \rightsquigarrow Q_\rho][\varrho \rightsquigarrow Q_\varrho] \models \varphi[\rho \rightsquigarrow Q_\rho][\varrho \rightsquigarrow Q_\varrho]$$

for $\rho, \varrho \in \{a, b\}$, $\rho \neq \varrho$ since alphabet disjointness with respect to the terms $P[\rho \rightsquigarrow Q_\rho]$ and Q_ϱ does not hold.

However it is possible to extract an interesting fragment of the Modal Mu-Calculus for which (under some particular restrictions) it is possible to guarantee the validity of the assertion

$$P \models \varphi \Rightarrow P[a \rightsquigarrow Q] \models \varphi[a \rightsquigarrow Q].$$

Furtheron, a fragment for which the assertion

$$P \models \varphi \Leftarrow P[a \rightsquigarrow Q] \models \varphi[a \rightsquigarrow Q]$$

holds will be presented.

In Section 2 we present a $TCSP$-like process calculus which contains an operator for syntactic action refinement. The concept of logical refinement for the Modal Mu-Calculus is defined in Section 3. Section 4 provides the link between those two refinement concepts and contains the main results as well as their applications to system verification. A summary, a short discussion on related papers and work in progress are given in Section 5. Section 6 contains some standard definitions.

---

[2]The alphabet of a process term P is the set of actions which occur in P.

[3]One argues that this restriction avoids the undesired mixture of different levels of abstraction [11]. Moreover as atomic actions are merely 'action names' it is possible to rename them without losing any generality.

2. SYNTACTIC ACTION REFINEMENT IN THE SYSTEM MODEL

In this section we fix the framework used to model reactive systems. Let $Act := \{a, b, \ldots\}$ be a set of actions and $Var_{Act} := \{v_1, v_2, \ldots\}$ be a set of distinguished action variables[4] such that $Act \cap Var_{Act} = \emptyset$. We let $\alpha, \beta, \gamma, \ldots$ range over the set $\mathcal{A} := Act \cup Var_{Act}$, the elements of which are called *atomic performances* or simply *performances*. Furtheron we let $Idf := \{x, y, \ldots\}$ be a fixed set of *identifiers*. As usual the process expression 0 is used to denote a process that is unable to perform any atomic performance.

Let $R\Delta$ be the language of process terms generated by the grammar

$$Q ::= \alpha \mid (Q + Q) \mid (Q; Q) \mid Q[\alpha \rightsquigarrow Q].$$

Let $R\Sigma$ be the language of process terms generated by the grammar

$$P ::= 0 \mid \alpha \mid x \mid (P + P) \mid (P; P) \mid (P \|_A P) \mid fix(x = P) \mid P[\alpha \rightsquigarrow Q]$$

where $Q \in R\Delta$ and $A \subseteq \mathcal{A}$. Let Σ, Δ be the languages of process expressions generated by the grammars for $R\Sigma$, $R\Delta$ respectively, without the rule $P ::= P[\alpha \rightsquigarrow Q]$.

An identifier x is *guarded* in a term $P \in R\Sigma$ iff each free occurrence of x only occurs in subexpressions F where F lies in a subexpression $(E; F)$ such that $E \notin \surd$, i.e. E can execute a performance (see Appendix A1). A term $P \in R\Sigma$ is called *guarded* iff in each subexpression $fix(x = Q)$ of P the identifier x is guarded in Q. We let $GR\Sigma$ ($G\Sigma$) be the set of all guarded $R\Sigma$ (Σ)-expressions.

The function $\xi : R\Sigma \rightarrow 2^{\mathcal{A}}$ that gives the set of performances occurring in a process expression is defined by induction on the structure of $R\Sigma$, e.g. $\xi(\alpha) := \{\alpha\}$ and $\xi((P_1 \|_A P_2)) := \xi(P_1) \cup \xi(P_2)$. The set of synchronisation performances of a process expression $P \in R\Sigma$ is given by the function $\chi : R\Sigma \rightarrow 2^{\mathcal{A}}$, e.g. $\chi(\alpha) := \emptyset$ and $\chi((P_1 \|_A P_2)) := \chi(P_1) \cup \chi(P_2) \cup A$.

A term $P_1 \in R\Sigma$ is called *ξ-disjoint* from a term $P_2 \in R\Sigma$ iff $\xi(P_1) \cap \xi(P_2) = \emptyset$. A term $P_1 \in R\Sigma$ is called *$\chi\xi$-disjoint*[5] from a term $P_2 \in R\Sigma$ iff $\chi(P_1) \cap \xi(P_2) = \emptyset$.

A term $P_1 \in R\Sigma$ is called *alphabet-disjoint* from a term $P_2 \in R\Sigma$ iff it is ξ-disjoint and $\chi\xi$-disjoint from P_2. A term $P \in R\Sigma$ is called *uniquely synchronized* iff for all terms $(P_1 \|_A P_2)$ that occur in P, $A = \chi(P_i)$ holds for $i = 1, 2$.

To give a meaning to refined terms $P[\alpha \rightsquigarrow Q]$, we make use of a *reduction function red* : $R\Sigma \rightarrow \Sigma$ that removes all occurrences of refinement

[4] Action variables will be used later to define the refinement of formulas (cf. section 3).
[5] Please note, that $\chi\xi$-disjointness is no symmetric relation on the set $R\Sigma$.

operators in a process expression illustrated by Figure 3. The function *red* uses syntactic substitution $\cdot\{Q/\alpha\}$ (see Appendix A2).

Let $P, P_1, P_2 \in R\Sigma$ and $Q \in R\Delta$ be process expressions. The function $red : R\Sigma \to \Sigma$ is defined as follows:

$red(*) := *$ for $* \in \{0\} \cup Idf \cup \mathcal{A}$ $\quad red(fix(x = P)) := fix(x = red(P))$

$red((P_1 \; op \; P_2)) := (red(P_1) \; op \; red(P_2))$ where $op \in \{+, ;, \|_A\}$

$red(P[\alpha \rightsquigarrow Q]) := (red(P))\{red(Q)/\alpha\}$

Figure 3 The Reduction Function

The operational semantics of the language Σ is given as usual (see Appendix A3). The semantics of a process expression P is a *labelled transition system with termination* ($LTST$) i.e. a tuple $\mathcal{T}(P) = (P, \Sigma, \mathcal{A}, \to, \surd)$[6]. Since the terms $P[\alpha \rightsquigarrow Q]$ and $red(P[\alpha \rightsquigarrow Q])$ are supposed to behave identically[7], we define $\mathcal{T}(P) := \mathcal{T}(red(P))$ to supply semantics for terms $P \in R\Sigma$ (cf. [1]). In what follows we sometimes identify the term P with the transition system $\mathcal{T}(P)$ if the context avoids ambiguity. The fact that the parallel operator cannot occur in terms $Q \in R\Delta$ is no severe restriction: due to the absence of the fix operator in all terms $Q \in R\Delta$ any such term induces a *finite state system* whence it is always possible to construct for every term Q in the language[8] $R\Delta^+$ a term $\tilde{Q} \in \Delta$ such that $\mathcal{T}(Q) \equiv_b \mathcal{T}(\tilde{Q})$ ($\equiv_b$ denotes strong bisimulation equivalence). If $Q \in R\Delta^+$ is deadlock free, the meaning of 'replacing' performances α in some term P by Q is the same as for the replacement of every α in P by $\tilde{Q}$. The exclusion of the empty process term 0 means that we disallow 'forgetful refinement'[9]. As the refinement of a (terminating) performance α by some infinite behaviour violates the intuition [12], no expression of the form $fix(x = P)$ is allowed to occur in a term $Q \in R\Delta$. Let $LTST_{fin} := \{\mathcal{T}(P) | P \in R\Sigma$ and the set of reachable states of $\mathcal{T}(P)$ is finite$\}$. The set of reach-

[6] $P \in \Sigma$ is the initial state and $\to \subseteq \Sigma \times \mathcal{A} \times \Sigma$ is the set of transitions, determined via the rules of the operational semantics. $\surd$ is the set of terminated states as defined in the Appendix.

[7] The syntactic difference between the terms $P[\alpha \rightsquigarrow Q]$ and $red(P[\alpha \rightsquigarrow Q])$ is that in the former information about α is scattered over several levels of abstraction whereas only one level of abstraction exists in the latter.

[8] $R\Delta^+$ is the language generated by the grammar for $R\Delta$ augmented with the rule $Q ::= (Q \|_A Q))$.

[9] Such refinements cannot be explained by a change in the level of abstraction [20] and are usually avoided (see e.g. [1]).

able states of $\mathcal{T}(P)$ consists of all states which can be reached from the state $red(P)$.

3. SYNTACTIC ACTION REFINEMENT IN THE SPECIFICATION LOGIC

Let $\mu\mathcal{L}$ be the (negation-free form of the) *Modal Mu-Calculus* [14] generated by the grammar

$$\Phi ::= \top \mid \bot \mid Z \mid (\Phi \wedge \Phi) \mid (\Phi \vee \Phi) \mid [\alpha]\Phi \mid \langle\alpha\rangle\Phi \mid \nu Z.\Phi \mid \mu Z.\Phi$$

where α ranges over $\mathcal{A}$ and Z over a fixed set Var of *variables*. Let $R\mu\mathcal{L}$ be the language generated by the above grammar with the additional rule $\Phi ::= \Phi[\alpha \rightsquigarrow Q]$ where $Q \in R\Delta$. Let $R\mu\mathcal{L}_{\langle\cdot\rangle}$ ($R\mu\mathcal{L}_{[\cdot]}$) be the language generated by the grammar for $R\mu\mathcal{L}$ without the rule $\Phi ::= [\alpha]\Phi$ ($\Phi ::= \langle\alpha\rangle\Phi$ resp.). We let σ range over the set $\{\mu, \nu\}$. A *fixed point formula* has the form $\sigma Z.\varphi$ in which σZ *binds* free occurrences of Z in φ. A variable Z is called *free* iff it is not bound. A $R\mu\mathcal{L}$-formula φ is called *closed* iff every variable Z which occurs in φ is bound. A $R\mu\mathcal{L}$-formula φ is called *guarded* iff every occurrence of a variable Z in φ lies in the scope of a modality $[\alpha]$ or $\langle\alpha\rangle$. For a language $L \subseteq R\mu\mathcal{L}$ we define the closed and guarded fragment of L by $CGL := \{\varphi \in L \mid \varphi \text{ is closed and guarded}\}$. The function $\xi : R\mu\mathcal{L} \to 2^{\mathcal{A}}$ gives the set of performances that occur in a formula. A formula $\varphi \in R\mu\mathcal{L}$ is called *ξ-disjoint* from a term $P \in R\Sigma$ iff $\xi(\varphi) \cap \xi(P) = \emptyset$.

We now give the concept of logical substitution by which we are able to define the reduction of formulas (see Figure 4):

Let $Q, Q_1, Q_2 \in \Delta$ and $\varphi, \varphi_1, \varphi_2 \in \mu\mathcal{L}$ and $n, m \in \mathbb{N}$. The operation of logical substitution, $(\varphi)\{\alpha \rightsquigarrow Q\}$ is defined as follows:

$$(*)\{\alpha \rightsquigarrow Q\} := * \quad \text{if } * \in \{\top, \bot\} \cup Var$$

$$((\varphi_1 \odot \varphi_2))\{\alpha \rightsquigarrow Q\} := ((\varphi_1)\{\alpha \rightsquigarrow Q\} \odot (\varphi_2)\{\alpha \rightsquigarrow Q\}) \quad \text{if } \odot \in \{\wedge, \vee\}$$

$$(\Delta_\beta \varphi)\{\alpha \rightsquigarrow Q\} := \Delta_\beta (\varphi)\{\alpha \rightsquigarrow Q\} \quad \text{if } \alpha \neq \beta$$

$$(\Delta_\alpha \varphi)\{\alpha \rightsquigarrow Q\} :=$$

$$\begin{cases} \Delta_\beta (\varphi)\{\alpha \rightsquigarrow Q\} & \text{if } Q = \beta \\ ((\Delta_{v_n}(\varphi)\{\alpha \rightsquigarrow Q\})\{v_n \rightsquigarrow Q_1\} \wedge (\Delta_{v_m}(\varphi)\{\alpha \rightsquigarrow Q\})\{v_m \rightsquigarrow Q_2\}) & \text{if } Q = (Q_1 + Q_2) \\ (\Delta_{v_n}(\Delta_{v_m}(\varphi)\{\alpha \rightsquigarrow Q\})\{v_m \rightsquigarrow Q_2\})\{v_n \rightsquigarrow Q_1\} & \text{if } Q = (Q_1; Q_2) \end{cases}$$

$$(\sigma Z.\varphi)\{\alpha \rightsquigarrow Q\} := \sigma Z.(\varphi)\{\alpha \rightsquigarrow Q\}$$

where in each clause Δ_γ means throughout either $\langle\gamma\rangle$ or $[\gamma]$ for all γ. For

Let $Q \in R\Delta$ be a process expression and $\varphi, \varphi_1, \varphi_2 \in R\mu\mathcal{L}$ be formulas. We define the logical reduction function $\mathcal{R}ed : R\mu\mathcal{L} \to \mu\mathcal{L}$ as follows:

$$\mathcal{R}ed(*) := * \quad \text{if } * \in \{\top, \bot\} \cup Var \qquad \mathcal{R}ed(\varphi[\alpha \rightsquigarrow Q]) := (\mathcal{R}ed(\varphi))\{\alpha \rightsquigarrow red(Q)\}$$

$$\mathcal{R}ed((\varphi_1 \odot \varphi_2)) := (\mathcal{R}ed(\varphi_1) \odot \mathcal{R}ed(\varphi_2)) \quad \text{if } \odot \in \{\wedge, \vee\}$$

$$\mathcal{R}ed([\beta]\varphi) := [\beta]\mathcal{R}ed(\varphi)\ , \quad \mathcal{R}ed(\langle\beta\rangle\varphi) := \langle\beta\rangle\mathcal{R}ed(\varphi)$$

$$\mathcal{R}ed(\sigma Z.\varphi) := \sigma Z.\mathcal{R}ed(\varphi)$$

Figure 4 The Logical Reduction Function

the introduction of variables v_n, v_m we require $v_n, v_m \notin \xi([\alpha]\varphi) \cup \xi(Q)$ and $n \neq m$ ($v_n, v_m \notin \xi(\langle\alpha\rangle\varphi) \cup \xi(Q)$ respectively).

To supply semantics for refined formulas $\varphi[\alpha \rightsquigarrow Q]$, we extend the standard satisfaction relation (see Sect. 6, Def. A4) which is defined relative to valuation functions[10] $\vartheta : Var \to 2^{R\Sigma}$ by means of the clause $P \models_\vartheta \varphi[\alpha \rightsquigarrow Q]$ iff $P \models_\vartheta \mathcal{R}ed(\varphi[\alpha \rightsquigarrow Q])$. We say P *satisfies* φ (with respect to ϑ) iff $P \models_\vartheta \varphi$. For a closed $R\mu\mathcal{L}$-formula φ we simply write $P \models \varphi$. For a fixed point formula $\sigma Z.\varphi$ we observe that least and greatest fixed points (denoted μ, ν resp.) always exist by the results of [19].

4. SAFETY FOR SYNTACTIC ACTION REFINEMENT AND ABSTRACTION

The first result shows that the concept of simultaneous syntactic action refinement is safe provided a condition of alphabet disjointness is met.

Theorem 1 *Let $P \in GR\Sigma$ be a process term, such that $\mathcal{T}(P) \in LTST_{fin}$ and $\varphi \in CGR\mu\mathcal{L}$ be a formula. Further let $Q \in R\Delta$, such that P is alphabet-disjoint from Q and φ is ξ-disjoint from Q. Then $P \models \varphi \Leftrightarrow P[\alpha \rightsquigarrow Q] \models \varphi[\alpha \rightsquigarrow Q]$.*

Proof Idea:
The proof of Theorem 1 is by induction on the structure of $\varphi \in CG\mu\mathcal{L}$ and a subsidiary induction on the structure of $Q \in R\Delta$. We first observe that $\mathcal{T}(P[\alpha \rightsquigarrow Q])$ is finite. The fact that an approximation over at most

[10]The customary updating notation is used: $\vartheta[\mathcal{E}/Z]$ is the valuation ϑ' which agrees with ϑ on all variables $Z \in Var$ except Z, and $\vartheta'(Z) = \mathcal{E}$.

$|\mathcal{T}_r(P)|_S$ steps [11] suffices to capture the actual fixed point on the system P can thus be used in the induction. In order to show Theorem 1, a serial of lemmata which relate the behaviour of a term P with the behaviour of $P[\alpha \rightsquigarrow Q]$ have to be proved.

Alphabet-disjointness is commonly used in approaches to syntactic action refinement. In order to meet the condition of Theorem 1, renaming of performances can often be applied successfully[12]. However alphabet-disjointness is too strong a condition in a situation as the second one described in the introduction. Alphabet-disjointness is necessary in Theorem 1 as can be seen by the following simple example. Consider the process expression $P = b$ and the formula $\varphi = [c]\langle d\rangle\top$. We have $P \models \varphi$ but $red(P[b \rightsquigarrow c]) \not\models \mathcal{R}ed(\varphi[b \rightsquigarrow c])$. Note that we have $\xi(\varphi) \cap \xi(Q) \neq \emptyset$, i.e. φ is not ξ-disjoint from Q. It is clear that dropping the condition of alphabet disjointness of the process terms P and Q only makes sense in conjunction with abandoning the ξ-disjointness of φ from Q. Dropping the latter restriction however leads to fundamental problems: Without it, repeated SSAR can transform an originally satisfiable formula into an unsatisfiable one[13]. The above examples show that we cannot hope for a result like Theorem 1 for any fragment $L \subseteq R\mu\mathcal{L}$ in which it is allowed to compose formulas $\varphi \in L$ containing both types of modalities, i.e. $\langle\alpha\rangle$ and $[\alpha]$ without accepting any restrictions on alphabet disjointness. This is the reason why we consider the logics $R\mu\mathcal{L}_{\langle\cdot\rangle}$ and $R\mu\mathcal{L}_{[\cdot]}$ where only one modality type can occur in the formulas. The logic $R\mu\mathcal{L}_{[\cdot]}$ can be used to express interesting properties of reactive systems, like e.g. the unless-property 'φ remains true in every computation unless ψ holds' or safety properties such as 'φ never holds again whenever ψ has become true'. Moreover, $R\mu\mathcal{L}_{[\cdot]}$ can be used to express liveness-properties under fairness and cyclic-properties (see [18]).

$R\mu\mathcal{L}_{\langle\cdot\rangle}$-formulas can be used to formalize properties like e.g. 'there exists a computation sequence of P in which φ holds infinitely often' or 'there exists a computation sequence of P along which φ is always attainable.' With techniques similar to those applied for Theorem 1 we obtain the following.

Theorem 2 *Let $P \in GR\Sigma$ be a uniquely synchronized process term, such that $\mathcal{T}(P) \in LTST_{fin}$ and $\varphi \in CGR\mu\mathcal{L}_{\langle\cdot\rangle}$ be a formula. Further let $Q \in R\Delta$, such that P is $\chi\xi$-disjoint from Q. If $(\alpha \notin \chi(P) \vee \xi(\varphi) \subseteq \chi(P))$ then $P \models \varphi \Rightarrow P[\alpha \rightsquigarrow Q] \models \varphi[\alpha \rightsquigarrow Q]$.*

---

[11] $|\mathcal{T}(P)|_S$ denotes the number of states of $\mathcal{T}(P)$.

[12] Cf. the assembly line example of the introduction where no renaming was necessary.

[13] Let $\varphi = (\langle\alpha_1\rangle\top \wedge [\alpha_2]\bot)$. Clearly, φ is satisfiable. On the other hand, the formula $\varphi[\alpha_1 \rightsquigarrow \alpha][\alpha_2 \rightsquigarrow \alpha]$ is not satisfiable.

Theorem 3 *Let $P \in GR\Sigma$ be a uniquely synchronized process term, such that $\mathcal{T}(P) \in LTST_{fin}$ and $\varphi \in CGR\mu\mathcal{L}_{[.]}$ be a formula. Further let $Q \in R\Delta$, such that P is $\chi\xi$-disjoint from Q. If $(\alpha \notin \chi(P) \vee \xi(\varphi) \subseteq \chi(P))$ then $P \models \varphi \Leftarrow P[\alpha \rightsquigarrow Q] \models \varphi[\alpha \rightsquigarrow Q]$.*

As an example, we consider a ('lock-step') solution of a two process mutual exclusion problem.

Example: The process terms P_α and P_β are given by □

$$fix(x = (((\alpha_1; (\alpha; \alpha_2)) + (\beta_1; (\beta; \beta_2))); x))$$

and

$$fix(y = (((\beta_1; (\beta; \beta_2)) + (\alpha_1; (\alpha; \alpha_2))); y))$$

respectively. It is easy to see that the process $ME_{\alpha\beta} = (P_\alpha \|_{\{\alpha_i, \beta_i, \alpha, \beta\}} P_\beta)$ where $i = 1, 2$ enters the (abstract) *critical sections* α and β in mutual exclusion. For $A \subseteq \mathcal{A}$ we let $\diamond_A \varphi$ abbreviate the formula $\bigvee_{\alpha \in A} \langle\alpha\rangle\varphi$ and $\Box_A \varphi$ abrreviate the formula $\bigwedge_{\alpha \in A} [\alpha]\varphi$. We have that

$$ME_{\alpha\beta} \models \varphi = \nu Y.(\mu Z.(\langle\alpha\rangle\top \vee \diamond_{\xi(ME_{\alpha\beta})} Z) \wedge \diamond_{\xi(ME_{\alpha\beta})} Y),$$

i.e. there exists a computation sequence of $ME_{\alpha\beta}$ along which it is always possible to reach a state where the performance α can be executed. Now let $Q = (\gamma_1 + \gamma_2)$. Then we have that

$$ME_{\alpha\beta}[\alpha \rightsquigarrow Q][\beta \rightsquigarrow Q] \models \phi[\alpha \rightsquigarrow Q][\beta \rightsquigarrow Q]$$

via Theorem 2 which says that $ME_{\alpha\beta}[\alpha \rightsquigarrow Q][\beta \rightsquigarrow Q]$ can execute a computation sequence along which it is always possible to reach a state where all performances of Q can be executed. This holds, though another performance (β in our case) is also refined by Q. □

Theorem 2 and Theorem 3 can both be established without the need to consider restrictions of alphabet disjointness whence we might use the two logics $CGR\mu\mathcal{L}_{\langle.\rangle}$ and $CGR\mu\mathcal{L}_{[.]}$ in the stepwise development of systems as demonstrated in the introduction.

Amongst the applications of Theorem 1 to the verification of reactive systems (as described in the introduction), the concept of a priori-verification might be the most interesting one. There, every property φ of a system P which can be expressed in $R\mu\mathcal{L}$ 'carries over' (in its refined form $\varphi[\alpha \rightsquigarrow Q]$) to the refined system $P[\alpha \rightsquigarrow Q]$, in particular this holds for all safety and liveness properties. Strong safety properties (which involve the box modality $[\alpha]$) might not be carried over (in the above sense) when dropping the restriction of alphabet disjointness. Theorem 2 shows however, that it is still possible to verify systems 'a priori' with respect to properties expressible in the logic $R\mu\mathcal{L}_{\langle\cdot\rangle}$, without the need to ensure alphabet disjointness of the considered process terms and formulas. In essence, these properties are 'existential' properties e.g. weak safety and liveness properties (according to [18]).

In its contrapositive form, Theorem 3 can be used to 'debug' a (concrete) reactive system by means of debugging an abstract system (where the abstraction is based on syntactic action refinement between those systems): If $P \not\models \varphi \in R\mu\mathcal{L}_{[\cdot]}$ then $P[\alpha \rightsquigarrow Q] \not\models \varphi[\alpha \rightsquigarrow Q]$. In the case we cannot disprove φ for P, no information about satisfaction of $\varphi[\alpha \rightsquigarrow Q]$ by $P[\alpha \rightsquigarrow Q]$ can be inferred. However, Theorem 3 might also be used to support model checking techniques for systems that otherwise would remain unfeasable due to the size of their state spaces: If P is such a system then no information at all about satisfaction with respect to any property φ can be established by means of model checking techniques. In particular we could not show $P \not\models \varphi$. However, if we could establish appropriate abstractions P_s and φ_s, i.e. $P = P_s[\alpha_1 \rightsquigarrow Q_1]\dots[\alpha_n \rightsquigarrow Q_n]$ and $\varphi = \varphi_s[\alpha_1 \rightsquigarrow Q_1]\dots[\alpha_n \rightsquigarrow Q_n]$ then P_s might well become manageable by a model checker since the state space of P_s might be exponentially smaller then the state space of P due to the well known state explosion problem[14]. Then we could apply the model checker to prove $P_s \not\models \varphi_s$ and conclude $P \not\models \varphi$ via Theorem 3. Various interesting properties can be expressed in $R\mu\mathcal{L}_{[\cdot]}$ (and therefore be used in the debug-procedure described above), in particular strong safety properties of a system P like e.g. $\nu Z.(\varphi \wedge \Box_{\xi(P)} Z)$ meaning 'φ holds in every state of P'.

5. CONCLUSION AND RELATED WORK

We showed that it is possible to establish the validity of the assertion $P \models \varphi$ iff $P[a \rightsquigarrow Q] \models \varphi[a \rightsquigarrow Q]$ $(*)$ where P and Q are terms from appropriate *TCSP*-like languages, φ is a *Modal Mu-Calculus*-formula and $\cdot[a \rightsquigarrow Q]$ denotes a refinement operator both on process terms and formulas. As we explained in the introduction, assertion $(*)$ can be used to simplify the verification task. Furthermore, the logical meaning of syntactic action refinement on process terms becomes evident. Most interestingly however is the method of a priori verification supplied by assertion $(*)$: Provided we have $P \models \varphi$, the specification refinement $\varphi[a \rightsquigarrow Q]$ automatically yields an a priori correct process $P[a \rightsquigarrow Q]$ via assertion $(*)$, i.e. $P[a \rightsquigarrow Q] \models \varphi[a \rightsquigarrow Q]$.

[14] A linear reduction of the number of performances in a term $P \in R\Sigma$ can entail an exponential reduction of the number of reachable states of $\mathcal{T}(P)$. Please note that model checking algorithms are based on the investigation of the involved system state spaces, regardless whether those spaces are represented explicitly or implicitly (e.g. using *BDDs* [3]).

We demonstrated that assertion (*) does not hold in general but under the condition of alphabet-disjointness. We showed the validity of the assertions $P \models \varphi \Rightarrow P[\alpha \rightsquigarrow Q] \models \varphi[\alpha \rightsquigarrow Q]$ (†) and $P \models \varphi \Leftarrow P[\alpha \rightsquigarrow Q] \models \varphi[\alpha \rightsquigarrow Q]$ (‡) where $P \in GR\Sigma$ is uniquely synchronized and $\varphi \in CGR\mu\mathcal{L}_{\langle . \rangle}$ in assertion (†), $\varphi \in CGR\mu\mathcal{L}_{[.]}$ in assertion (‡). The important point is that the restriction of alphabet disjointness is not necessary to prove the correctness of the two assertions above. Consequently, assertion (*), (†) and (‡) show that the expressive power of the specification logic determines the applicability of SSAR to the verification of reactive systems. A future topic of research is to further investigate this 'expressiveness-applicability' tradeoff.

We explained that SSAR can also be conceived as a concept of abstraction simplifying the verification of reactive systems. In this sense our approach is related to the method of *abstract interpretations* [6]. There, one can conclude that properties hold for a (concret) system $\mathcal{C}$ if they hold for a (abstract) system $\mathcal{A}$ where the state spaces of $\mathcal{A}$ and $\mathcal{C}$ are related with each other via appropriate abstractions thereby reducing the costs of *model checking* techniques (see e.g. [4, 8]). There however, the formula under consideration remains 'unchanged' whereas it is changed (through refinement) in our approach. In [13], *semantic action refinement* for *synchronisation structures* is used to induce refinement functions for formulas of the considered specification logic whereas we use *syntactic action refinement* for process terms and formulas. In the above paper a *linear time* logic is used whereas our approach is based on the *branching time* Modal Mu-Calculus.

The concept of syntactic action refinement can easily be used by a system developer and the transition system semantics used is closely related to the well known concept of automata. The Modal Mu-Calculus is often conceived as the generic process-logic (see e.g. [8]). We thus believe that our framework satisfies the requirement that formal methods should be easy to comprehend to be of any use in practice [5].

Work is in progress that extends the above results: We study the consequences of introducing the 'hiding'-operator (see e.g. [17]) to the process algebra $R\Sigma$. Further, we investigate how the reduction of formulas can be determined efficiently. Investigations concerning infinite state systems are a future topic as well as investigations to what extend the presented abstraction technique can be automatized. We are currently performing some case studies to determine the practical applicability of our results.

6. AUXILIARY DEFINITIONS

A1 Terminated States:

To evaluate the semantics of the operator ';' it is common to use a special predicate $\surd$: Let $\surd \subseteq R\Sigma$ be the least set which contains the term 0 and is closed under the rules $(P_1 \in \surd \wedge P_2 \in \surd) \Rightarrow (P_1\ op\ P_2) \in \surd$ where $op \in \{\|_A, +, ;\}$ and $(P \in \surd) \Rightarrow fix(x = P) \in \surd$ and $(P \in \surd) \Rightarrow P[\alpha \rightsquigarrow Q] \in \surd$.

A2 Syntactic Substitution:

Let $P, P_1, P_2 \in \Sigma$ and $Q \in \Delta$ be process expressions. Syntactic substitution, denoted $(P)\{Q/\alpha\}$ is defined as follows:

$$(*)\{Q/\alpha\} := * \text{ where } * \in \{0\} \cup Idf \qquad (\alpha)\{Q/\beta\} := \begin{cases} Q & \text{if } \alpha = \beta \\ \alpha & \text{otherwise} \end{cases}$$

$$((P_1\ op\ P_2))\{Q/\alpha\} := ((P_1)\{Q/\alpha\}\ op\ (P_2)\{Q/\alpha\}) \text{ where } op \in \{+, ;\}$$

$$((P_1 \|_A P_2))\{Q/\alpha\} := \begin{cases} ((P_1)\{Q/\alpha\} \|_{A\setminus\{\alpha\}\cup\xi(Q)} (P_2)\{Q/\alpha\}) & \text{if } \alpha \in A \\ ((P_1)\{Q/\alpha\} \|_A (P_2)\{Q/\alpha\}) & \text{if } \alpha \notin A \end{cases}$$

$$(fix(x = P))\{Q/\alpha\} := fix(x = P\{Q/\alpha\})$$

A3 Operational Semantics:

Let $P, Q \in \Sigma$ be process expressions.

$$\frac{}{\alpha \xrightarrow{\alpha} 0} \qquad \frac{P \xrightarrow{\alpha} P'}{(P+Q) \xrightarrow{\alpha} P'} \qquad \frac{Q \xrightarrow{\alpha} Q'}{(P+Q) \xrightarrow{\alpha} Q'}$$

$$\frac{Q \xrightarrow{\alpha} Q'}{(P;Q) \xrightarrow{\alpha} Q'} \text{ if } P \in \surd \qquad \frac{P \xrightarrow{\alpha} P'}{(P;Q) \xrightarrow{\alpha} (P';Q)}$$

$$\frac{P \xrightarrow{\alpha} P'}{(P\|_A Q) \xrightarrow{\alpha} (P'\|_A Q)} \text{ if } \alpha \notin A \qquad \frac{Q \xrightarrow{\alpha} Q'}{(P\|_A Q) \xrightarrow{\alpha} (P\|_A Q')} \text{ if } \alpha \notin A$$

$$\frac{P[fix(x=P)/x] \xrightarrow{\alpha} Q}{fix(x=P) \xrightarrow{\alpha} Q} \qquad \frac{P \xrightarrow{\alpha} P' \quad Q \xrightarrow{\alpha} Q'}{(P\|_A Q) \xrightarrow{\alpha} (P'\|_A Q')} \text{ if } \alpha \in A$$

A4 Extended Satisfaction Relation:

Let $P \in R\Sigma$, $Q \in R\Delta$, $\varphi, \psi \in R\mu\mathcal{L}$ and $Z \in Var$.

$$P \models_\vartheta \top, \qquad P \not\models_\vartheta \bot\ , \qquad P \models_\vartheta Z \text{ iff } P \in \vartheta(Z)$$

$P \models_\vartheta (\varphi \wedge \psi)$ iff $P \models_\vartheta \varphi$ and $P \models_\vartheta \psi$

$P \models_\vartheta (\varphi \vee \psi)$ iff $P \models_\vartheta \varphi$ or $P \models_\vartheta \psi$

$P \models_\vartheta [\alpha]\varphi$ iff $P \in \{E \in R\Sigma | \forall E' \in R\Sigma(E \xrightarrow{\alpha} E' \Rightarrow E' \models_\vartheta \varphi)\}$

$P \models_\vartheta \langle\alpha\rangle\varphi$ iff $P \in \{E \in R\Sigma | \exists E' \in R\Sigma(E \xrightarrow{\alpha} E'$ and $E' \models_\vartheta \varphi)\}$

$P \models_\vartheta \mu Z.\varphi$ iff $P \in \bigcap \{\mathcal{E} \subseteq R\Sigma | \{E \in R\Sigma | E \models_{\vartheta[\mathcal{E}/Z]} \varphi\} \subseteq \mathcal{E}\}$

$P \models_\vartheta \nu Z.\varphi$ iff $P \in \bigcup \{\mathcal{E} \subseteq R\Sigma | \mathcal{E} \subseteq \{E \in R\Sigma | E \models_{\vartheta[\mathcal{E}/Z]} \varphi\}\}$

$P \models_\vartheta \varphi[\alpha \rightsquigarrow Q]$ iff $P \models_\vartheta \mathcal{R}ed(\varphi[\alpha \rightsquigarrow Q])$

References

[1] Aceto, L. and Hennessy, M. (1991). Adding action refinement to a finite process algebra. *Lecture Notes in Computer Science*, 510:506–519.

[2] Brookes, S. D., Hoare, C. A. R., and Roscoe, A. W. (1984). A theory of communicating sequential processes. *Journal of the ACM*, 31(3):560–599.

[3] Bryant, R. (1986). Graph-Based Algorithms for Boolean Function Manipulation. *IEEE Transactions on Computers*, C-35(8):677–691.

[4] Clarke, E., Grumberg, D., and Long, D. (1994). Model Checking and Abstraction. *ACM Transactions on Programming Languages and Systems*, 16(5):1512–1542.

[5] Clarke, E. M. and Wing, J. M. (1996). Formal methods: State of the art and future directions. Technical Report CMU-CS-96-178. ftp://reports.adm.cs.cmu.edu/usr/anon/1996/CMU-CS-96-178.ps.

[6] Cousot, P. and Cousot, R. (1992). Abstract interpretation frameworks. *Journal of Logic and Computation*, 2(4):511–547.

[7] Dam, M. (1994). CTL* and ECTL* as fragments of the modal μ-calculus. *Theoretical Computer Science*, 126(1):77–96.

[8] Dams, D., Gerth, R., and Grumberg, O. (1997). Abstract interpretation of reactive systems. *ACM Transactions on Programming Languages and Systems*, 19(2):253–291.

[9] Emerson, E. A. (1990). Temporal and Modal Logic. In J. van Leeuwen, editor, *Handbook of Theoretical Computer Science*, volume B, pages 996–1072, Amsterdam. Elsevier Science Publishers.

[10] Emerson, E. A. and Lei, C. L. (1986). Efficient model checking in fragments of the propositional μ-calculus. In *Symposium on Logic in Computer Science (LICS '86)*, pages 267–278, Washington, D.C., USA. IEEE Computer Society Press.

[11] Goltz, U., Gorrieri, R., and Rensink, A. (1994). On syntactic and semantic action refinement. *Lecture Notes in Computer Science*, 789:385–404.

[12] Gorrieri, R. and Rensink, A. (1999). Action refinement. Technical Report UBLCS-99-9, University of Bologna (Italy). Department of Computer Science.

[13] Huhn, M. (1996). Action refinement and property inheritance in systems of sequential agents. In Montanari, U. and Sassone, V., editors, *CONCUR '96: Concurrency Theory, 7th International Conference*, volume 1119 of *Lecture Notes in Computer Science*, pages 639–654, Pisa, Italy. Springer-Verlag.

[14] Kozen, D. (1983). Results on the propositional mu -calculus. *Theoretical Computer Science*, 27(3):333–354.

[15] Majster-Cederbaum, M. and Salger, F. (1999). A verification technique based on syntactic action refinement in a TCSP-like process algebra and the Hennessy-Milner-Logic. In *Advances in Computing Science-ASIAN'99*, volume 1742 of *LNCS*, pages 379–380. Springer.

[16] Milner, R. (1980). *A Calculus of Communicating Systems.* Springer, Berlin, 1 edition.

[17] Olderog, R. (1986). TCSP: Theory of communicating sequential processes. In *Advances in Petri Nets 1987, ed. Grzegorz Rozenberg, LNCS 266; Petri Nets: Central Models and Their Properties, Advances in Petri Nets 1986, LNCS 254-255, 1987, LNCS 188 (1984), LNCS 340 (1988), LNCS 483 (1991).*

[18] Stirling, C. (1996). Modal and temporal logics for processes. *Lecture Notes in Computer Science*, 1043:149–237.

[19] Tarski, A. (1955). A lattice-theoretic fixpoint theorem and its applications. *Pacific Journal of Mathematics*, 5:285–309.

[20] van Glabbeek, R. and Goltz, U. (1989). Equivalence notions for concurrent systems and refinement of actions. In Kreczmar, A. and Mirkowska, G., editors, *Proceedings of the Conference on Mathematical Foundations of Computer Science*, volume 379 of *LNCS*, pages 237–248, Berlin. Springer.

Part Two

TEST GENERATION

FROM RULE-BASED TO AUTOMATA-BASED TESTING

Kousha Etessami, Mihalis Yannakakis

Bell Labs

Murray Hill, NJ 07974

kousha@research.bell-labs.com, mihalis@research.bell-labs.com

Abstract

Rule based languages have been used extensively to provide a declarative description of causal relationships between events and conditions in a wide variety of systems. On the other hand, automata (finite state machine) based models provide a detailed operational specification of how the system state evolves over time. While rules are a convenient declarative mechanism, a finite automaton is more flexible and far more easily analyzed than a collection of rules with potentially complicated semantics.

In this paper, we address this dichotomy in the context of large scale testing. We describe a (probabilistic) rule based system, which has been developed and used over many years as the testing model of a large telephone switch. One important drawback of the rule-based system is that it is inflexible in terms of the test generation mechanisms it supports, essentially only allowing the generation of random tests.

We have designed and implemented a translation from the rule-based language to a probabilistic finite automaton model, and applied it to the switch model. Our translation makes use of several automata-theoretic algorithms to keep the size of the resulting automata manageable.

The automaton model allows much more flexible and targeted test generation. We have designed and implemented an assortment of targeted test generation algorithms which are applied to the probabilistic automaton.

Keywords: System testing, rule based systems, finite automata.

1. INTRODUCTION

For large complex software systems, ad-hoc testing methods can not be trusted to assure adequate quality. In order to test a system rigorously and systematically, it is necessary to have some kind of model which reflects, if not the structure of the system, at least the legiti-

Tommaso Bolognesi and Diego Latella (Eds.), *Formal Methods for Distributed System Development.*

mate behavior expected of the system. A notable example of a large complex software system is a telephone switch along with the apparatus surrounding it. The code for these systems typically reaches well into millions of lines, with many interacting components.

Finite state machines (of various types) have often been used to model communication protocols and features for testing and verification purposes ([3, 4, 9, 5, 11]), and commercial testing tools based on them are already in the marketplace. A different kind of modeling mechanism, which has seen widespread use in other areas, is a *rule-based* model. Sets of "if ... then ..." rules are used to specify that if a certain combination of facts/events/actions hold/occur, then a certain other combination of facts/events/actions must also hold/occur. Depending on the context, rules are interpreted either in a passive manner as constraints (properties) of the system, or in an active manner, as directions for action. Rules are used for example in active databases (see, e.g., [12], for a description of a number of such systems), in expert systems and knowledge bases ([10]), in workflows ([7]), in policy description languages in various contexts (firewalls, routing, network management), requirement properties (for example, telephony feature properties ([6])). Rules can provide a high level description of the causal relationships that must hold in a system in a declarative and localized manner. This often makes them very convenient for users, who only need to isolate individual causal relationships, rather than describe the global behavior of the system in a monolithic fashion, as would be done by, say, a global finite state machine (automaton) model.

However, the descriptive convenience gained by using rule bases comes at a cost: the global behavior of the interacting rules is difficult to predict or to analyze automatically. In the context of testing, this translates to the fact that it is difficult to generate flexible and targeted tests with desired properties based on the constraints specified by the rules.

This is where our work enters the picture, by providing a bridge between the rule-based and automata-based worlds for large scale testing, applied to a real world commercial system. The work was motivated by the testing of a large telephone switch. The switch was developed and has been evolving since the early 1980's, has millions of lines of code, and supports over 400 features. In order to test the switch, in the early 90's the testers developed a rule-based modeling language and tool called TGAP (for Test Generation and Analysis Platform). The TGAP system and the test model for the switch have continued to evolve and grow over the years. The TGAP model is a high level, functional model of the whole system which reflects the expected behavior of the system from the point of view of various participants in communication scenarios, as well as the underlying provisioning information (types of trunks, devices etc.). Participants include, for example, a caller, a callee, an operator, and so on, as well as administrative and setup operations.

The specification mechanisms of TGAP, in rough outline, consist of (hierarchical, probabilistic) *graphs* for all the participants, and of a large set of rules (in the form of "if .. then .." conditions) that prescribe precisely how behaviors of one user restrict those of another in valid phone scenarios. The detailed syntax and semantics of the TGAP model will be provided in the next section. These graphs plus rules together define the set of all legal scenarios in the system. The TGAP tool uses the probabilities provided to take a random walk on the graphs, eliminating illegal scenarios and backtracking until it finds a legal scenario, which it then outputs. This process is repeated as often as necessary in order to obtain a sufficient number of test scenarios. The scenarios are then translated into a home-grown test-execution language, via another tool, the *Converter*, that we do not address in this paper; the outcome of the Converter describes in detail a sequence of actions that are to be performed on an actual phone system and their expected outcomes.

As it stands, the testing platform suffers from several weaknesses stemming from its rule based design. The first of these is that, the only way to access the information in the test model is to generate a random scenario via the random walk outlined above. This limits significantly the tester's ability to perform *targeted* tests that exercise particular features of the system. Since exhaustive testing is impossible given the enormous number of possible scenarios (in fact, infinitely many when cycles are allowed in the graphs), testers need to be selective, using their knowledge to focus on important and/or problematic components. Automated analysis of the behavioral model can also help in designing tests that meet particular coverage criteria.

We describe a translation from the rule-based system to a probabilistic finite state automaton model. Once we established that such a translation is possible even in principle, it became clear that because of the large number of interacting rules in the TGAP model for the switch, there was a great potential for state explosion. The biggest challenge we faced in our translation was to keep the size of the resulting automata manageable. We designed a new incremental scheme in which we alternate between building up the automaton and doing minimization. Along the way, we made interesting use of several automata theoretic algorithms. The result is that the translation has already been used to successfully translate the model for the switch into a probabilistic finite state machine that is feasible, only 230,000 states and 480,000 transitions.

Once an automaton for the test scenarios became available, it enabled many forms of analysis and targeted test generation. We developed a host of algorithms to apply to the probabilistic automaton for a number of tasks such as: Generate the n most probable scenarios, to optimize coverage for a given number of scenarios. Compute statistics on the profile of the model and its coverage, for example count the number of possible scenarios Restrict the model to a subsystem or feature of interest specified in terms of keywords or patterns, and compute targeted scenarios that optimize desired test criteria. These algorithms are critical in order for testers to be able to optimize and target their tests to get adequate coverage and achieve desired testing goals.

2. SYNTAX AND SEMANTICS OF TGAP

As described in the introduction, a TGAP model consists of hierarchical *graphs* and *rules.* There is a set of users, such as caller, callee, operator, administrators etc. Each user has an associated user type, and for each type a hierarchical graph captures the possible behaviors for that type of user. The nodes of the graph are annotated with *keywords* that describe at a high level the events at the user. Rules dictate how behaviors of one user in a call constrain the behaviors of other users. They also dictate how call provisioning constraints, such as the kind of phone device a user has, limit the features accessible to that user.

The aim of graphs + rules is to define a set of valid *scenarios.* Informally, a scenario is a description of the precise sequence of events that occurs at each user in one use of the system, along with peripheral information such as what phone devices each user is using. Formally, a scenario is represented as a sequence of keywords from an alphabet Σ, separated by the special delimiter keyword "$\longrightarrow$", i.e., a string from $(\longrightarrow (\Sigma^+))^+$. The sequence immediately after each "$\longrightarrow$" represents the behavior witnessed by one user in the scenario, and is called a *transaction.* Note that in this formalism, the executions of different users are listed sequentially one after the other, rather than in the usual interleaved time order. One consequence of this is a succinctness in representation: a scenario represents really a potentially exponential set of possible interleavings of the individual user transactions. As we mentioned in the introduction, a TGAP scenario is processed by a Converter, which interleaves the transactions in some order (and fills in the details).

Now we describe the syntax of graphs and rules, and we then put the two together to describe how their combined semantics defines a set of valid scenarios.

Hierarchical graphs

Given an alphabet Σ of *keywords*, a TGAP (hierarchical) *graph* over Σ is a tuple $H = (G_1, G_2, \ldots, G_n)$ of *pure graphs*, where each pure graph, $G_i \doteq \langle W_i = V_i \cup \{g_1^i, \ldots, g_{m_i}^i\}, E_i, l_i, b_i, p_i, s_i \rangle$, consists of

1 A set of nodes W_i, partitioned into ordinary nodes V_i, and some special graph nodes $\{g_1^i, \ldots, g_{m_i}^i\}$.

2 An edge relation $E_i \subseteq V_i \times W_i$, forms a directed graph on $W_i \times W_i$, where the special graph nodes $\{g_1^i, \ldots, g_{m_i}^i\}$ are sinks (i.e., have out-degree 0). There is a single source node, s_i (i.e., a node with no incoming edge), which is the start node of the graph G_i.

3 A labeling function $l : V_i \rightarrow \Sigma$ that assigns a keyword to each ordinary node.

4 A labeling $b_i : \{g_1^i, \ldots, g_{m_i}^i\} \mapsto \{i+1, \ldots, n\}$ that assigns to each special graph node in G_i an index $j > i$ for another pure graph G_j. Reuse is allowed, i.e., many special nodes may be mapped to the same graph G_j.

5 A probability function $p_i : E_i \rightarrow [0, 1]$ that assigns a probability to each edge such that for each non-sink $v \in V_i$

$$\sum_{\{(v,w) | (v,w) \in E_i\}} p((v, w)) = 1 \qquad (1)$$

Now, each graph can be converted to an ordinary directed graph using the following *flattening* process:

Let $G'_n := G_n$.
For $i = (n-1)$ down to 1 do:
For each edge $(v, g_k^i) \in E_i$, let $j = b_i(g_k^i)$, and substitute $(v, g_k^i) \in E_i$ by the edge (v, s'_j), where s'_j is the start node of G'_j. Let G'_i be the resulting flattened graph.

Note that the final result, G'_1, is a labeled flat graph with consistent node probabilities, meaning equation 1 holds for each non-sink node v. Note also that in this case the flattening to a directed graph does not blow up the size, i.e., the size of the final flat graph G'_1 is linear in the size of the hierarchical graph (the sum of the sizes of the pure graphs G_i). This is due to the limited use that is made of the hierarchy, since only sinks can be expanded with other graphs. Conciseness over graphs would be gained if internal nodes, rather than sinks, were also allowed to be expanded (see, e.g., ([1])). The main reason for using hierarchy in the TGAP platform is that it allows the graphs to be presented with GUIs in a clean partitioned way. We remark that in the tool's terminology, the hierarchical graph is termed a *tree* (although cycles are allowed in the tool and 'tree' is defined in the man pages as a 'cyclic directed graph'), partly because for the main switch system which has been modeled the hierarchical graph is actually a hierarchical tree without cycles. (For trees, hierarchy even only at sinks gives an exponential succinctness over flat trees, though they are equivalent in expressiveness to directed acyclic graphs.) We take some advantage of the acyclicity in our implementation, but our algorithms remain valid or can easily be modified to the more general cyclic setting.

Given a flattened graph, a sequence of labels along a path from start to sink node will be called a *transaction*. *Scenarios* will be built up out of transactions from different user graphs as prescribed by the rules.

Rules

TGAP rules are quite complicated. Rather than provide the formal syntax of all rules, we will describe the important details of all essential rules. The rules specify how to build a scenario: how to start a scenario, i.e., with which user and what conditions on the first transaction; how to continue a partial scenario, i.e., conditions on the next user and the next transaction depending on the history; and when to terminate a scenario, i.e., conditions on whether a scenario is complete or has to continue. There are five essential types of rules:

1. START_USER: This rule describes which user starts the scenario, with given probabilities. There is only one such rule. It takes the form: START_USER { user_1 prob_1 ... user_k prob_k }

2. FIRST_TRANS: These rules describe constraints, given by right-patterns, on what may appear in the first transaction of a scenario. The format of the rule is: FIRST_TRANS right-pattern

3. MORE_TRANS: The format is: left-pattern MORE_TRANS
The meaning of the rule is "Any partial scenario which satisfies the left-pattern is not complete, i.e., more transactions are required in this scenario."

4. NEXT_USER: left-pattern NEXT_USER { user_1 prob_1 ... user_k prob_k }. This rule means that, for those partial scenarios satisfying the left hand side, and requiring more transactions, the next transaction must come from the given user types user_i, with the given probabilities prob_i, respectively.

5. NEXT_TRANS: left-pattern NEXT_TRANS right-pattern
The meaning of these rules is that, if a partial scenario satisfies the left

side, and requires more transactions, then the next transaction must satisfy the right-pattern. When a partial scenario satisfies the left side of several NEXT_TRANS rules, all of these rules apply, in effect creating a conjunction of the right hand sides.

We will not formally define the format of left-patterns and right-patterns. They are particular forms of regular expressions, augmented with extra expressive features such as specifying the number of transactions, negations, and so on.

There is another rule which allows us to define aliases to denote any of a number of keywords: aliasname ALIAS keyword$_1$ ... keyword$_k$. Thus, whenever the word aliasname appears, it is to be interpreted as a disjunction $\bigvee_i$ keyword$_i$. Aliases can be defined in terms of prior aliases.

How graphs and rules are put together

As we have seen, a TGAP model is given by a sequence $\mathcal{H} = \langle H_1 \dots H_r \rangle$ of hierarchical graphs, one corresponding to each user type, as well as a sequence $\mathcal{R} = \langle R_1, \dots R_d \rangle$ of rules. For the rest of this section, view the hierarchical graphs in *flattened* form, $H_i^!$. In order to simplify our treatment here, we will first ignore probabilities, and only consider a scenario valid if it can be generated by the model $(\mathcal{H}, \mathcal{R})$ with non-zero probability.

Given $(\mathcal{H}, \mathcal{R})$, a string $w \in (\rightarrow \Sigma^+)^+$ consisting of a sequence of transactions $w_1 \dots w_m$, where each $w_i \in (\rightarrow \Sigma^+)$, is called a *valid scenario* if it satisfies the following criteria:

1. The first transaction w_1 in w must exist as a source-sink path with non-zero probability in the graph H_i, such that the user type user$_i$ is selected with non-zero probability in the START_USER rule.
2. w_1 must also satisfy the right-patterns given by all FIRST_TRANS rules.
3. For $i < m$, $w_1 \dots w_i$ must satisfy some MORE_TRANS rule's left-pattern.
4. For $i < m$, w_{i+1} must exist as a source-sink path in a graph H_i such that user$_i$ is selected with non-zero probability in the *first* NEXT_USER rule in $\mathcal{R}$ whose left-pattern is satisfied by $w_1 \dots w_i$.
5. For $i < m$, all the NEXT_TRANS rules whose left-patterns are satisfied by $w_1 \dots w_i$, must have right-patterns that are satisfied by w_{i+1}.
6. There must be no MORE_TRANS rule whose left-pattern is satisfied by w.

What is needed in order to complete the above description is to describe how probabilities affect the generation of partial scenarios along the way. Although this is not difficult to do, the details of such a description would take us too far afield, so we only mention that one can proceed in a depth-first fashion to construct the scenario, picking children according to edge probabilities, and backtracking when a rule is violated.

3. (GRAPHS + RULES) → PROBABILISTIC AUTOMATA

Although the graphs + rules model is complex and has many details, it is no more powerful in expressive power than a probabilistic finite state machine (automaton) model. The following theorem establishes this for finite automata, again ignoring probabilities for now for simplicity.

Theorem 1 *For a TGAP model $\mathcal{M} = (\mathcal{H}, \mathcal{R})$, the set of valid scenarios consistent with $\mathcal{M}$ is a regular language. That is, we can construct a finite automaton $A_{\mathcal{M}}$ that accepts exactly the scenarios generated by $\mathcal{M}$.*

Proof. We only sketch the proof. The basic idea is that, since left-patterns and right-patterns in rules are given by variants of regular expressions, each rule can be represented by one or two deterministic finite automata, depending on whether the rule has both a left and a right side, and the entire model can be represented as a kind of "product automaton" of all the rule automata along with the user graphs.

In more detail, consider for example a NEXT_TRANS rule

R =*left-pattern* NEXT_TRANS *right-pattern*

The rule can be represented by two coupled automata A_{left} and A_{right}, where A_{left} monitors the input, and goes to a special state when the *left-pattern* is satisfied. A_{right}, on the other hand, monitors the state of A_{left} and begins processing *right-pattern* as soon as A_{left} enters the special state.

The automaton $A_{\mathcal{M}}$, is built up from these component automata as follows. A state of $A_{\mathcal{M}}$ is given by a tuple $(q, i, a_1, a_2, \ldots)$, consisting of the current node q in some user graph, the index i of the current transaction, and the state a_j of each rule automaton. The transitions of $A_{\mathcal{M}}$ reflect the processing of the input, in lock step (synchronously), by all the rule automata, as we advance down some user graph. Transitions on the delimiter symbol $\rightarrow$ depend on being at a terminal node of a user graph and appropriate accepting states of the relevant rule automata. Acceptance is determined when we have reached a sink node on a user graph, and all rule automata are in an "agreeable" state (including, MORE_TRANS automata, which must indicate that no more transactions are required). ■

The difficulty with the above proof is that, even ignoring probabilities, it results in a hopelessly huge automaton for any reasonable model. To give a sense of the problem, consider the model for the large switch which has been our focus in this project. It has 29 users (there are 11 user types), most user graphs contain up to 100,000 nodes after flattening, and roughly 500 rules, some consisting of patterns with complicated regular expressions (including negation). Even assuming, very optimistically, that each rule could be modeled by an automaton of at most 10 states, the construction in the above proof would build a global automaton $A_{\mathcal{M}}$ with 10^{506} states!! So, yes, it is a *finite* automaton, but just barely so. It was a priori not at all clear what the size of such an equivalent automaton would be, and furthermore if it were small, whether and how one could construct it without blowing up the size in the process. This required a number of optimizations which we outline below.

Our translation produces a probabilistic automaton $A = \langle S, \delta, l, start \rangle$ where S is a set of states, $\delta \subseteq S \times [0,1] \times S$ is a transition relation with associated probabilities for each edge, such that for all v

$$\sum_{\{(v,p_{(v,w)},w) | w \in S\}} p_{(v,w)} = 1.0$$

$l : S \rightarrow \Sigma \cup \{\longrightarrow\}$ is a labeling that assigns a keyword to each node. (For convenience, keywords label states rather than edges in these automata.)

Given the model $(\mathcal{H}, \mathcal{R})$ we obtain the probabilistic automaton $A_{\mathcal{M}}$ according to the following general outline:

1 For each rule, with general form "If exp1 then exp2", convert both exp1 and exp2 to (non-deterministic) automata A_{exp1} and A_{exp1}.

2 Incrementally apply each rule, in their prescribed sequential order, to the partial automaton built thus far, as follows

> intersect the current automaton with A_{exp1}, and on those paths that satisfy A_{exp1}, take the product with A_{exp2}, removing the paths not satisfying exp2.

3 minimize the result after each rule is applied.

This process involves different actions for different rule types, so what we've described only serves as a rough guideline. In essence, the algorithm consists of a process of repeated "producting" and minimization of an automaton. Each of the steps outlined requires us to take several measures to insure that the automata do not get too large. We sketch below some the measures. For most rules, Step 2 is further broken down into substeps depending on the rule and the patterns that appear in it. Since many of the rules have complicated expressions with many patterns, instead of constructing one automaton for the whole expression *exp*1 (or *exp*2), which can be quite large, we break the expression down into suitable subexpressions, and apply the producting followed by minimization routine for each subexpression.

Furthermore, when converting rule expressions to automata, we will need to address the problems we discussed arising from the use of negation and aliases in the expressions in rules. Two techniques we use for this purpose are: first we do not directly determinize the expression automata; we leave them nondeterministic and resolve nondeterminism on the fly while producting with the system automaton. Second, we do not explicitly produce transitions of the automata for all input symbols (there are just too many keywords). Rather we specify them implicitly, using formulas as labels for rule automata. Thus, rather than ordinary automata, we will convert exp1 and exp2 in a rule into *special automata.* Let $B(\Sigma)$ denote the set of boolean formulas over the alphabet Σ (in fact, DNF formulas suffice for our purposes, but we give the more general definition). A special automaton has the form $A_{exp_i} = \langle S', \delta, l, F, start' \rangle$ where, S' is the set of states, $\delta \subseteq S' \times S'$, $F \subseteq S'$ is a set of final states, but this time the labeling function $l : S' \rightarrow B(\Sigma)$, assigns to each state a *boolean formula* over the keywords, rather than one particular keyword.

The interpretation of such an automaton is as follows. A string $w = w_1 \dots w_k$ over Σ is in the language $L(A_{exp_i})$ of the automaton iff there is a sequence of states $s_0, s_1, \dots, s_k$, beginning with the start state $start'$ $= s_0$, such that keyword w_i satisfies $l(s_i)$, in the sense that if we view w_i as a assignment of truth values to Σ which assigns `true` to w_i itself and *false* to every other keyword, $l(s_i)$ is satisfied by this assignment.

Finally, we product A_{exp_i} with the current scenario automaton A. Again, we have to be very careful if we want to maintain feasibility, and we also have to maintain probabilities properly. A is very large, so we do not wish to incur the worst case $|A| \times |A_{exp_i}|$ blowup, which is typically unnecessary. To do this, our producting algorithm produces states "on-the-fly", in a way reminiscent of on-the-fly algorithms for state exploration in model checking (see, e.g., [5]). Second, because A_{exp_i} is non-deterministic, we need to do something akin to subset construction while our producting is proceeding, in order to obtain a final result in the deterministic form of A.

The basic outline of the algorithm for producting the system automaton with a special expression automaton, which proceeds via depth-first exploration of the product, is as follows:

automaton_producting_algorithm {
    Initialize stack $Z \leftarrow [(start, \{start'\})]$.

```
(we assume start and start' are a consistent pair)
 While Z is not empty do
  remove (s,X) from atop stack Z
  foreach child t of s  do
       find  the set X' of next-states of X in A_exp
       which are  consistent with t, meaning the label of t
       satisfies the formula labeling each x' ∈ X'.
         if (t, X') is not already in hash table  then
           add (t, X') to hash table,
           place (t, X') atop stack Z,
           and create the edge ((s, X), p_(s,t), (t, X'))
}
```

Note that in the algorithm the probability of a new edge is inherited from the graphs, thus maintaining the sequence of weight on a given scenario. But the process no longer assures that probabilities will sum to 1, so we will need to do some rescaling. There is a subtle point here regarding the correct way to do this rescaling; we omit the details.

Finally, after each producting, we run a minimization algorithm on the resulting automaton A' in order to get rid of redundancies that might have been created in the process of producting. Note that our minimization algorithm will need to preserve probabilities. What we use is a bisimulation reduction for probabilistic automata (see, e.g., [2]). We describe the algorithm for arbitrary automata, although we use the acyclicity of the switch automaton to make the algorithm run more efficiently (in fact in linear time).

```
state_minimization_for_probabilistic_automata {
  Initialize: ∀ states s, color(s) := l(s).
  (initially "color" the states according to the label of the state)
   Repeat
       ∀ states s,
         new_color(s) := (color(s), {(c, p_c) | p_c = Σ_{t|color(t)=c} p_(s,t)})
       color := new_color
       (rename colors, using sorting, to represent them concisely)
   Until the total number of colors does not increase
}
```

The resulting automaton has one state for each color class, and its transition relation can be read off from the final coloring.

4. TRANSLATION IMPLEMENTATION, AND THE SWITCH MODEL

A complete translator from TGAP models into probabilistic automata has been implemented. The implementation was carried out in the programming language ML, using roughly 7000 lines of code. We have used the translator to translate the large model for the switch. The dimensions of this model were as follows: There are 266 keywords, plus roughly 100 aliases. The largest flattened graph contains, approximately 100,000 nodes (this gets reduced dramatically, to 4700 states, after minimization). There are approximately 500 rules.

Our translation produced an automaton having only 230,000 states and 480,000 transitions. This is well within the realm of feasibility, as many of the test generation algorithms we apply to this automaton require only linear time. The one drawback to the translation is that it is currently very slow. However, since all of the analysis and test generation is performed directly on the resulting automaton, translation is an off-line cost that is only incurred once, performed prior to testing when the model is completed. Hence, it is a one-time penalty that is reasonable to pay.

5. TARGETED TEST GENERATION ALGORITHMS

The ruled-based TGAP system provided only random sampling as a test generation capability. The user requests a desired number N of scenarios, and the system attempts to generate N random scenarios using the user models, the associated probabilities in the operational profile, and the rules. We say only "attempts" because the system may not (and often does not) succeed in producing N scenarios, and may return a smaller number. The reason is that the system tries to construct the scenarios incrementally walking through the graph models and applying the rules. As a consequence it is possible that a partial scenario cannot be extended to a complete legal scenario, because of the rules; when the scenario fails to complete, the TGAP system backtracks and tries another route. After a fixed number of failures, the system terminates the search. Note that it is possible that there are not even N legal scenarios, for example if one adds rules in an effort to narrow down the scenarios in order to restrict to a subsystem of interest. However, there is no way to tell directly from the rules whether this is the case.

Moving to an automaton based model enables a number of testing capabilities with better test selection, analysis of coverage, targeting to subsystems of interest, and enforcement of desired test criteria. We have developed and implemented several algorithms and applied them to the switch automaton. We summarize below some of the tasks addressed, and our experience in the application to the switch model, without going into the details of the algorithms, which will be given elsewhere.

Counting scenarios. This is useful when a user asks for N scenarios, (1) to determine whether there exist that many scenarios, without searching in the dark, and (2) to determine which fraction of the total are accounted for by the N scenarios. Of course, the number of scenarios is finite only if the automaton is acyclic, as is the case with the switch automaton. The number of legal scenarios in this test model is more than 10^{20} (the precise number is 905,772,615,646,147,548,919); Obviously, it is impossible to generate that many items. However, note that we can count the scenarios without explicitly enumerating them. The algorithm runs in linear time in the size of the automaton and takes just a few seconds for the switch model.

Sampling scenarios. That is, generate N random scenarios according to the probabilities in the model. This is the only kind of testing that the TGAP system allows. Having the automaton makes this problem very easy; the algorithm runs much faster because it does not have to match and apply the rules online at runtime, and does not have to worry about dead ends, i.e. work on partial scenarios that don't lead anywhere. A random scenario corresponds to a random walk on the Markov chain represented by the automaton. Thus, one can generate thousands of random scenarios in seconds. Besides, producing the scenarios, we can also compute their associated probabilities, and thus calculate the fraction of the probability space that is covered by the produced test set.

Most probable scenarios. That is, generate the N scenarios with the highest probability. The rationale for this kind of test generation is to maximize the coverage of the operational profile (as measured by the probabilities) with a limited number of tests. Given the size of the model, obviously we need to pay attention to the efficiency. A naive approach takes time $O(N \cdot input)$. Our algorithm takes essentially linear time in the input (the size of the automaton) and the output (the N tests). The algorithm also runs fast in practice, comparable to the simple random sampling. For the switch automaton, on a time shared 250MHz MIPS processor, after reading the input and setting up the data structures (9 secs.) it generates the maximum probability scenarios in order at the rate of more than 3,000 scenarios per second. Again, besides the scenarios themselves we can also compute the associated probabilities to analyse the coverage, obtaining a profile of the probability distribution of the scenario space. The following table shows some data for the switch automaton; the second column gives the probability of the Nth test and the third column the coverage (cumulative probability) of the first N tests. We computed statistics for random testing, and noted that it requires 8-10 times as many tests to achieve the same coverage as the maximum probability method.

test	probability	coverage
100	0.00049	13 %
500	0.00013	22 %
1000	5.54e-05	26 %
5000	1.18e-05	35 %
10000	5.68e-06	39 %
50000	1.12e-06	48 %
100000	5.57e-07	52 %

Often one wants to target tests to specific subsystems or features of interest. The keywords are used for this purpose; some keywords may be irrelevant to the subsystem, while others represent activities of the feature or subsystem, thus, are relevant and are expected to appear often in scenarios of the subsystem. For example, to test functionalities related to call centers, testers compiled a list of special 'relevant' keywords that involve call center activities. The objective then is to generate relevant scenarios that exercise the parts of the system that have relevant keywords.

Most relevant scenarios. For a given list of relevant keywords, we can assign to each scenario (path through the automaton) a relevance score: the number of its nodes with relevant keywords. More generally, we can assign weights to the keywords according to their relative importance, and the relevance score is then the sum of the weights of the scenario's keywords. For a given number N of desired scenarios, we can generate the N scenarios with highest relevance score. The algorithm runs essentially in linear time in the input and output as in the case of probabilities and is also fast in practice.

Minimum cover. Find a small number of scenarios that cover all the relevant keywords, i.e., each relevant keyword appears at least once. This problem was addressed in [8], where algorithms were proposed to find a near minimum set (finding the minimum set is NP-hard). For example, 28 tests suffice to cover all keywords of the whole switch automaton, and 5 tests cover the special keywords in the call center list. Similarly, we can find a small test set that covers all the nodes of the automaton or the original user graphs that are labeled with relevant keywords.

Balanced relevant cover. Suppose we can afford to run N tests (which may well be more than the minimum needed for simple coverage). We could find the N scenarios with highest relevance score but it is possible that all of them involve only a subset of the special keywords. Usually, it is desirable to cover all the relevant keywords, and furthermore to cover them multiple times in a roughly balanced manner. We have developed an algorithm for this purpose; it is based on the above algorithm for computing the most relevant scenarios but it uses a dynamic reweighting of the keywords through the computation to bias the generation towards a balanced coverage.

Pattern-based Restriction. A more refined way to characterize subsystems and features is by using regular expression patterns, similar to the rules. The patterns can be used to restrict the automaton, that is, we can combine the system automaton with the pattern automaton to produce a new automaton that characterizes the subsystem. Any of the above algorithms then can be applied to the resulting automaton to produce desired targeted tests for the subsystem.

6. CONCLUSION

We have described an efficient translation from a rule-based behavioral testing model to a probabilistic automaton. Automata are a more familiar and well-studied model which provides much more flexibility in test generation. The motivating application for this work was the testing of a major switch.

Our goal was to enable the application of state-of-the-art automata based testing algorithms while leveraging the existing rule-based model that had been developed over many years. Although feasibility of the translation from a rule-based to automata-based model was a big open question given the scale of the problem, the experiment turned out favorably, and we were able to generate a reasonable size equivalent automaton. We also developed several efficient algorithms for targeted test generation and coverage analysis which went well beyond anything that was possible with the rule-based system.

As we mentioned in the introduction, rule-based models are used widely to specify systems in a wide variety of domains. On the other hand, automata-based models are more convenient for verification and testing purposes; indeed most of the automated techniques in conformance testing and model checking are based on automata (finite state machines). Bridging the rule and automata paradigms, as we have done in this paper, will bring these testing techniques to bear on a variety of other domains.

References

[1] Alur, R. and Yannakakis, M. (1998). Model checking of hierarchical state machines. In *Proc. 6th ACM Symp. on Foundations of Software Engineering*, pages 175–188.

[2] Baier, C., Engelen, B., and Majster-Cederbaum, M. (2000). Deciding bisimlarity and similarity for probabilistic processes. *Journal of Computer and System Sciences*, 60(1):187–231.

[3] Beizer, B. (1990). *Software Testing Techniques.* Thompson.

[4] Brinksma, E. (1988). A theory for the derivation of tests. In *Proc. IFIP WG6.1 8th PSTV.*

[5] Holzmann, G. J. (1990). *Design and Validation of Protocols.* Prentice-Hall.

[6] Holzmann, G. J. and Smith, M. H. (1999). A practical method for the verification of event driven systems. In *Pro.c Intl. Conf. on Software Engineering*, pages 597–608.

[7] Hull, R., Kumar, B., Llirbat, F., Simon, E., Su, J., and Zhou, G. (1999). Declarative workflows that support easy modification and dynamic browsing. In *Proc. WACC.*

[8] Lee, D. and Yannakakis, M. (1996a). Optimization problems from feature testing of communication protocols. In *Proc. Intl. Conf. on Network Protocols*, pages 66–75.

[9] Lee, D. and Yannakakis, M. (1996b). Principles and methods of testing finite state machines - a survey. *Proc. of IEEE*, 84(8):1090–1123.

[10] Ullman, J. D. (1988). *Principles of Database and Knowledge-base systems.* Computer Science Press.

[11] Vardi, M. Y. and Wolper, P. (1986). An automata-theoretic approach to automatic program verification. In Kozen, D., editor, *First IEEE Symp. on Logic in Computer Science (LICS)*, pages 322–331.

[12] Widom, J. and S. Ceri, E. (1996). *Active Database Systems: Triggers and Rules for Advanced Database Processing.* Morgan Kaufmann.

INTEGRATED SYSTEM INTEROPERABILITY TESTING WITH APPLICATIONS TO VOIP

Nancy Griffeth, Ruibing Hao, David Lee*, Rakesh K. Sinha

Bell Laboratories

Murray Hill, New Jersey

* *Bell Laboratories Research China*

Beijing, China

{nancyg,rbhao,leedavid,rks1}@lucent.com

Abstract This work has been motivated by the need to test interoperability of systems carrying voice calls over the IP network. Voice over IP (VoIP) systems must be integrated and interoperate with the existing Public Switched Telephone Network (PSTN) before they will be widely adopted. Standards have been developed to address this problem, but unfortunately different standards bodies and commercial consortiums have defined different standards. Furthermore, the prevailing commercial standard (H.323) is incomplete, complex, and presents the implementers with "vendors latitudes". As a result, there is no guarantee that the integrated VoIP systems would interoperate properly even if the implementations are all H.323-compliant. Interoperability testing has become indispensable.

We discuss criteria that cover all the required patterns of "interoperating behavior." We want to test all such system interoperations. On the other hand, test execution in real environment is expensive, and we want to minimize the number of tests while maintaining the coverage.

We present a general method for automatic generation of test cases, which cover all the required system interoperations and contain a minimal number of tests. We study data structures and efficient test generation algorithms, which take time proportional to the total test case size. We report experimental results on VoIP systems.

Keywords: VoIP, H.323 protocol, integrated system, interoperability testing, coverage, redundancy

Tommaso Bolognesi and Diego Latella (Eds.), *Formal Methods for Distributed System Development.*

1. INTRODUCTION

With the rapid growth of Internet, more and more new protocols are proposed and new products are developed, such as Voice over IP(VoIP) and Virtual Private Network(VPN). When two or more entities in separate communicating systems are integrated and need to interact with each other to perform a certain task the capability to operate as desired is called interoperability, which is considered an essential aspect of correctness of integrated systems. However, products from different vendors or even from the same vendor often do not interoperate properly. Two main causes of non-interoperation are ambiguity of protocol specification and a vendor's proprietary extensions. Interoperability testing is done to check the interoperations among integrated system implementations.

The research work on interoperability testing can be roughly classified into two categories: general concepts and experiences of interoperability testing, and systematic generation of interoperability test suites [7].

Most of the early work [6, 4, 14] (beginning of 90s) belongs to the first category. Majority of the recent research work is related to interoperability test suite derivation. One approach for interoperability test generation is to apply conformance test generation techniques on composed finite state machines, which are constructed from several components systems via a reachability analysis. A finite set of test cases is then selected to test the interoperability. Rafiq et al. [13] was one of the first papers in this field. It proposes the above test generation procedure and also an interoperability test architecture. Kang and Kim [9] proposed a technique for symmetric protocol control portion interoperability testing. However, for practical systems, this system composition approach often leads to state explosion [12, 11, 5, 8]. Furthermore, we may not have complete information on all the interoperating systems. For instance, in VoIP systems, we can model end user behavior and H.323 protocols, but we do not have a specification of the communication system, which includes the IP network and is too complex to model.

Our work has been motivated by the need to test interoperability of systems carrying voice calls over the IP network. Voice over IP (VoIP) systems must be integrated and interoperate with the existing Public Switched Telephone Network (PSTN) before they will be widely adopted. We model the system behaviors by extended finite state machines (EFSM), and propose a testing architecture for interoperability testing. Based on the experiences of domain experts, we discuss various criteria that cover the required patterns of interoperations. A key idea in developing the coverage criteria is that interoperability errors are introduced only when the integrated systems are "interacting" with each other. Based on this observation, our test generation only requires information about part of the integrated communication systems; it is impossible to obtain, model and analyze the whole systems in practice. For instance, for VoIP interoperability testing, we only need information about end user required behaviors and H.323 interfaces. Consequently, it does not require the modeling and construction of the composition of all the integrated systems, including the IP network, and avoids the state explosion problem while checking all the required interoperations.

We design efficient data structures and test generation algorithms, which guarantee desired coverage and contain a minimal number of tests.

These algorithm are part of a software tool, ITIS, developed at Bell Laboratories. With this tool, we have generated interoperability test cases for: End users vs. the rest of the communication system; and End Users and H.323 vs. the rest of the communication system. For clarity, we only report the experimental results of interoperability testing of the end users behaviors vs the rest of the systems.

The remainder of this paper is organized as follows. After introducing a formal model, we present the interoperability test generation architecture and techniques in Section 2. Section 3 contains more details of the test generation algorithms, which incorporate a range of coverage and redundancy criteria. Experimental results on VoIP are reported in Section 4.

Formal Model

Communicating system behaviors can often be modeled by extended finite state machines, which are finite state machines extended with variables. We denote a finite set of variables by a vector: $\vec{x} = (x_1, \ldots, x_k)$. A predicate on variable values $P(\vec{x})$ returns FALSE or TRUE; Given a function $A(\vec{x})$, an action is an assignment: $\vec{x} := A(\vec{x})$. Each transition t is a 6-tuple: $t = (s_t, q_t, a_t, o_t, P_t, A_t)$, where s_t, q_t, a_t, and o_t, P_t, A_t are the start state, end state, input, output, predicate, and action respectively. In state s_t, upon input a_t, if $\vec{x}$ is valid for P_t, i.e., $P_t(\vec{x}) = \text{TRUE}$, then the machine follows the transition t, outputs o_t, changes the current variable values by action $\vec{x} := A_t(\vec{x})$, and moves to state q_t.

Each combination of a state and variable values is called a *configuration*. We can construct a reachability graph of configurations starting from the initial configuration. If each variable has a finite number of values (Boolean variables for instance), then there are a finite number of configurations and the reachability graph is an equivalent (ordinary) FSM with configurations as states [10]. Therefore, an EFSM with finite variable domains is a compact representation of an FSM.

2. INTEROPERABILITY TEST GENERATION

When we do interoperability testing, we are concerned only with those failures that occur when integrated systems are interoperating. For instance, consider Voice over IP systems in Figure 1. Two H.323 end-users want to communicate. They each access the communication system through an H.323 gateway, using H.323 protocol. We may want to test the interoperability between H.323 end users and the rest of the systems, i.e., the H.323 gateways and the communication system.

Figure 1 Model of interoperating systems

For interoperability testing, we would like to test all possible system interoperations. For instance, in the case of interoperability of end users and the rest of the system, we might want to test all the interoperations between the end users vs. the rest of the system. For this test, assume that we can access the system through end user behaviors only. We model the joint behaviors of the two end users by an EFSM. Ideally, we want to check every input sequence to the end user EFSM so that all its interoperations with the rest of the system are tested. It is an impossible task. We want to remove the redundant tests (redundancy will be defined in the sequel) and, hopefully, reduce to a small number of tests.

In general, we want to test interoperations of system A and B. We have complete information about A but not B or choose not to model B, either because the information is not available or is too complex. We can only access the systems through the interface with A. Based on our information about A, we want to construct a set of non-redundant tests such that all the interoperations between A and B are tested. Our goals are: (1) *Completeness*: all the interoperations between the two systems are tested; and (2) *Non-Redundancy*: all the redundant tests are removed to minimize the total number of tests.

Note that interoperability testing is different than conformance testing; even a *checking sequence* [10] of system A may not be sufficient due to the following reasons. A checking sequence guarantees structural isomorphism between the implementation and specification machines under the *assumption* that the implementation machine has no more states than that of the specification machine. It may not be true for the expected behavior of an integrated system, in which the number of states is unknown. Furthermore, we have no information or only partial information of system B. Therefore, for interoperability testing, we need different tests to ensure the coverage of all the interoperations between A and B, given only information about A.

There are various techniques to remove redundancy. For instance, in the configuration of Figure 1, a key idea in developing the test cases is that interoperability errors will be introduced only when the gateways are actually talking to each other about a call. Thus we can ignore local activities involved in the protocol, such as obtaining information from an end-user. We label as "white" the transitions that can be ignored and label all other transitions as "black." A white transition is purely local, that is, it reads the state of only one gateway and writes that same state. A black transition involves both gateways, because it reads both states, or reads one state and writes the other.

This is illustrated in the call flow for a simple telephone call in Figure 2. The transition Off-hook from Idle state to Dialing state is white, because it involves only the originating gateway and the end user. The Dial transition from Dialing to Calling is black, because the state is changed in both gateways. Similarly, the Off-hook transition which connects the call changes the state in both gateways from Calling to Connected.

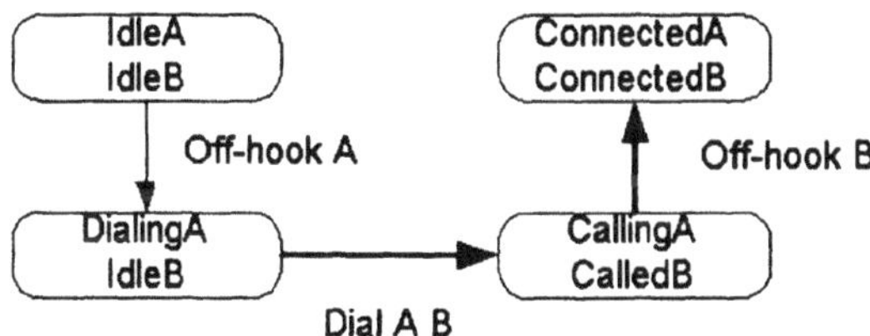

Figure 2 Partial call flow of a telephone call

The test cases that we generate focus only on black edges. The purpose of white edges is to interconnect the sequence of black edges.

2.1. TEST GENERATION STRATEGY

We want to test interoperability of system A and B, assuming we have complete information about A. We can model it by an EFSM and construct its reachability graph, which is a directed graph, a transition diagram [10]. Throughout the rest of the paper, we will be using graph terminology. For example, we will denote the initial state as the "source node", transitions as "edges", etc.

Ideally, we want to cover (test) all possible execution sequences, often called scenarios. Therefore, covering all possible execution sequences requires that we cover all the branches and all the possible paths. Consequently, we cover all the interoperations from A to B, and this would be our interoperability coverage criteria. However, it is impossible; in general, transition diagrams often contain cycles and, therefore, will have infinitely many distinct paths. We need to remove the redundant paths, if possible, and reduce the number of tests to a manageable size.

Our test generation process contains three steps:

step 1 Generate all possible acyclic paths, i.e. paths without any repeated vertices.

step 2 Generate all possible simple cycles, i.e. cycles that do not contain any smaller cycles.

step 3 "Combine" the paths (from step 1) and simple cycles (from step 2) to generate the final set of paths.

The number of paths and cycles generated in the first two steps is finite. In a later section, we will describe various criteria for removing the redundant cycles and paths. The goal of the "combine" process in step 3 is to generate a finite number of paths that cover all necessary scenarios. This is done as follows:

> First we include all the acyclic paths. Then we generate additional cyclic paths as follows. For each acyclic path P, find all the simple cycles that share a node with P. Let these cycles be $C_1, C_2, \ldots, C_k$ and let v_i, $1 \leq i \leq k$, be a node common to C_i and P. Then generate a new (cyclic) path by replacing node v_i in P with cycle C_i.

Now we elaborate on steps 1 and 2. We first find all the strongly connected components (SCC) of the transition graph. This has two advantages: we know that any cycle is completely contained inside an SCC. So we can perform step 2 by looking at each SCC in turn and finding all simple cycles within this SCC. The second advantage is that we can "shrink" each SCC into a node and obtain a Directed Acyclic Graph (DAG), i.e., a graph without any cycles [3]. This naturally translates into a two phase algorithm for step 1. First, generate all acyclic paths on the resulting DAG and then replace each SCC on any given path with a set of acyclic paths within the SCC.

2.2. NEXT-TRANSITION-TREE

We describe a simple data structure, next-transition-tree, that will be convenient for both steps 1 and 2. The data structure can be defined for any graph but in our application, the graph will always be an SCC.

For any node v, next-transition-tree(v) stores all acyclic paths from v to other vertices in its SCC. The tree has v as its root and its height is equal to the number of nodes in the SCC containing v. The children of v are all the nodes in its SCC that have a direct edge from v. In general, the children of any node u are all the nodes in its SCC that have a direct edge from node u. Please note that a node may appear multiple times in this tree. (Precisely speaking we should say that each node has a label where the labels are not necessarily unique. The root node has label v. A node labeled u has as many children as the outdegree of u in its SCC and these children are labeled with the corresponding nodes in the SCC.)

We do not need a next-transition-tree for each node in the graph. We need it only for a few of the nodes; the details are in Section 4. For ease of presentation, we assume that we maintain a separate next-transition-tree for each node. The actual implementation will have lots of shared pieces among next-transition-trees belonging to the same SCC.

2.3. GENERATING ALL SIMPLE CYCLES

We outline how to generate all simple cycles within an SCC. Let T be the next-transition-tree of an arbitrary node in this SCC.

We process each root-to-leaf path of T as follows: if this path contains (exactly) two instances of the same node then the path segment connecting these two instances corresponds to a simple cycle. It can be shown that this procedure gives all possible cycles. (Proof Sketch: let $C = v_1 v_2 v_3 \cdots v_r v_1$ be any simple cycle. Find a node v_i of C that is closest to the root in T. If there are multiple nodes all closest to the root, pick any one of them. v_i must contain an edge to v_{i+2}, which in turn must contain an edge to v_{i+1} and so on. Inductively, one can show that the subtree of T rooted at v_i contains the path $v_i v_{i+1} \cdots v_r v_1 v_2 v_{i-1}$.) However, the same cycle may appear more than once. So we need to maintain a set data structure of all cycles generated. Before inserting a new cycle, we check whether it is already present or not.

3. GENERATING ALL ACYCLIC PATHS

As mentioned earlier, we first find all the strongly connected components (SCC) of the transition graph. Then we construct a *component graph* where each node represents an SCC and an edge from node u to node v represents an edge (in the original transition graph) from a node belonging to u's SCC to a node belonging to v's SCC. This component graph is a Directed Acyclic Graph (DAG), i.e., a graph without any cycles [3].

We first generate all acyclic paths in the DAG and then replace each SCC on any given path with a set of acyclic paths within the SCC.

Redundancy. There are a finite number of distinct paths (tests) in a DAG. However, we do not want to test all of them; Our goal is to generate a minimal number of tests, which satisfies the coverage criterion and does not contain any *redundant* tests.

The tests need to start from the start state, so we only consider paths starting from the source node (node without any incoming transitions). In addition, the following is a very popular *redundancy* criterion in practice.

(R1) *Proper prefixes are redundant and can be discarded.*

From the redundancy criterion above, we only consider paths that terminate at a sink node (node without any outgoing transitions). Any path not terminating at a sink node will be a proper prefix of a path that has been extended to a sink node. Combining the above two, we only need to generate paths that start from the source node and terminate at one of the sink nodes.

Figure 3 gives an algorithm for generating all source to sink paths in a DAG.

PATHS-IN-DAG

Input. a DAG G with one source s and multiple sinks.
Output. all source–sink paths in G.

1 topologically sort nodes in G: $s = v_0, v_1, \ldots, v_{n-1}$;
2 **for** $i = n-1, \ldots, 0$
3   if v_i is a sink node then
4         $p(v_i) = \{\Lambda\}$; /* a singleton set of empty path */
5   else
6   let outgoing edges from v_i be: $w_1, \ldots, w_r$;
7         $p(v_i) = \bigcup_{j=1}^{r} (v_i, w_j) p(w_j)$;
8 **return** $p(v_0)$

Figure 3 Generating all source–sink paths in a DAG

Generating All Acyclic Paths in a DAG. The algorithm (Figure 3) is bottom-up, starting from one of the the sink nodes $t = v_{n-1}$. When processing a node v_i, we examine all its outgoing edges (v_i, w_j) where the paths from w_j to sink nodes have been computed. At Line 7, we concatenate edge (v_i, w_j) to each path computed at w_j and collect it at node v_i. After processing the source node $s = v_0$, we have obtained all the paths, which are non-redundant and have a complete coverage.

A topological sort takes time proportional to the number of edges. We account for the examination and concatenation of each edge, which is processed only once at Line 7, and the total cost is proportional to the total path lengths:

Proposition 1 *The algorithm in Figure 3 constructs all source–sink paths. Its time and space requirement is linear in the output (the total lengths of all the constructed tests).*

Proposition 2 *For machines with a reset to source s, the constructed test sequences are a checking sequence.*

Generating All Acyclic Paths within an SCC. Up to now we have discussed test (path) generation for DAGs where nodes can be SCC. For each edge connecting two SCC's on the path, we need to replace it with an edge in the original graph. In general, it will be possible to replace the edge between SCC's with one of the several possible edges in the original graph, and we will obtain a separate path for each choice of the replacement edge. We also need to replace each SCC node with a set of acyclic paths within this SCC. Suppose the incoming and outgoing edges of this SCC node are on nodes u and v (respectively) of this SCC. Then we need to replace this SCC node with all possible acyclic paths from u to v in this SCC. These paths can be generated from next-transition-tree(u).

Figure 4 summarizes the steps in generating all acyclic paths and Figure 5 gives the algorithm for generating all possible tests.

Remark: Here we are replacing any SCC with all possible acyclic paths. This provides exhaustive coverage but generates lots of tests. In certain applications, we may not need to cover the SCC as thoroughly. There are two other options commonly used in practice.

Chinese Postman Tour. For an SCC, we want to test each edge (transition) at least once and we want to minimize the test sequence length. Such a path is called *Chinese Postman Tour* [2].

Checking Sequence. A more thorough coverage is provided by a checking sequence, which guarantees the structural isomorphism of the implementation and specification machines [10]. The length of the test sequence is longer than that of a Chinese Postman Tour.

3.1. REDUNDANCY CRITERIA

The algorithm outlined in the previous sections (Figure 4) provides exhaustive coverage. However, the number of tests generated is enormous even for systems of moderate size.

EXHAUSTIVE-COVERAGE

Input. transition graph G with a designated source state.
Output. set of all acyclic paths in G.
1 find all SCC of G;
2 for each SCC
3 for each node in this SCC
4 construct next-generation-tree;
5 shrink each SCC of G to obtain a DAG G';
6 apply PATHS-IN-DAG on G' to obtain the set of paths P'
7 for each path p' in P'
8 replace nodes and edges in p' with paths and nodes in G to obtain the set P of acyclic paths;

Figure 4 Generating all acyclic paths for exhaustive coverage

EXHAUSTIVE-TEST-GENERATION

Input. An EFSM of the system.
Output. minimal set of tests providing exhaustive coverage.
1 construct an equivalent FSM and its transition diagram G.
2 apply EXHAUSTIVE-COVERAGE on G to obtain acyclic paths P.
3 obtain the set C of all simple cycles.
4 combine the sets P and C (as described in Section 2.1).

Figure 5 Generating all tests for exhaustive coverage

As stated in the introduction, a key insight is that that interoperability errors will be introduced only when the gateways are actually talking to each other about a call. We label the transitions as either "white" (local activity) or "black" (involves both gateways).

Thus for interoperability test generation, we do not need to cover white transitions. This motivates additional redundancy criteria.

(R2) *Remove all-white test sequences.*

(R3) *Let (u, v) be the first black edge in the test sequence. Then replace the path from the source node to node u with the shortest path between the source node and node u.*

(R4) *Let (u, v) be the last black edge in the test sequence. Then replace the path from node v to the sink node with the shortest path between node v and the sink node.*

(R5) *If there is a sequence of white edges, in which each party separately reaches the idle state, terminate the sequence at the last black edge and then use a shortest path to the state "IdleA,IdleB" at the end of the sequence.*

The first three criteria reflect our belief that white transitions are not relevant for interoperability testing. Their only use is to connect (relevant) black transitions. For example, $R3$ states that the sequence of transitions before the first black transition is not relevant to us, so we might as well replace that part with the shortest all-white sequence. Redundancy criterion $R4$ is dual of $R3$. The last criterion recognizes that if both states reach the "idle" state through a sequence of non-relevant transitions then whatever happens in the rest of the sequence will be tested at the beginning of another generated sequence.

Generating Test Sequences according to $R1$-$R5$. We give an algorithm (Figure 6) for generating all acyclic paths satisfying these criteria.

The algorithm starts with a graph G and generates another graph G' such that the set of source–sink paths in G' is the same as the set of source–sink paths in G with criteria $R2$-$R5$.

The redundancy criterion $R3$ says that the path before the first black transition should be replaced with the shortest all-white path.

Lines 1-4 considers all possible black transitions that can be the first in any sequence and adds a marker 'super-edge' for the shortest all-white path from the source node. Line 9 makes sure that a black transition follows a super-edge. The marker 'super-edge' gets replaced on line 10 by the shortest all-white path. This takes care of redundancy criterion $R3$.

Lines 5-8 perform an analogous function for sink nodes and the last black transitions in the sequence, taking care of redundancy criterion $R4$.

Criterion $R2$ is satisfied because line 9 makes sure that there is at least one black transition.

Instead of generating all paths and then processing them according to criterion $R5$ (as suggested on line 11) the actual implementation makes sure that the incremental construction of paths in procedure EXHAUSTIVE-COVERAGE never generates a path in violation of $R5$.

Proposition 3 *The algorithm in Figure 6 constructs all acyclic paths according to criteria $R1$-$R5$.*

3.2. ADDITIONAL REDUNDANCY CRITERIA

On most practical systems, we expect algorithm ADEQUATE-COVERAGE (Figure 6) to generates a considerably smaller set of test sequences than

ADEQUATE-COVERAGE

Input. transition graph G with a designated source state.
Output. complete set of paths according to criteria $R1$-$R5$.

```
1  for each node v that has an outgoing black edge
2      if there is an all-white path from the source node to v then
3          add a super-edge from the source node to v;
4  delete all outgoing edges (not super-edges) from the source node;
5  for each node v that has an incoming black edge
6      if there is an all-white path from v to a sink node then
7          add a super-edge from v to this sink node;
8  delete all incoming edges (not super-edges) to the sink node;
9  generate all acyclic paths in the resulting graph by
   EXHAUSTIVE-COVERAGE with the modification that no
   super-edge should follow or precede a white edge;
10 replace each super-edge with the shortest all-white path in the
   original graph; /* this may introduce some loops */
11 process each path to get rid of loops;
12 process each path according to redundancy criterion R5;
```

Figure 6 Generating paths for adequate coverage (according to $R1$-$R5$)

the naive algorithm (Figure 4). Unfortunately even such a smaller set may be too large for manual test execution.

In this section, we suggest a set of more restrictive criteria for test generation that concentrates only on black transitions.

(R6) *Generate acyclic paths that consist only of black edges, except that the prefix from the source node and the suffix to the sink node is allowed to contain white edges.*

(R7) *Generate simple cycles consisting only of black edges.*

Generating Test Sequences according to *R6*-*R7*. We give an algorithm BASIC-COVERAGE (Figure 7) for generating the test-set according to $R6$ and $R7$. The algorithm starts with a graph G and transforms into graph G' such that the set of source–sink paths in G' is the same as the set of source–sink paths in G with criterion $R6$. (We do not discuss criterion $R7$ but it is easy to modify the procedure to generate all simple cycles to satisfy $R7$.)

The intuition of the algorithm is to delete all white edges except those needed to reach black transitions from the source node (or from the black transitions to the sink nodes). Lines 1-3 maintains a marker 'super-edge' for each source node to black transition path. On line 10, we replace this marker with the shortest path.

Lines 4-6 perform an analogous function for sink nodes.

```
BASIC-COVERAGE

Input. transition graph G with a designated source state.
Output. test generation according to R6.
1  for each node v that has an outgoing black edge
2       if there is a path from the source node to v then
3            add a super-edge from the source node to v;
4  for each node v that has an incoming black edge
5       if there is a path from v to a sink node then
6            add a super-edge from v to this sink node;
7  delete all white edges from the graph;
   /* the only remaining edges are super-edges or black edges */
8  generate all acyclic paths in the resulting graph by
   EXHAUSTIVE-COVERAGE with the modification that no
   super-edge should follow or precede a white edge;
9  replace each super-edge with the shortest path in the original
   graph /* this may introduce some loops */
10 process each path to get rid of loops;
```

Figure 7 Generating all acyclic paths for basic coverage(according to $R6$)

Proposition 4 *The algorithm in Figure 7 constructs all acyclic paths according to criterion $R6$.*

Discussion: We feel that both algorithms BASIC-COVERAGE and ADEQUATE-COVERAGE are interesting in their own rights. If automated test execution is available then ADEQUATE-COVERAGE is the preferred algorithm. If on the other hand, test cases have to be executed by hand then BASIC-COVERAGE is likely to produce a manageable set of test sequences. Alternatively, one can always start with BASIC-COVERAGE since it captures the most critical interoperability behavior. If the system passes tests generated by BASIC-COVERAGE, one can move to a broader coverage provided through ADEQUATE-COVERAGE.

4. ITIS: AN INTEROPERABILITY TEST GENERATION SOFTWARE TOOL

The algorithms described in earlier sections are part of the ITIS (Interoperability Testing Intelligent System) project at Lucent Bell Laboratories. ITIS is a software tool for automated interoperability test generation. It is inexpensive and completely portable. ITIS is written in ANSI C and Tcl/Tk. We first describe its architecture and then report experimental results.

4.1. ARCHITECTURE OF ITIS

ITIS has a GUI for user input and for displaying test sequences. The workflow of ITIS is shown in Figure 8.

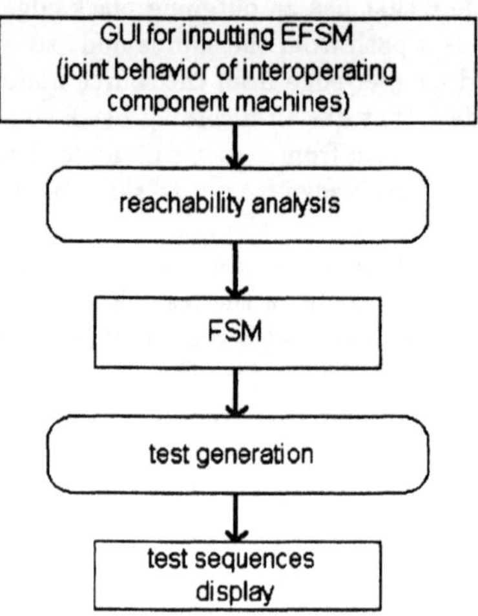

Figure 8 The workflow ITIS for interoperability test generation

The input to ITIS is an Extended Finite State Machine description of the system behavior, of which we have complete information. ITIS will first do reachability analysis to convert the EFSM into an FSM, and then use different algorithms discussed in the above sections to generate test sequences.

4.2. EXPERIMENTS

We have used ITIS to generate the interoperability test cases for End-user VoIP testing.

The EFSM for this protocol has 21 states and 68 transitions, among them 24 transitions are colored as black, others white. The shrunk DAG generated by ITIS from this EFSM contains three SCC nodes; only one of them contains more than one states. The number of generated test sequence by applying different algorithms is shown in the following table. (A longer version of this paer, that includes detailed description of EFSM and the experiments, is available as a Bell labs TM.)

The coverage of the 22 test sequences generated by the algorithm BASIC-COVERAGE is complete with respect to the requirements for the system modeled by the EFSM. The requirements appear in Bellcore's LATA Switching Systems Generic Requirements [1], which includes hundreds of requirements on the behavior of a telephone switching system.

Algorithms	*loop-free paths*	*loops*	*final tests*
EXHAUSTIVE-COVERAGE	950	424	1752
ADEQUATE-COVERAGE	508	424	908
BASIC-COVERAGE	16	4	22

These include many, such as trunking requirements and power requirements, that are not relevant to Internet telephony and are not modeled by the EFSM. However, there are 40 requirements that are applicable to Internet telephony. All of these are tested by the 22 test sequences generated by BASIC-COVERAGE.

5. CONCLUSION

Heterogeneity is one of the prominent features of networking systems, and interoperability is ubiquitous and has become a major hurdle for system reliability and quality of service. Interoperability testing is indispensable for the integration of reactive systems. Conventional conformance testing techniques do not apply since we want to test the system interfaces but not to check implementations vs. specifications.

We propose several algorithms for automatic generation of interoperability testing cases, incorporating a range of coverage and redundancy criteria. The tests generated are complete and non-redundant. They are applied to VoIP interoperability testing, and provide promising results.

References

[1] LATA switching systems generic requirements: Call processing. Technical Report GR-505-CORE, Bellcore, December 1997.

[2] A. V. Aho, A. T. Dahbura, D. Lee, and M. U. Uyar. An optimization technique for protocol conformance test generation based on UIO sequences and rural chinese postman tours. *IEEE Trans. on Communication*, 39(11):1604–15, 1991.

[3] A.V. Aho, J. E. Hopcroft, and J. D. Ullman. *The Design and Analysis of Computer Algorithms.* Addison-Wesley, 1974.

[4] G. Bonnes. IBM OSI interoperability verification services. In *IFIP TC6 WG6.1 The 3rd International Workshop on Protocol Test System*, 1990.

[5] R. Castanet and O. Kone. Deriving coordinated testers for interoperability. In O. Rafiq, editor, *Protocol Test Systems, VI(C-19)*,

pages 331–345. Elsevier Science Publisher B. V.(North-Holland), 1994.

[6] J. Gadre, Rohre C, C. Summers, and S. Symington. A COS study of OSI interoperability. *Computer Standards and Interfaces*, 9(3):217–237, 1990.

[7] R. Hao. *Protocol Conformance and Interoperability Testing based on Formal Methods.* PhD thesis, Tsinghua University, P.R.China, 1997.

[8] S. Kang and M. Kim. Test sequence generation for adaptive interoperability testing. In *Proceeding of Protocol Test Systems, VIII*, pages 187–200, 1995.

[9] S. Kang and M. Kim. Interoperability test suite derivation for symmetric communication protocols. In *Proceeding of FORTE/PSTV'97*, 1997.

[10] D. Lee and M. Yannakakis. Principles and methods of testing finite state machines - a survey. *The Proceedings of IEEE*, 84(8):1089–1123, August 1996.

[11] G. Luo, G. Bochmann, and A. Petrenko. Test selection based on communicating nondeterministic finite-state machines using a generalized wp-method. *IEEE Transactions on S.E.*, 20(2):149–162, 1994.

[12] N.Arakawa and T. Soneoka. A test case generation method for concurrent programs. In R.J. Heijink J. Kroon and E. Brinksma, editors, *Protocol Test Systems, IV*, pages 95–106. Elsevier Science Publisher B. V.(North-Holland), 1992.

[13] O Rafiq and R. Castanet. From conformance testing to interoperability testing. In *The 3rd International Workshop on Protocol Test Systems*, 1990.

[14] G. S. Vermeer and H. Blik. Interoperability testing: Basis for the acceptance of communicating systems. In *Protocol Test Systems, VI(C-19)*. Elsevier Science Publisher B. V.(North-Holland), 1994.

ON TEST DERIVATION FROM PARTIAL SPECIFICATIONS

Alex Petrenko[a] and Nina Yevtushenko[b]
[a] *CRIM, Centre de recherche informatique de Montréal, 550 Sherbrooke West, Suite 100, Montreal, H3A 1B9 Canada, E-mail: petrenko@crim.ca*
[b] *Tomsk State University, 36 Lenin Str., Tomsk, 634050, Russia, E-mail: yevtushenko.RFF@elefot.tsu.ru*

Abstract The paper addresses the problem of test derivation from partially defined specifications. A specification is modeled by an Input/Output FSM such that transitions from some states on some inputs are not specified (a partial FSM). Tests have to be derived for a weak conformance relation between FSMs as a conformance relation. The paper further elaborates the state-counting approach by providing an insight into the structure of tests with fault coverage for partial deterministic machines and by offering risk-free optimizations which reduce the length of resulting tests. Based on this approach, a method for deriving tests with fault coverage guarantee (checking experiments) is proposed. It is demonstrated that the method is superior to other test derivation methods for deterministic state machines.

Keywords: I/O FSMs, partially defined machines, test derivation, weak conformance testing, fault detection, checking experiments.

1. INTRODUCTION

There are not many papers on test derivation that deal with partial specifications, in general, and with partial Input/Output FSMs, in particular. One of the reasons is that a classical state identification approach pioneered in [7] does not extend to an arbitrary partially defined FSM, as its states may no longer be distinguishable using only "core" or specified transitions. The

Tommaso Bolognesi and Diego Latella (Eds.), *Formal Methods for Distributed System Development*.

core transitions have exclusively to be used to check for a *weak* conformance, as opposed to the *strong* conformance (equivalence) used for complete machines. A pragmatic solution that sometimes works is to complete a partial specification in one way or another. A complete specification can then be minimized so that the state identification approach becomes applicable; however, some conforming implementations may now be recognized as nonconforming as already argued, e.g., in [14], [20]. Note that in [20] it is assumed that a partial machine is reduced, which seems to be only a special case. In a more general case, a partial machine may not be reduced and another approach is required. As a solution to the problem, we attempted to develop such an approach, called a state-counting approach, in a number of papers [21], [22], [13], [17], [16]. This approach relies on counting states traversed by test sequences to decide when a particular sequence could be trimmed. It subsumes the classical state identification approach of Hennie [7] and his followers, see, e.g., [19], [2], [4], [11], [18].

The goal of this paper is to further advance the state-counting approach by providing an insight into the structure of tests with fault coverage for partial (deterministic) machines with compatible, quasi-equivalent, and distinguishable states and by offering risk-free optimizations which reduce the length of resulting tests.

The rest of the paper is organized as follows. In Section 2, we give necessary basic notions. In Section 3, we discuss the structure of tests for various types of FSMs, which may have compatible, quasi-equivalent and distinguishable states and propose a method for test derivation from partial FSMs with complete fault coverage. We also discuss the complexity of test suites obtained with the proposed method. In Section 4, we summarize the contributions of the paper and indicate possible future work.

2. BASIC DEFINITIONS AND NOTATIONS

A finite state machine (FSM), often simply called a machine throughout this paper, is an initialized deterministic Mealy machine which can be formally defined as follows. An FSM A is a 7-tuple $(S, s_0, X, Y, D_A, \delta, \lambda)$, where S is a finite set of states with the initial state s_0, X is a finite set of inputs, Y is a finite set of outputs, D_A is a subset of $S \times X$, called a *specification domain*, and δ and λ are a transition function δ: $D_A \rightarrow S$ and an output function λ: $D_A \rightarrow Y$, respectively.

An FSM A is said to be *completely* specified (a complete FSM) if $D_A = S \times X$. We will omit the specification domain D_A in the case of complete machines. If D_A is a proper subset of $S \times X$ then A is called a *partially* specified machine (a partial FSM). Note that, sometimes [6], the functions δ and λ are defined over different specification domains, however, to simplify our discussions we will not consider here such cases. Moreover, we note that, for example, in typical protocol applications, both functions of protocol machines are defined over the same domain.

Given a string $\alpha = x_1..x_k$ of the set X^* of all possible input sequences, α is said to be an *acceptable* input sequence at state $s \in S$ if there exist states s_1, ..., s_k, s_{k+1}, where $s_1=s$, such that $(s_i,x_i) \in D_A$ and $\delta(s_i,x_i) = s_{i+1}$ for any $i = 1$, ..., k. In words, α is an acceptable input sequence at state s if the behavior of A for the sequence α is defined. We use $X_A^{\cdot}(s)$ to denote the set of all acceptable input sequences for state s and $X_A^{\cdot}$ for the state s_0, i.e. for A. The set $X_A^{\cdot}(s)$ may be empty for state s, but hereinafter we assume that the set $X_A^{\cdot}$ is not empty.

We extend the transition and output functions from input symbols to acceptable input sequences, including the empty word ε, as usual. Let $\delta(s, \varepsilon) = s$ and $\lambda(s, \varepsilon) = \varepsilon$ for any $s \in S$. Suppose that β is an acceptable input sequence for state s and $\delta(s, \beta) = s'$ and $\lambda(s, \beta) = \gamma$. Then for any $x \in X$ such that $(s', x) \in D_A$ we define $\delta(s, \beta x) = \delta(s', x)$ and $\lambda(s, \beta x) = \gamma\lambda(s', x)$. For convenience, we use the same notations δ and λ for the extended functions, as well, since in our discussions, this does not imply any contradiction.

Given states $s, s' \in S$, a sequence $\alpha \in X_A^{\cdot}(s)$ such that $\delta(s, \alpha) = s'$ is a *transfer* sequence from s to s'. For any state s, the empty sequence ε is a transfer sequence from s to s. A transfer sequence, if we do not specify from which state it is, starts, by default, from the initial one of a given FSM.

We say that the sequence α, applied at state s, *traverses* state s' if there exists a non-empty prefix β of α such that β is a transfer sequence from s to s'. We also say that a set of sequences *traverses* a given state if it contains at least one sequence that traverses state. We consider here only connected machines. FSM $A = (S, s_0, X, Y, D_A, \delta, \lambda)$ is said to be *connected* if all its states are reachable from the initial state, i.e., for any state $s \in S$, there exists a transfer sequence $\alpha \in X_A^{\cdot}$ from s_0 to s.

A *state cover* of FSM A with n states is defined as a set of n transfer sequences that take the machine A from the initial state to every state. An input sequence γ is said to *cover* a transition from state s with input x if it can be represented as a concatenation $\gamma = \alpha x \beta$ such that α is a transfer sequence to s. We also say that a given set of sequences covers a given transition if it has at least one sequence that covers the transition.

Given state s of an FSM $A = (S, s_0, X, Y, D_A, \delta, \lambda)$ and state t of an FSM $B = (T, t_0, X, Y, D_B, \Delta, \Lambda)$, states s and t are said to be *compatible* if either $X_A^{\cdot}(s) \cap X_B^{\cdot}(t) = \varnothing$ or $\lambda(s, \alpha) = \Lambda(t, \alpha)$ for any input sequence $\alpha \in X_A^{\cdot}(s) \cap X_B^{\cdot}(t)$. Otherwise, i.e., when there exists an input sequence $\alpha \in X_A^{\cdot}(s) \cap$

$X_B(t)$ such that $\lambda(s, \alpha) \neq \Lambda(t, \alpha)$, states s and t are said to be *distinguishable* (by α), written $s \_ t$ ($s \__{\alpha} t$). Given a set $W \subseteq X_A(s) \cap X_B(t)$, we also use $s \__W t$ to denote the fact that there exists $\alpha \in W$ such that $s \__{\alpha} t$. In the case of a complete FSM, two distinguishable states can already be distinguished by input sequence α of length $|\alpha| \leq n - 1$, where n is the number of states in the given machine. In the case of a partial machine, the corresponding bound is $n(n - 1)/2$ [14]. An FSM is *reduced* if any two of its states are distinguishable; a machine with compatible states is said to be *unreduced*.

An unreduced FSM may have quasi-equivalent or equivalent states. Given two compatible states, s of an FSM $A = (S, s_0, X, Y, D_A, \delta, \lambda)$ and t of an FSM $B = (T, t_0, X, Y, D_B, \Delta, \Lambda)$, state t of B is said to be *quasi-equivalent* to state s of A, written $t \_ s$, if $X_B(t) \supseteq X_A(s)$. The quasi-equivalence relation, introduced in [5], is also called weak conformance in [20] and other works. If states s and t are quasi-equivalent to each other, we refer to these states as to *equivalent* states, written $s \_ t$.

Given FSMs A and B over the same input alphabet, B is said to be *quasi-equivalent to A*, written $B \_ A$, if the initial state of FSM B is quasi-equivalent to the initial state of FSM A. The two machines are *distinguishable*, written $B \_ A$, if their initial states are distinguishable.

If a reduced FSM B is quasi-equivalent to A then it is called a *reduced form* of A. Methods for constructing a reduced form from a partial machine are more involved than that for complete machines. The reason is that, in general, the relation of state compatibility is not transitive for partial FSMs, i.e. an appropriate state of an implementation FSM may be quasi-equivalent to different compatible states of a specification FSM. This feature explains why test derivation from partial machines is a more difficult problem than that for complete machines. We use the quasi-equivalence relation (weak conformance) for deriving tests directly from a given FSM without first minimizing it.

Let X be the input alphabet of a (partial or complete) specification FSM A. A *test suite* for FSM A is a finite set of acceptable input sequences (tests) of A. We thus assume that a reliable reset is available in any potential implementation machine to bring it back to the initial state to resume testing. We assume also that all implementations are represented by a set $\Im(X)$ of complete FSMs defined over the input alphabet X, the set $\Im(X)$ is often called a *fault domain*. A test suite TS is said to be *complete* for A w.r.t. the fault domain $\Im(X)$ if, for all $B \in \Im(X)$, $B \_ A$ implies $B \__{TS} A$. Let $\Im_m(X)$ be a universal set of all complete FSMs with m states over the input alphabet X. A test suite is said to be *m*-complete if it is complete w.r.t. the fault domain $\Im_m(X)$. Clearly, *m*-complete test suite is also *k*-complete for any $k < m$.

3. TEST DERIVATION

3.1 Complete Test Suite and Distinguishing Machine

Let A be a given specification FSM, partially or completely defined, and B be a completely defined implementation FSM, $B \in \Im_m(X)$. To verify whether or not the FSM B is quasi-equivalent to A, we construct a designated FSM, called a distinguishing machine of A and B, whose states are pairs of states of A and B. Its initial state is the pair of initial states of the two machines, the remaining states are determined by performing a reachability analysis.

Formally, we define the distinguishing machine of FSM $A = (S, s_0, X, Y, D_A, \delta, \lambda)$ and FSM $B = (T, t_0, X, Y', \Delta, \Lambda)$ as an FSM with the state set $Q \subseteq S \times T$, the input set X, the output set $Y \cup \{fail\}$, $fail \notin (Y \cup Y')$, the transition function ψ, the output function φ, and the initial state (s_0, t_0), using the following rules:

- $q_0 = (s_0, t_0)$,
- $\psi((s, t), x) = (\delta(s, x), \Delta(s, x))$, if $(s, x) \in D_A$, otherwise, $\psi((s, t), x)$ is undefined;
- $\varphi((s, t), x) = \lambda(s, x)$ if $\lambda(s, x) = \Lambda(t, x)$, and
- $\varphi((s, t), x) = fail$ if $\lambda(s, x) \neq \Lambda(t, x)$;

such that the state set Q is the smallest set obtained by application of the above rules. The machine $(Q, q_0, X, Y \cup \{fail\}, D_{A\oplus B}, \psi, \varphi)$, where $D_{A\oplus B} = \{((s, t), x) \mid (s, t) \in Q \wedge (s, x) \in D_A\}$, is called the *distinguishing machine* (DM) of A and B, and is denoted by $A\oplus B$.

Clearly, the set Q includes only states reachable from the initial state, i.e., DM is an initially connected machine. Note that similar notions of FSM products are used in other work, see, e.g., [1], [3], [16], [9].

If FSM B is distinguishable from A, then any input sequence that, applied at the initial state of DM $A\oplus B$, covers a fail-transition (one with the output *fail*), distinguishes B from A and can be used as a test case to expose faults modeled by B. The DM enjoys several other properties useful for test generation, some of them are collected in the following proposition.

Proposition 1. Given two FSMs, $A = (S, s_0, X, Y, D_A, \delta, \lambda)$ with n states and $B = (T, t_0, X, Y', \Delta, \Lambda)$ with m states, let $A\oplus B = (Q, q_0, X, Y \cup \{fail\}, D_{A\oplus B}, \psi, \varphi)$ be DM of A and B. Then

(a) $X_{A\oplus B}(s, t) = X_A(s)$, for all $s \in S$ and $t \in T$ such that $(s, t) \in Q$.

(b) $|Q| \leq nm$.

□

Proposition 1(a) says that any input sequence acceptable in DM is also acceptable in the specification machine A and can be used as a test. The following proposition, which is an immediate corollary to Proposition 1(b), establishes an upper bound on tests for partial FSMs.

Corollary 1. For any FSM A with n states and the set X_A of acceptable input sequences, there exists a subset of the set $\{\alpha \in X_A \mid |\alpha| = nm\} = X_A^{nm}$ that is an m-complete test suite.

❑

Example

Consider the specification FSM A with n states and implementation FSM B with m states shown in Figure 1 [21]. DM has the only fail-transition out of state (s_n, t_m). The shortest input sequence that covers this transition from the initial state is $(1^{n-1}2)^m$, its length is nm. Therefore, the bound nm is indeed tight for partial machines.

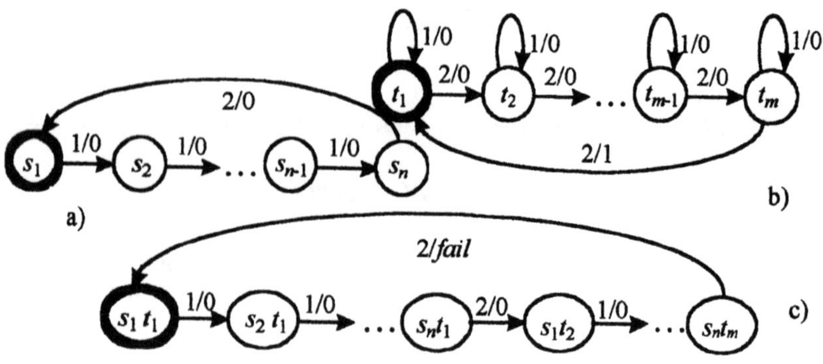

Figure 1. An example of machines distinguished by a sequence of length nm: a) FSM A, b) FSM B, c) DM $A \oplus B$

If the set of acceptable sequences is finite (when, for example, the machine has no cycling behavior) then it is a complete test suite regardless of the bound on the number of states in implementation machines. A systematic method is required to determine complete test suites for FSMs with cyclic behavior. We first develop a method for several important classes of unreduced FSMs and then extend it to an arbitrary deterministic FSM, defined partially or completely.

3.2 Deriving Tests for Special Classes of Unreduced FSMs

We will describe a method for test derivation that accepts any deterministic I/O FSM, completely or partially specified, which may have compatible, equivalent, quasi-equivalent and distinguishable states. By combination, one can define many FSM classes, which differ in relations between the states of any machine from the class. Here we consider the following classes:

- FSMs such that each two states are compatible, but neither quasi-equivalent nor distinguishable.
- FSMs such that each two states are either compatible or quasi-equivalent, but not distinguishable.
- A general type of FSMs such that each two states are compatible, quasi-equivalent, or distinguishable. This class includes reduced machines and completely defined machines.

3.2.1 FSMs Without Distinguishable or Quasi-Equivalent States

Given an implementation FSM $B \in \Im_m(X)$ such that $B \_ A$, a test suite to be m-complete should have a sequence that covers a fail-transition in DM $A \oplus B$. It is known that the length of a shortest input sequence that covers a transition of any FSM does not exceed the number of its states. On the other hand, states of distinguishing machines consist of states of FSMs A and B, therefore any state of A constitutes at most m states of DM. It means that a transfer sequence traversing the initial state of A m times is cyclic in DM and can be replaced by a shorter (acyclic) one. A transfer sequence is said to be *acyclic* if any two prefixes of it take FSM into different states. These considerations lead us to the following statements.

Lemma 1. Given a state cover V of FSM A and state q of DM $A \oplus B$, there exist a sequence $\alpha \in V$ and a transfer sequence $\alpha\beta \in X_{A\square B}$ to state q such that β, applied at state $\delta(s_0, \alpha)$, traverses any state of FSM A at most m -1 times.

Proof. Let $\alpha\beta \in X_{A\square B}$ be a transfer sequence to state q in DM $A \oplus B$ such that β, applied at state $\psi(q_0, \alpha)$ is an acyclic transfer sequence. Then each state of FSM A is traversed by β at most m times. Suppose there exists state s of A traversed by β, applied at state $\delta(s_0, \alpha)$, exactly m times. Let q_1, ..., q_m be the sequence of states $A \oplus B$ with the component state s traversed by β, applied at state $\psi(q_0, \alpha)$. For β to be acyclic in $A \oplus B$ it is required that all component states of FSM B in q_1, ..., q_m be distinct states, denoted t_1, ..., t_m, where $t_m = t$. Consider a transfer sequence $\omega \in V$ to state s in FSM A. Then the corresponding state $\Delta(t_0, \omega)$ of FSM B must be one of the states $t_1, \ldots, t_m$, since FSM B has at most m states, i.e., there exists a transfer sequence $\omega\beta' \in X_{A\square B}$ to state q in DM $A \oplus B$, such that $\omega \in V$ and β', applied at state $\delta(s_0, \omega)$, traverses state s of FSM A at most m-1 times.

□

The lemma allows us to determine the maximal length of input sequences covering each transition in an arbitrary DM $A \oplus B$ to obtain a complete test suite.

Proposition 2. Let V be a state cover of A and $E = \bigcup_{\alpha_i \in V} \alpha_i L_i \subseteq X_A^*$ be

the state q with an emanating fail-transition. Let x be the input that labels the fail-transition. By Lemma 1, there exists a transfer sequence $\alpha_i\beta \in X^*_{A\square B}$ to state q such that $\alpha_i \in V$, $\delta(s_0, \alpha_i) = s_i$, and β traverses any state of FSM A at most m -1 times. Then the sequence β is a proper prefix of some sequence in the set L_i. Hence, the sequence βx is a prefix of some sequence in the set L_i. Thus, $B \__E A$.

□

The use of a prefix-closed state cover facilitates merging sequences in sets $\alpha_i L_i$ and may thus reduce tests. However, there might be even no need to consider each transfer sequence from a state cover, as follows from the proof of Lemma 1.

Corollary 2. Given an m-complete test suite E for A from Proposition 2, let $s_j \in S$ be state of A, different from the initial state, such that $s_j = \delta(s_0, \alpha_j)$ and for all $\alpha_i \in V$, s_j is traversed less than m times by any input sequence $\beta \in L_i$, then the set $E \setminus \{\alpha_j L_j\}$ is an m-complete test suite for A.

□

In a general case, the length of sequences in the resulting test suite cannot be further reduced if states of the given machine are neither distinguishable nor quasi-equivalent. FSM A in Figure 1 serves as an example of such machines. For other classes of machines, which have quasi-equivalent or distinguishable states, shorter (and fewer) sequences than in $X_A^{''''}$ may constitute a complete test suite. In the next section, we consider how quasi-equivalent states can be taken into account to reduce a complete test suite.

3.2.2 FSMs without Distinguishable States

Assume now that a given machine has quasi-equivalent states, but no two states are distinguishable. The existence of quasi-equivalent states may shorten transfer sequences in distinguishing machines, compared to Lemma 1. In fact, let (s', t) and (s, t) be states of DM $A \oplus B$ such that $s' \_ s$, i.e., $(s', t) \_ (s, t)$. Any transfer sequence to state q that first traverses (s', t) and then $(s,$

t) can be replaced by a shorter sequence that takes the machine to a state that is quasi-equivalent to state q, but avoids state (s, t). Indeed, successors of quasi-equivalent states for any acceptable input sequence are also quasi-equivalent states. This observation motivates the following definitions.

A transfer sequence $\nu \in X_{A\square B}$ is said to be *acyclic in DM A⊕B* if, for any two non-empty prefixes α and β of ν, $|\alpha| < |\beta|$, it holds that if $\psi(q_0, \alpha) = (s, t)$, $\psi(q_0, \beta) = (s', t')$, and $s' \_ s$, then $t \neq t'$.

We introduce a designated function, denoted #, which for given states s, $p \in S$ and input sequence $\alpha \in X_A^{'}(s)$, returns a maximal length of a sequence of states, traversed by α, applied at state s, each of which is quasi-equivalent to all subsequent states in the sequence, including state p. Formally, $\#_s(p, \alpha) = l$, where l is the maximal number of prefixes $\alpha_1, \ldots, \alpha_l$ of α such that $0 < |\alpha_1| < \ldots < |\alpha_l|$, and $\delta(s, \alpha_1) \_ \ldots \_ \delta(s, \alpha_l) \_ p$. $|\alpha| \geq \#_s(p, \alpha) \geq 0$ and $\#_s(p, \alpha) = 0$ if α is an empty sequence or none of the states traversed by a non-empty sequence α is quasi-equivalent to p.

We also introduce the concept of a core of an FSM. Given FSM A, we determine a minimal set of states of A such that contains the initial state and, for each state $s \in S$, a state quasi-equivalent to s, called a *core* of FSM A. (A set is minimal (maximal) with respect to the inclusion ordering.) In the case, when the machine A has no quasi-equivalent states, the core coincides with the state set S. A subset $K \subseteq X_A^{'}$ is said to be a *core cover* of A if, for each state in the core of A, it has exactly one transfer sequence.

Example

Consider the partial FSM A shown in Figure 2. States 1 and 3 are compatible states. These states have no common acceptable input sequence. We also have $2 \_ 3$. The core of FSM A includes states 1 and 2. The machine has the core cover $K = \{\varepsilon, 1\}$, where ε is an empty sequence.

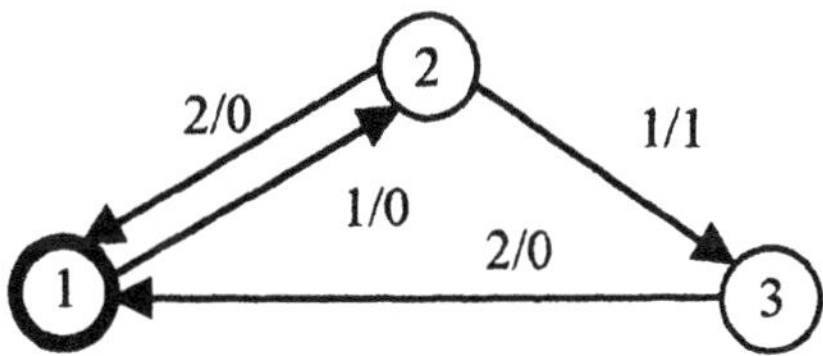

Figure 2. The partial FSM *A*.

Using the above notions, Lemma 1 can be generalized to the case of quasi-equivalent states as follows.

Lemma 2. Given a core cover K of FSM A and state q of DM $A⊕B$, there exist a sequence $\alpha \in K$ and a transfer sequence $\alpha\beta \in X_{A\square B}$ to state $q' \_ q$ in DM $A⊕B$, such that $\#_{\delta(s_0,\alpha)}(s, \beta) \leq m - 1$ for all $s \in S$.

Proof. Let $\nu\beta \in X_{A\square B}$ be a transfer sequence to state q in DM $A⊕B$ such that β, applied at state $\psi(q_0, \nu)$ is an acyclic transfer sequence in DM

$A \oplus B$, i.e., $\#_{\delta(s_0,v)}(s, \beta) \leq m$ for all $s \in S$. Assume that there exists state $s \in S$ such that $\#_{\delta(s_0,v)}(s, \beta) = m$, while $\#_{\delta(s_0,v)}(s', \beta) < m$ for all $s' \in S$, $s' \neq s$. We have m prefixes $\beta_1, ..., \beta_m$ of β such that $0 < |\beta_1| < ... < |\beta_m|$ and $\delta(s_0, v\beta_1) \_ ... \_ \delta(s_0, v\beta_m) \_ s$. The sequence β is acyclic, i.e., the corresponding m states of FSM B are distinct. Thus, we have m states of DM $A \oplus B$ $(\delta(s_0, v\beta_1), \Delta(t_0, v\beta_1)), ..., (\delta(s_0, v\beta_m), \Delta(t_0, v\beta_m))$, such that $\delta(s_0, v\beta_1) \_ ... \_ (\delta(s_0, v\beta_m) \_ s$ and $\Delta(t_0, v\beta_i) = \Delta(t_0, v\beta_j)$ implies $i = j$. By definition of a core cover, there exists $\alpha_i \in K$ such that $\delta(s_0, \alpha_i) = s_i$ and $s_i \_ \delta(s_0, v\beta_1) \_ ... \_ (\delta(s_0, v\beta_m) \_ s$. The sequence α_i takes DM $A \oplus B$ into state $\psi(q_0, \alpha_i) = (s_i, \Delta(t_0, \alpha_i))$ such that $\Delta(t_0, \alpha_i) = \Delta(t_0, v\beta_j)$ and $\delta(s_0, \alpha_i) \_ \delta(s_0, v\beta_j)$ for some non-empty prefix β_j of β, as FSM B has at most m states. Then $\psi(q_0, \alpha_i) \_ \psi(q_0, v\beta_j)$ and the transfer sequence $v\beta$ to state q can be represented as a concatenation $v\beta_j\beta'$, such that $\#_{\delta(s_0,v\beta_j)}(s, \beta') \leq m - 1$ for all $s \in S$. $\psi(q_0, \alpha_i) \_ \psi(q_0, v\beta_j)$ implies that $\psi(q_0, \alpha_i\beta') \_ \psi(q_0, v\beta_j\beta') = q$. Thus, there exists a transfer sequence $\alpha_i\beta'$ to a state quasi-equivalent to the given state q, such that $\#_{\delta(s_0,\alpha_i)}(s, \beta') \leq m - 1$ for all $s \in S$.

❑

The lemma leads us to the following statement that includes Proposition 2 as a special case.

Proposition 3. Let K be a core cover of A and $F = \bigcup_{\alpha_i \in K} \alpha_i M_i \subseteq X_A^*$ be the set of input sequences such that $\beta \in M_i$ iff $\beta \in X_A^*(\delta(s_0, \alpha_i))$, $\#_{\delta(s_0,\alpha_i)}(\delta(s_0, \alpha_i\beta), \beta) = m$ and $\#_{\delta(s_0,\alpha_i)}(s, \beta) < m$ for all $s \neq \delta(s_0, \alpha_i\beta)$. The set F is an m-complete test suite of FSM A.

Proof. Let $B \in \Im_m(X)$ and $B \_ A$. If $B \_ A$ then in DM $A \oplus B$, there exists state q with an emanating fail-transition. Let x be the input that labels the fail-transition. By Lemma 2, there exists a transfer sequence $\alpha_i\beta \in X_{A \square B}$ to state $q' \_ q$, such that $\alpha_i \in K$ and $\#_{\delta(s_0,\alpha_i)}(s, \beta) \leq m - 1$ for all $s \in S$. Then the sequence β is a proper prefix of some sequence in the set M_i. $q' \_ q$ implies that $X_{A \square B}(q') \supseteq X_{A \square B}(q)$ and x induces a fail-transition from the state q'. Hence, the input x is also acceptable at state q' and sequence βx is a prefix of some sequence in the set M_i that covers a fail-transition. Thus, $B \__F A$.

❑

Similar to Corollary 2, we have the following.

Corollary 3. Given an m-complete test suite F from Proposition 3, let $\alpha_j \in K$ be a transfer sequence from a core cover of A such that $s_j = \delta(s_0, \alpha_j)$, $s_j \neq s_0$ and for any $\alpha_i \in V$, $i \neq j$, any $p \in S$, $s_j \_ p$, it holds that $\#_{\delta(s_0,\alpha_i)}(p, \beta) < m$ for all $\beta \in M_i$, then the set $F \setminus \{\alpha_j M_j\}$ is an m-complete test suite.

3.3 Deriving Tests for General Type of FSMs

Now we allow for the specification FSM A to have any type of states, compatible, quasi-equivalent or distinguishable.

Given FSM A, we determine distinguishable states of FSM A, using standard algorithms for state minimization of partial FSMs [5], [8], [6], [10]. We will use maximal sets of mutually distinguishable states. These sets can be determined as cliques of the so-called state distinguishability graph, where two nodes representing states have an edge if these states are distinguishable.

Given a transfer sequence $\alpha\beta \in X_A$, let $D_{\beta\text{after}\alpha}$ denote a maximal set of mutually distinguishable states of A, such that at least one of them is traversed by β, applied at state $\delta(s_0,\alpha)$. Let $n_{\max}$ denote the maximal number of mutually distinguishable states of A. Hereinafter, we assume that m, the number of states in an implementation machine, is not less than $n_{\max}$. Distinguishable states of FSM A may create distinguishable states in DM $A\oplus B$. In particular, states of DM $A\oplus B$ (s, t) and (s', t) are said to be *conflicting* if s _ s' [16].

To generalize Proposition 3 for the case of conflicting states in DM $A\oplus B$, we need the following notation. Given state $p \in S$, an input sequence $\beta \in X_A^*(p)$ and a set of mutually distinguishable states $D \subseteq S$, let $\#_p(D, \beta)$ denote $\square_{s\{ D}\#_p(s, \beta)$. Let $\#_p(D, \beta) = l$. If $l > 0$ then by definition of the function #, β has exactly l prefixes $\beta_1, ..., \beta_l$, $0 < |\beta_1| < ... < |\beta_l|$ with the following property. For each $s \in D$, such that $\#_p(D \setminus \{s\}, \beta) < l$ and for each i, $i \leq l$, if $\delta(s, \beta_i)$ _ s and $\delta(s, \beta_j)$ _ s for some $j < i$, then $\delta(s, \beta_j)$ _ $\delta(s, \beta_i)$. We use $P(\beta, D)$ to denote the set $\{\beta_1, ..., \beta_l\}$.

Proposition 4. Let K be a core cover of FSM A and $G = \bigcup_{\alpha_i l K}\alpha_i N_i \subseteq X_A^*$ be a set of input sequences, where N_i comprises each shortest sequence $\beta \in X_A^*(\delta(s_0, \alpha_i))$ such that $\#_{\delta(s_0,\alpha_i)}(D, \beta) = m - |D| + 1$ holds for an appropriate set D of mutually distinguishable states of A. The following statements hold.

1. If FSM A has no distinguishable states then the set G covers a fail-transition in each DM $A\oplus B$, such that $B \in \Im_m(X)$ and B _ A.
2. Let $B \in \Im_m(X)$, B _ A and B $_G$ A. Then there exist $\alpha_i \in K$, $\beta \in N_i$ and $D_{\beta\text{after}\alpha_i}$ such that $|D_{\beta\text{after}\alpha_i}| > 1$ and $\#_{\delta(s_0,\alpha_i)}(D_{\beta\text{after}\alpha_i}, \beta) = m - |D_{\beta\text{after}\alpha_i}| + 1$, moreover, in DM $A\oplus B$, the set of states $\{\psi(q_0, \alpha) \mid (\alpha \in K \wedge \exists\, s \in D_{\beta\text{after}\alpha_i}: \delta(s_0, \alpha)$ _ $s) \vee \exists\, \gamma \in P(\beta, D_{\beta\text{after}\alpha_i}) : \alpha = \alpha_i\gamma\}$ contains two conflicting states.

Proof. 1. If FSM A has no distinguishable states then the set G coincides with the set F in Proposition 3, i.e., G is an m-complete test suite.

2. We have to demonstrate that for an FSM B that has passed the test suite G, but is distinguishable from A, the set of states in FSM $A \oplus B$ traversed by an input sequence from some N_i, applied at the state $\psi(s_0 t_0, \alpha_i)$, together with the set of states where sequences of the core cover K take the $A \oplus B$ from the initial state, has two conflicting states. Moreover, in these two conflicting states, states of FSM A are quasi-equivalent to states of the set D that bounds the input sequence.

Let B _ A and β be a shortest sequence such that the sequence $\alpha_i \beta \in X_{A \square B}$, $\alpha_i \in K$, is a transfer sequence in DM $A \oplus B$ to state q with a fail-transition. By definition of the set G, there exists a set $D_{\beta \mathrm{after} \alpha_i}$ such that $\#_{\delta(s_0, \alpha_i)}(D_{\beta \mathrm{after} \alpha_i}, \beta) \geq m - |D_{\beta \mathrm{after} \alpha_i}| + 1$, since B _$_G$ A. Let ν be the shortest prefix of β such that:

1. $\#_{\delta(s_0, \alpha_i)}(D_\nu, \nu) = m - |D_\nu| + 1$ for some set D_ν of mutually distinguishable states of A,
2. ν is an acyclic sequence from the state $\psi(q_0, \alpha_i)$, moreover, for any state (s, t) traversed by ν and any state $(s', t') = \psi(q_0, \mu)$, $\mu \in K$, it holds that s' _ $s \Rightarrow t' \neq t$.

We have the set $P(\nu, D_\nu) = \{\nu_1, \ldots, \nu_{m-|D_\nu|+1}\}$ of $m - |D_\nu| + 1$ prefixes of ν such that $0 < |\nu_1| < \ldots < |\nu_{m-|D_\nu|+1}|$. If the corresponding $m - |D_\nu| + 1$ states of FSM B, $\Delta(t_0, \alpha_i \nu_1), \ldots, \Delta(t_0, \alpha_i \nu_{m-|D_\nu|+1})$, are not distinct then, by definition of an acyclic sequence, the sequence ν traverses two conflicting states (s, t) and (s', t), where s, s' are quasi-equivalent to two different states of the set D_ν, i.e., $|D_\nu| > 1$. Assume, therefore, that the states $\Delta(t_0, \alpha_i \nu_1), \ldots, \Delta(t_0, \alpha_i \nu_{m-|D_\nu|+1})$ are distinct. Let K_ν be a subset of the core cover K of A that has exactly $|D_\nu|$ transfer sequences such that for any $j = 1, \ldots, m - |D_\nu| + 1$ there exists $\gamma \in K_\nu$ for which $\delta(s_0, \gamma)$ _ $\delta(s_0, \alpha_i \nu_j)$ holds. The corresponding $|D_\nu|$ states of FSM B are distinct, if the core cover K does not traverse conflicting states. Then the two subsets of states of FSM B, $\{\Delta(t_0, \alpha_i \nu_1), \ldots, \Delta(t_0, \alpha_i \nu_{m-|D_\nu|+1})\}$ and $\{\Delta(t_0, \gamma) \mid \gamma \in K_\nu\}$ intersect, as, together, they have $m - |D_\nu| + 1 + |D_\nu| = m + 1$ states, while B has only m states. Thus, there exist a sequence $\mu \in K_\nu$ and non-empty prefix ν_j of ν such that $\Delta(t_0, \mu) = \Delta(t_0, \alpha_i \nu_j)$, while $\delta(s_0, \mu)$ and $\Delta(t_0, \alpha_i \nu_j)$ are quasi-equivalent to some states in the set D_ν. If they are quasi-equivalent to the same state then $\delta(s_0, \mu)$ _ $\delta(s_0, \alpha_i \nu_j)$ by definition of the set K_ν. The case $\delta(s_0, \mu)$ _ $\delta(s_0, \alpha_i \nu_j)$ contradicts to property (2) of the sequence ν. Thus, $\delta(s_0, \mu)$ and $\delta(s_0, \alpha_i \nu_j)$ are quasi-equivalent to two different states of the set D_ν. Hence, $|D_\nu| > 1$ and the set of states $\{\psi(q_0, \alpha) \mid \alpha \in K \wedge \exists\, s \in D_\nu :$ $\delta(s_0, \alpha)$ _ $s \vee \exists\, \gamma \in P(\nu, D_\nu) : \alpha = \alpha_i \gamma\}$ contains two conflicting states $[\delta(s_0, \mu), \Delta(t_0, \mu)]$ and $[\delta(s_0, \alpha_i \nu_j), \Delta(t_0, \alpha_i \nu_j)]$, where $\delta(s_0, \mu)$ and $\delta(s_0, \alpha_i \nu_j)$ are quasi-equivalent to two different states of the set D_ν.

❑

The set G from Proposition 4 can easily be extended to cover fail-transitions missed by G, using the following property of distinguishing machines.

Lemma 3. Let (s', t) and (s, t) be conflicting states of $A \oplus B$ such that $s' \__{\alpha}$ s for some input sequence α, then the input sequence α covers a fail-transition in DM $A \oplus B$ from (s', t) or (s, t).

❑

The presence of distinguishable states in FSM A allows us to decrease the length (and therefore, the number) of sequences needed to traverse states (cf. Proposition 3 and 4) at a price of including additional input sequences distinguishing states in A. The gain increases with the number of distinguishable states. Informally speaking, the more distinguishable states FSM A has, the more the structure of complete test suites resembles that implied by existing test derivation methods for reduced machines. To demonstrate the point more formally, we determine the set G of Proposition 4 for the case of completely defined reduced FSM A.

The core cover K of such a machine coincides with a state cover V of the given machine. All n states of A are mutually distinguishable, thus $|D_{\beta \text{after} \alpha}| = n$ for any $\alpha\beta \in X_A$ with non-empty β. Then each set N_i is nothing more but the set of all input sequences of the length up to $m - n + 1$ acceptable in state $\delta(s_0, \alpha_i)$, i.e. $N_i = X^{m-n+1} \cap X_A(\delta(s_0, \alpha_i))$. If A is completely defined then $N_i = X^{m-n+1}$ for all $\alpha_i \in V$. Thus, $G = VX^{m-n+1}$. Let W denote a characterization set of FSM A. For any two distinguishable states, it contains a sequence that distinguishes them. According to Lemma 3, the sequences of W, concatenated to each sequence in the set G, cover any remaining fail-transition. The set $VX^{m-n+1}W$ is exactly the test suite derived by the W-method [19], [2], which is m-complete. A similar observation applies to the test derivation method developed for reduced partial FSMs in [20]. This indicates that the method we are developing includes the existing methods as special cases.

We are now ready to present the detailed description of the method.

Input. FSM A; the list of pairs of distinguishable states and for each pair (s, s'), an input sequence $\gamma(s, s')$ that distinguishes s and s'; all maximal sets of mutually distinguishable states; and an integer m equal or greater than the maximal number of mutually distinguishable states.

Output. An m-complete test suite for FSM A.

The method includes the following steps.

1. Determine a core cover K of A.
2. For each $\alpha_i \in K$, determine the set of all shortest input sequences $N_i \subset X_A(\delta(s_0, \alpha_i))$ such that $\#_{\delta(s_0,\alpha_i)}(D_{\beta \text{after} \alpha}, \beta) = m - |D_{\beta \text{after} \alpha_i}| + 1$ and $\delta(s_0, \alpha_i\beta)$

$\in D_{\beta after \alpha_i}$ for each sequence $\beta \in N_i$. Let $D(N_i)$ denote the union of sets $D_{\beta after \alpha_i}$ over all $\beta \in N_i$.

3. $\alpha_1 := \varepsilon$. $R := \{\alpha_1\}$.
4. For each N_i, where $\alpha_i \in R$, and for each state $s \in D(N_i)$, determine $\alpha \in K$ such that $\delta(s_0, \alpha)$ _ s and include α in R. Repeat this step until $R = K$ or no more sequences from K are included into R.
5. Determine the set $\{\alpha_i\gamma(\delta(s_0, \alpha_i), \delta(s_0, \alpha_j)) \mid \alpha_i, \alpha_j \in R\} = TS_1$.
6. Determine the set $\{\alpha_i\kappa\gamma(s, s') \mid \alpha_i \in R \wedge \beta \in N_i \wedge \kappa \in P(\beta, D_{\beta after \alpha_i})\} \wedge s, s' \in D_{\beta after \alpha_i}\} = TS_2$.
7. Determine $TS_1 \cup TS_2 = TS$, that is an m-complete test suite (all sequences that are prefixes of other sequences can be removed).

❑

We provide a few comments on the steps of this method. Determining a core of a given FSM in Step 1 is a typical combinatorial, set cover problem, i. e., find a minimal number of states that "cover" all the states of a given machine. We say that one state "covers" another state if the former is quasi-equivalent to the latter. Once a core is determined, a core cover can be obtained from a spanning tree similar to the one used to find a state cover, see e.g. [Chow78]. Step 2 is a key part of the method. It corresponds to constructing the set of traversal sequences VX^{m-n+1} in the W-, Wp-methods and other methods for reduced machines. The step can be performed as follows. We start with an empty input sequence and grow it in all the possible ways (obtaining acceptable sequences) to satisfy the condition $\#_{\delta(s_0,\alpha_i)}(D_{\beta after \alpha}, \beta) = m - |D_{\beta after \alpha_i}| + 1$. Step 4 attempts to minimize the number of sets of traversal sequences applied after transfer sequences from the core cover (see Corollaries 2 and 3), and thus, can always be omitted, by assuming $R = K$.

The set TS_1 ensures that if sequences in the core cover (or the set R, to be more precise) traverse conflicting states in a particular DM $A \oplus B$ then a fail-transition is covered by a sequence from TS_1. Similarly, the set TS_2 guarantees that if conflicting states are traversed by a sequence in N_i then a fail-transition is covered. Note that if FSM A has no distinguishable states then $TS_1 = R$ and $TS_2 = G = \bigcup_{\alpha_i \in R} \alpha_i N_i \supseteq R$, i.e., Steps 5-7 can be skipped (see Proposition 3).

Example

We derive a complete test suite for the partial FSM A shown in Figure 2, assuming that the number of states in any implementation machine does not exceed 3, i.e., $m \leq 3$. The core cover $K = \{\varepsilon, 1\}$. We have 1 _ 2 and $\gamma(1, 2) = 1$, as the machine in response to input 1 produces output 0 in state 1 and 1 in state 2. (1, 2) is the only pair of distinguishable states, hence, we have two maximal sets of mutually distinguishable states - $D_1 = \{1, 2\}$ and $D_2 = \{3\}$.

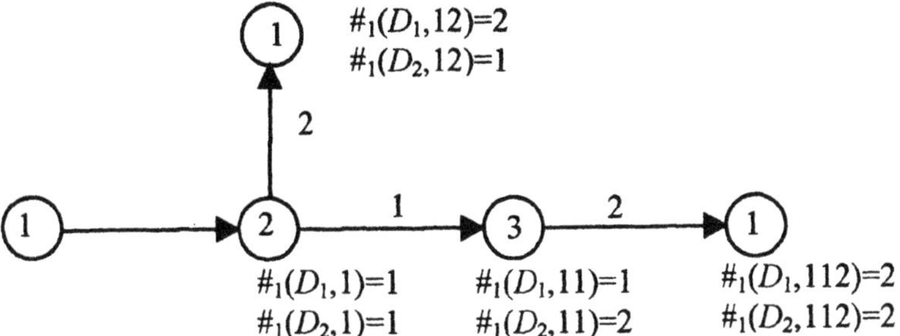

Figure 3. Constructing the set N_1 of traversal sequences.

We determine traversal sequences. In Step 2, let $\alpha_1 = \varepsilon$. The process of determining the set of input sequences N_1 is illustrated in Figure 3. A sequence β in N_1 has to be trimmed as soon as $\#_1(D_\beta, \beta) = m - |D_\beta| + 1$, this means when either $\#_1(D_1, \beta) = m - |D_1| + 1 = 3 - 2 + 1 = 2$ or $\#_1(D_2, \beta) = m - |D_2| + 1 = 3 - 1 + 1 = 3$. As an example, the sequence 11 cannot be terminated for neither of the two conditions is satisfied. However, for the sequence 112, we have $\#_1(D_1, 112) = 2$, so it is not extended any farther. We obtain $N_1 = \{112, 12\}$.

Similarly, the set $N_2 = \{121, 21\}$ is determined for $\alpha_2 = 1$. Figure 4 illustrates the process.

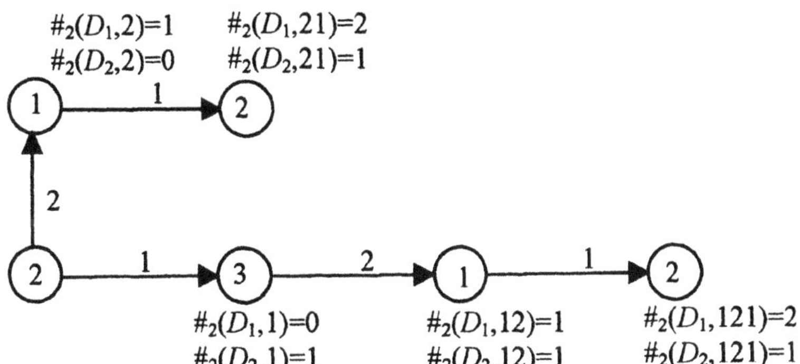

Figure 4. Constructing the set N_2 of traversal sequences.

In Step 3 and 4, we have $R = K = \{\varepsilon, 1\}$. Executing Step 5 yields $TS_1 = \{\alpha_i\gamma(\delta(s_0, \alpha_i), \delta(s_0, \alpha_j)) \mid \alpha_i, \alpha_j \in R\} = \{1, 11\}$. Step 6 - the set $TS_2 = \{\alpha_i\kappa\gamma(s, s') \mid \alpha_i \in R \wedge \beta \in N_i \wedge \kappa \in P(\beta, D_{\beta\text{after}\alpha_i})\} \wedge s, s' \in D_{\beta\text{after}\alpha_i}\} = \{11, 1121, 11211, 121, 1211\} = \{11211, 1211\}$. The test suite is $TS = TS_1 \cup TS_2 = \{11211, 1211\}$, it is m-complete for $m \leq 3$.

Proposition 5. The set of input sequences obtained with the above method is an m-complete test suite for a given FSM A.

❑

The statement comes as a corollary to Proposition 4 and Lemma 3.

Finally, we discuss the complexity of test suites obtained with the proposed method. As Figure 1 illustrates, there exist pathological FSMs for which the smallest *m*-complete test suite may coincide with the set X_A^{mnm}. On the other hand, in the class of completely defined reduced FSMs, the resulting test suite coincides with the set $VX^{m-n+1}W$ (the W-method). The size of tests for FSMs with distinguishable states depends on the choice of distinguishing sequences $\gamma(s, s')$ (constituting a characterization set W), for these sequences are the input to our method. "Optimal" distinguishing sequences should yield a test suite of a minimal complexity. The problem of determining such sequences remains open for all existing test derivation methods with fault coverage, our method included. Treating distinguishing sequences as the input to our method, we separate this combinatorial problem from all the other issues of deriving tests for partial machines, for which we offer the solutions in this paper.

4. CONCLUSION

We revisited in the paper the problem of test derivation with fault coverage from partially defined deterministic FSMs and further elaborated the so-called state-counting approach outlined in our earlier work.

The contributions of this paper can be summarized as follows. We investigated the role quasi-equivalent states play in the structure of tests with complete fault coverage. We demonstrated that in presence of quasi-equivalent states, the use of a state cover is definitely redundant. A subset of it is sufficient to obtain a complete test suite. This result allowed us to extend the notion of a state cover to a core cover for a given FSM. The use of this new characteristic of partial machines significantly reduces the number of traversal sequences necessary to cover transitions in implementation machines. The resulting save is indeed essential, since each transfer sequence in a state cover is concatenated with several traversal sequences which, in turn, are concatenated with state identification sequences (if they exist).

The rules for trimming traversal sequences were further refined, enforcing their earlier termination in the presence of quasi-equivalent states. We showed that the length of traversal sequences could be defined by counting the number of times quasi-equivalent states and not necessarily the same state (as was previously suggested) are traversed. We also demonstrated that there is no need to apply distinguishing sequences after each prefix of traversal sequences.

These new features of our method can significantly shorten the resulting test suite in the presence of quasi-equivalent and distinguishable states. Due to these non-trivial improvements, our method subsumes the existing methods; moreover, there are FSM classes where our method is superior to them, as it yields shorter tests.

Several research directions can be pursued from this work. The ideas presented in this paper may help improve the results of [17], [12], [16] for the nondeterministic case. Based on this approach it is also interesting to develop methods for deriving complete tests from partial specifications in the form of partially defined LTSs and I/O automata. Another potential direction is to try to construct a single checking sequence from the resulting complete test suite for a partial machine, similar to what was done in [18] for complete machines. Last but not least, in spite the fact that the proposed method provides a significant reduction in length of tests, the problem of finding a complete test suite of a minimal overall length for a given I/O FSM remains open and requires further research.

5. ACKNOWLEDGMENTS

This work was in part supported by the NSERC grant OGP0194381.

REFERENCES

[1] R. Alur, C. Courcoubetis, and M. Yannakakis, *Distinguishing tests for nondeterministic and probabilistic machines*, Proceedings of the 27th ACM Symposium on Theory of Computing, 1995, pp. 363-372.

[2] T. S. Chow, *Test software design modeled by finite state machines*, IEEE Transactions, SE-4, 3, 1978, pp. 178-187.

[4] S. Fujiwara, G. v. Bochmann, F. Khendek, M. Amalou, and A. Ghedamsi, *Test selection based on finite state models*, IEEE Transactions, SE-17, 6, 1991, pp. 591-603.

[3] J.-C. Fernandez, C. Jard, T. Jéron, and C. Viho, *An experiment in automatic generation of test suites for protocols with verification technology*, Science of Computer Programming, 1996.

[5] A. Gill, *Introduction to the theory of finite-state machines*, Mc Graw-Hill, New York, 1962.

[6] A. Grasselli and F. Luccio, *A method for minimizing the number of internal states in incompletely specified sequential networks*, IEEE Transactions, EC-14, 3, 1965, pp. 350-359.

[7] F. C. Hennie, *Fault detecting experiments for sequential circuits*, Proceedings of the IEEE 5th Ann. Symp. On Switching Circuit Theory and Logical Design, 1964, pp. 95-110.

[8] Z. Kohavi, *Switching and finite automata theory*, Mc Graw-Hill, New York, 1970.

[9] I. Koufareva, A. Petrenko, and N. Yevtushenko, *Test generation driven by user-defined fault models*, Proceedings of the Twelfth International Workshop on Testing of Communicating Systems, Hungary, 1999, pp. 215-233.

[10] D. Lee and M. Yannakakis, *Principles and methods of testing finite state machines, a survey*, Proceedings of the IEEE, vol. 84, 8, 1996, pp. 1090-1123.

[11] G. Luo, G. v. Bochmann, and A. Petrenko, *Test selection based on communicating nondeterministic finite state machines using a generalized Wp-method*, IEEE Transactions, SE-20, 2, 1994, pp. 149-162.

[12] G. Luo, A. Petrenko, and G. v. Bochmann, *Selecting test sequences for partially specified nondeterministic finite state machines*, Proceedings of the IFIP Seventh International Workshop on Protocol Test Systems, Japan, 1994, pp. 95-110.

[13] A. Petrenko, *Checking experiments with protocol machines*, Proceedings of the IFIP Fourth International Workshop on Protocol Test Systems, the Netherlands, 1991, pp. 83-94.

[14] A. Petrenko, G. v. Bochmann, and R. Dssouli, *Conformance relations and test derivation*, Invited Paper, Proceedings of the IFIP Sixth International Workshop on Protocol Test Systems, France, 1993, pp. 157-178.

[15] A. Petrenko, G. v. Bochmann, and M. Yao, *On fault coverage of tests for finite state specifications*, Computer Networks and ISDN Systems, 29, December 1996, pp. 81-106.

[16] A. Petrenko, N. Yevtushenko, and G. v. Bochmann, *Testing deterministic implementations from their nondeterministic specifications*, Proceedings of the IFIP Ninth International Workshop on Testing of Communicating Systems, Germany, 1996, pp. 125-140.

[17] A. Petrenko, N. Yevtushenko, A. Lebedev, and A. Das, *Nondeterministic state machines in protocol conformance testing*, Proceedings of the IFIP Sixth International Workshop on Protocol Test Systems, France, 1993, pp. 363-378.

[18] A. Rezaki and H. Ural, *Construction of checking sequences based on characterization sets*, Computer Communications, 18, December 1995, pp. 911-920.

[19] M. P. Vasilevskii, *Failure diagnosis of automata*, Cybernetics, Plenum Publishing Corporation, NY, No. 4, 1973, pp. 653-665.

[20] M. Yannakakis and D. Lee, *Testing finite state machines: fault detection*, Journal of Computer and System Sciences, 50, 1995, pp. 209-227.

[21] N. Yevtushenko and A. Petrenko, *Checking capabilities of multiple experiments*, Automatic Control and Computer Sciences, Allerton Press, Inc., USA, #3, 1989.

[22] N. Yevtushenko and A. Petrenko, *Test derivation method for an arbitrary deterministic automaton*, Automatic Control and Computer Sciences, Allerton Press, Inc., USA, #5, 1990.

Part Three

MODEL CHECKING – THEORY

COMPOSITIONALITY FOR IMPROVING MODEL CHECKING

Antonella Santone
Dipartimento di Ingegneria dell'Informazione, Università di Pisa, Italy
santone@iet.unipi.it

Abstract

Model checking is an automatic technique for verifying finite state systems: in this approach, properties are expressed in a temporal logic and systems are modelled as transition systems. A main problem of model checking is *state explosion*: very complex systems are often represented by transition systems with a prohibitive number of states. The primary cause of this problem is the parallel composition of interacting processes. Many techniques have been proposed to attack this problem, among them *compositional techniques*. These techniques reduce state explosion exploiting the natural decomposition of complex systems into processes. In this paper we present a formula-based compositional rule that allows us to deduce a property of a parallel composition of processes by checking it only on a component process.

Keywords: model checking, compositionality, temporal logic, state explosion.

1. INTRODUCTION

Model checking is an automatic technique for verifying finite state systems: in this approach, properties are expressed in a temporal logic and systems are modelled as transition systems. A model checker accepts two inputs, a transition system and a temporal formula, and returns "true" if the system satisfies the formula and "false" otherwise. Several efficient model checkers have been developed, see for example [5, 10]. A main problem of the model checking technique is *state explosion*: very complex systems are often represented by transition systems with a prohibitive number of states. The primary cause of this problem is the parallel composition of interacting processes. The problem occurs because the number of states of the global transition system is exponential in the number of component processes.

Tommaso Bolognesi and Diego Latella (Eds.), *Formal Methods for Distributed System Development.*

Many techniques have been proposed to attack this problem. While all these techniques cannot solve the problem in general, they provide significant improvements in performance. These techniques can be roughly divided into two categories. The works in the first category generate the whole transition system, corresponding to the concurrent system, but encode it *symbolically* to reduce the memory size [6, 18]. The works belonging to the second category avoid the generation of the whole transition system; local model checking [9, 23], on-the fly techniques [12, 14], abstractions [3, 7] and compositional reasoning [1, 8, 13, 15] belong to this category.

In our work we follow a compositional approach. Compositional techniques reduce state explosion exploiting the natural decomposition of complex systems into processes. Many finite state systems are composed of multiple processes running in parallel. The goal of this approach is to verify the processes individually and then piece together the results to conclude that the original system is correct.

In this paper we present a formula-based compositional rule that allows us to deduce a property of a parallel composition of processes by checking it only on a component process. Suppose that we want to show that a property φ holds for a system composed of two parallel processes P and Q, with synchronisation on the set of actions ω. Suppose that the transition system of the global system is too large to be handled by the existing verification environments. The first step of our approach consists of proving whether P (or equivalently Q) satisfies that property. This can be done by using standard model checkers, since the size of a single component is obviously less than that of the global system. If the first step succeeds, then we check that Q does not alter the satisfaction of φ when composed in parallel with P. This is done by checking a suitable relation between P and Q, based both on the formula φ and on the set ω of communication actions. If this second step succeeds too, then we can conclude that the original system satisfies φ, without generating the global transition system. This compositional method can significantly reduce the state explosion problem arising in the direct model checking method. Moreover it can be automated and it is completely transparent to the user. Other works based on compositional reasoning (see for example [15, 21]) require more user intervention, although they can achieve better results.

In this paper we use the LOTOS [4] language to specify concurrent and distributed systems, and a particular temporal logic, called *selective-μ-calculus*, to express properties. The selective-μ-calculus has been defined by the author and others in [2, 3]. We have chosen this logic since it is particularly suited to be used in a compositional verification approach. However the method we present is general, since it is applicable, with suitable modifications, with other logic, for example standard modal μ-calculus; moreover it can be used with different formal specification languages.

The paper is organised as follows. Some notions related to LOTOS are recalled in Section 2, while the selective-μ-calculus is introduced in Section 3. Our compositional approach is described in Section 4. Considerations are made in Section 5, while comparisons with related works are given in Section 6.

2. BASIC LOTOS

Basic LOTOS is the version of LOTOS [4] without value-passing; thus it is used to describe synchronisation aspects of the system. We assume that the reader is familiar with Basic LOTOS, which is widely used in the specification of concurrent and distributed systems, and so we recall only some main concepts. The reader can refer to [4] for further details. From now on we write LOTOS instead of Basic LOTOS. A LOTOS program is defined as:

```
process ProcName := P
        where ℰ
endproc
```

where `P` is a *process*, `ProcName := P` is a *process declaration* and $\mathcal{E}$ is a *process environment*, i.e. a set of process declarations. A process is the composition, by means of a set of operators, of a finite set $\mathcal{A} = \{i, a, b, ...\}$ of atomic *actions*. The action i is called the *unobservable action*. For the sake of simplicity, we do not consider all operators which are allowed in LOTOS. The simplified syntax of a LOTOS process is the following:

$$P ::= \texttt{stop} \mid \alpha; P \mid P[]P \mid P|[S]|P \mid \texttt{hide}\ S\ \texttt{in}\ P \mid P[f] \mid X$$

where X ranges over a set of process names, α ranges over $\mathcal{A}$, $S \subseteq \mathcal{A}-\{i\}$ and $f : \mathcal{A} \to \mathcal{A}$ is an action relabelling function, with the property that $f(i) = i$. We call $\mathcal{P}$ the processes generated from P. By **stop** we denote the empty process. The operators to build processes are *action prefix* $(a; P)$, *choice* $(P_1[]P_2)$, *parallel composition* $(P_1|[S]|P_2)$, *hiding* (**hide** S **in** P), *relabelling*, $(P[f])$ and *process instantiation* (X).

Given a set $\mathcal{E}$ of process declarations, the standard *operational semantics* is given by a relation $\longrightarrow_{\mathcal{E}} \subseteq \mathcal{P} \times \mathcal{A} \times \mathcal{P}$. $\longrightarrow_{\mathcal{E}}$ ($\longrightarrow$ for short) is the least relation defined by the rules in Table 1. In Table 1 the symmetric rules for choice and parallel composition are not shown.

$\alpha \in \mathcal{A}$, $l \in \mathcal{A} - \{i\}$

$$\textbf{pre} \quad \frac{}{\alpha; P \xrightarrow{\alpha} P} \qquad \textbf{choice} \quad \frac{P_1 \xrightarrow{\alpha} P_1'}{P_1 \,[]\, P_2 \xrightarrow{\alpha} P_1'}$$

$$\textbf{inst} \quad \frac{P \xrightarrow{\alpha} P'}{X \xrightarrow{\alpha} P'} \; X := P \in \mathcal{E} \qquad \textbf{rel} \quad \frac{P \xrightarrow{\alpha} P'}{P[f] \xrightarrow{f(\alpha)} P'[f]}$$

$$\textbf{par} \quad \frac{P_1 \xrightarrow{\alpha} P_1'}{P_1 \,|[S]|\, P_2 \xrightarrow{\alpha} P_1' \,|[S]|\, P_2} \; \alpha \notin S$$

$$\textbf{com} \quad \frac{P_1 \xrightarrow{\alpha} P_1', \; P_2 \xrightarrow{\alpha} P_2'}{P_1 \,|[S]|\, P_2 \xrightarrow{\alpha} P_1' \,|[S]|\, P_2'} \; \alpha \in S$$

$$\textbf{hide}_1 \quad \frac{P \xrightarrow{\alpha} P'}{\texttt{hide } S \texttt{ in } P \xrightarrow{\alpha} \texttt{hide } S \texttt{ in } P'} \; \alpha \notin S$$

$$\textbf{hide}_2 \quad \frac{P \xrightarrow{l} P'}{\texttt{hide } S \texttt{ in } P \xrightarrow{i} P'} \; l \in S$$

Table 1 Standard operational semantics of Basic LOTOS

If $\delta \in \mathcal{A}^*$ and $\delta = \alpha_1 \ldots \alpha_n, n \geq 1$, we write $P \xrightarrow{\delta} Q$ to mean $P \xrightarrow{\alpha_1} \cdots \xrightarrow{\alpha_n} Q$ and $\alpha_1, \ldots, \alpha_n$ is called a *finite computation* of P. For the empty sequence λ of actions we have $P \xrightarrow{\lambda} P$ for every $P \in \mathcal{P}$; clearly also λ is a finite computation of P.

A *labelled transition systems* $\mathcal{T}$ is a quadruple $(S, T, \longrightarrow, s_0)$, S is a set of states, T is a set of transition labels, $s_0 \in S$ is the initial state and $\longrightarrow \subseteq S \times T \times S$. The *standard transition system* for a LOTOS process P, denoted by $\mathcal{S}(P)$, is the transition system $(\mathcal{P}, \mathcal{A}, \longrightarrow, P)$. A LOTOS process P is finite, if the standard transition system for P is finite.

We now define a function that associates a process to a set of actions:

Definition 1 **($\mathcal{L}(P)$)** *Let P be a LOTOS process and $\mathcal{E}$ be a set of process declarations. The sort of P with $\mathcal{E}$ is the set $\mathcal{L}_\mathcal{E}(P) \subseteq \mathcal{A}$ defined as* *the least solution of the following recursive definition:*

$$
\begin{array}{lll}
\mathcal{L}_{\mathcal{E}}(\texttt{stop}) & = & \emptyset \\
\mathcal{L}_{\mathcal{E}}(\alpha.P) & = & \begin{cases} \mathcal{L}_{\mathcal{E}}(P) \cup \{\alpha\} & \textit{if } \alpha \neq i \\ \mathcal{L}_{\mathcal{E}}(P) & \textit{if } \alpha = i \end{cases} \\
\mathcal{L}_{\mathcal{E}}(P[]Q) & = & \mathcal{L}_{\mathcal{E}}(P|[S]|Q) \quad = \quad \mathcal{L}_{\mathcal{E}}(P) \cup \mathcal{L}_{\mathcal{E}}(Q) \\
\mathcal{L}_{\mathcal{E}}(P[f]) & = & \{f(\alpha) : \alpha \in \mathcal{L}_{\mathcal{E}}(P)\} \\
\mathcal{L}_{\mathcal{E}}(\texttt{hide } S \texttt{ in } P) & = & \mathcal{L}_{\mathcal{E}}(P) - S \\
\mathcal{L}_{\mathcal{E}}(X) & = & \mathcal{L}_{\mathcal{E}}(P) \quad \textit{if } X := P \in \mathcal{E}
\end{array}
$$

When clear from the context, we use $\mathcal{L}(P)$ in place of $\mathcal{L}_{\mathcal{E}}(P)$.

From now on, without loss of generality, we consider the parallel composition of two processes P and Q with synchronisation of the actions common to both of their sorts $(P\ |[\mathcal{L}(P) \cap \mathcal{L}(Q)]|\ Q)$.

3. SELECTIVE-μ-CALCULUS

The selective-μ-calculus, introduced by the author and others in [2, 3], is a branching temporal logic to express behavioural properties of systems. It is equi-expressive to μ-calculus [16, 22], but it differs from it in the definition of the modal operators. The basic characteristic of the selective-μ-calculus is that each formula guides the definition of a reduced transition system on which the formula can be equivalently checked.

The syntax of the selective-μ-calculus is the following, where K and R range over sets of actions, while Z ranges over a set of variables:

$$\varphi ::= \texttt{tt} \mid \texttt{ff} \mid Z \mid \varphi_1 \vee \varphi_2 \mid \varphi_1 \wedge \varphi_2 \mid [K]_R\,\varphi \mid \langle K\rangle_R\,\varphi \mid \nu Z.\varphi \mid \mu Z.\varphi$$

The satisfaction of a formula φ by a state s of a transition system, written $s \models \varphi$, is defined as follows:

each state satisfies $\texttt{tt}$ and no state satisfies $\texttt{ff}$; a state satisfies $\varphi_1 \vee \varphi_2$ ($\varphi_1 \wedge \varphi_2$) if it satisfies φ_1 or (and) φ_2. $[K]_R\,\varphi$ and $\langle K\rangle_R\,\varphi$ are the selective modal operators:

$[K]_R\,\varphi$ is satisfied by a state which, for every performance of a sequence of actions not belonging to $R \cup K$, followed by an action in K, evolves to a state obeying φ.

$\langle K\rangle_R\,\varphi$ is satisfied by a state which can evolve to a state obeying φ by performing a sequence of actions not belonging to $R \cup K$, followed by an action in K.

As in standard μ-calculus, a fix-point formula has the form $\mu Z.\varphi$ ($\nu Z.\varphi$) where μZ (νZ) *binds* free occurrences of Z in φ. An occurrence of Z is free if it is not within the scope of a binder μZ (νZ). A formula is *closed* if it contains no free variables. $\mu Z.\varphi$ is the least fix-point of the recursive equation $Z = \varphi$, while $\nu Z.\varphi$ is the greatest one.

A transition system $\mathcal{T}$ satisfies a formula φ, written $\mathcal{T} \models \varphi$, if and only if $P \models \varphi$, where P is the initial state of $\mathcal{T}$. A LOTOS process P satisfies φ if $\mathcal{S}(P)$ satisfies φ.

$$
\begin{array}{lll}
P \not\models \texttt{ff} & & \\
P \models \texttt{tt} & & \\
P \models \varphi \wedge \psi & \text{iff} & P \models \varphi \text{ and } P \models \psi \\
P \models \varphi \vee \psi & \text{iff} & P \models \varphi \text{ or } P \models \psi \\
P \models [K]_R\,\varphi & \text{iff} & \forall P'.\forall \alpha \in K.P \stackrel{\alpha}{\Longrightarrow}_{K \cup R} P' \text{ implies } P' \models \varphi \\
P \models \langle K \rangle_R\,\varphi & \text{iff} & \exists P'.\exists \alpha \in K.P \stackrel{\alpha}{\Longrightarrow}_{K \cup R} P' \text{ and } P' \models \varphi \\
P \models \nu Z.\varphi & \text{iff} & P \models \nu Z^n.\varphi \text{ for all } n \\
P \models \mu Z.\varphi & \text{iff} & P \models \mu Z^n.\varphi \text{ for some } n
\end{array}
$$

where, for each n, $\nu Z^n.\varphi$ and $\mu Z^n.\varphi$ are defined as:

$$
\begin{array}{ll}
\nu Z^0.\varphi = \texttt{tt} & \mu Z^0.\varphi = \texttt{ff} \\
\nu Z^{n+1}.\varphi = \varphi[\nu Z^n.\varphi/Z] & \mu Z^{n+1}.\varphi = \varphi[\mu Z^n.\varphi/Z]
\end{array}
$$

where the notation $\varphi[\psi/Z]$ indicates the substitution of ψ for every free occurrence of the variable Z in φ.

Table 2 Satisfaction of a closed formula by a state

The precise definition of the satisfaction of a closed formula φ by a state of a transition system $\mathcal{T} = (\mathcal{S}, \mathcal{A}, \longrightarrow, Q)$ is given in Table 2. It uses the transition relation $\Longrightarrow_I$, parametric with respect to $I \subseteq \mathcal{A}$, which ignores all non-interesting actions (i.e. those in $\mathcal{A} - I$).

Definition 2 ($\Longrightarrow_I$ relation) *Let P be a LOTOS process and $I \subseteq \mathcal{A}$, we define the relation $\Longrightarrow_I \subseteq \mathcal{P} \times I \times \mathcal{P}$ such that, for each $\alpha \in I$, $P, Q \in \mathcal{P}$:*

$$P \stackrel{\alpha}{\Longrightarrow}_I Q \quad \textit{iff} \quad P \stackrel{\delta\alpha}{\longrightarrow} Q, \textit{ where } \delta \in (\mathcal{A} - I)^*$$

By $P \overset{\alpha}{\Longrightarrow}_I Q$ we express the fact that it is possible to pass from P to Q by performing a (possibly empty) sequence of actions not belonging to I and then the action α in I. In the following, Q is called an *α-derivative of P with* $\Longrightarrow_I$ (or *α-derivative* for short). Note that $\Longrightarrow_{\mathcal{A}} = \longrightarrow$.

Example 3 *Consider the LOTOS program in Table 3 that describes a coffee and tea dispenser in which the user first reads the instructions (action* `read_instr`*) and then he/she inserts a coin (action* `coin`*) if the* `Dispenser` *is ready to accept a coin (action* `ready`*). Afterwards, the user decides (action* `deciding`*) either to collect immediately a cup of coffee (action* `coffee`*); or to insert another coin before collecting, in case, a cup of tea (action* `tea`*). However, if the user does not insert the second coin quickly, i.e. he/she inserts the second coin after a timeout, internal to the* `Dispenser`*, (action* `timeout`*), the user is left without neither tea nor coffee.*

```
process System := Dispenser |[coin,coffee,tea]| User
   where ℰ
endproc
```

$\mathcal{E}$ is composed of the following process declarations:

```
Dispenser := ready;coin;(timeout;(coin;Dispenser [] coffee;Dispenser)
                         []
                         coffee;Dispenser
                         []
                         coin;(tea;Dispenser [] Dispenser))

User := read_instr;coin;deciding;(coffee;User
                                  []
                                  coin;(User [] tea;User))
```

Table 3 A coffee and tea dispenser

$\mathcal{S}$(`System`*) has 13 states and 22 transitions, while $\mathcal{S}$(*`Dispenser`*) has 5 states and 9 transitions and $\mathcal{S}$(*`User`*) has 5 states and 7 transitions.*

To explain the use of the selective operators, we give some examples of selective-μ-calculus formulae, which are satisfied by `System`.

$\varphi_1 = \langle \mathtt{tea} \rangle_{\{\mathtt{timeout}\}}\, \mathtt{tt}$: *"it is possible to get a cup of tea if the second coin has been inserted quickly, i.e. the timeout has not occurred".*

$\varphi_2 = [\mathtt{coffee}]_{\{\mathtt{coin}\}}\, \mathtt{ff}$: *"a cup of coffee cannot be obtained if a coin has not been inserted".*

$\varphi_3 = \langle \texttt{coin} \rangle_{\{\texttt{coin}\}} \big(\langle \texttt{coffee} \rangle_{\emptyset} \texttt{tt} \wedge [\texttt{tea}]_{\{\texttt{coin}\}} \texttt{ff} \big)$: *"after the user has inserted a coin, it is possible to get a cup of coffee and a cup of tea cannot be collected if another coin has not been inserted"*.

4. COMPOSITIONAL MODEL CHECKING

Compositional model checking is a method for reducing the complexity of temporal logic model checking in systems composed of many parallel processes. The goal is to check a property on a single component of a system and then deduce that the original system satisfies that property. In general, a property satisfied by a process could be not preserved when that process is composed with another process. For example: $a; stop$ satisfies $\langle a \rangle_{\{b\}} \texttt{tt}$, while $a; stop \,|[a]|\, b; a; stop$ does not satisfy $\langle a \rangle_{\{b\}} \texttt{tt}$, since there exists no path from the initial state ending with a and not containing b. The problem is to find sufficient conditions under which a property of a process remains true in a parallel composition involving that process.

In this section we define a formula-based compositional rule that allows us to deduce a property of a parallel composition by checking it only on a parallel component. In particular, if we want to verify that a process of the form $P|[\omega]|Q$ has a property φ, we reduce this task to: 1) check whether P satisfies φ; and 2) check that Q does not alter the satisfaction of φ when composed in parallel with P, over the set of actions ω. To solve the first point we use the standard model checkers which we can found in all existing verification environments [5, 10]. To solve the second point we check whether P and Q are related by a suitable relation, based both on the formula φ and on the set of communication actions ω. A main point of this paper is the definition of this relation, which represents a sufficient condition for the preservation of the satisfaction of a formula.

For the sake of clarity, we first explore our approach considering selective-μ-calculus formulae containing only $\langle K \rangle_R$ modal operators, from now on referred to as $\langle\rangle s\mu$ formulae. We use the coffee and tea dispenser of Example 3 to explain our approach. Let us suppose that we want to check whether `System` satisfies the $\langle\rangle s\mu$ formula:

$$\varphi_1 = \langle \texttt{tea} \rangle_{\{\texttt{timeout}\}} \texttt{tt}.$$

It is easy to show that the process `Dispenser` satisfies φ_1. If we want to deduce that also the complete system (`Dispenser` $|[\omega]|$ `User`) satisfies φ_1 (with $\omega = \{\texttt{coin}, \texttt{coffee}, \texttt{tea}\}$) we must check that `User` does not alter the satisfaction of φ_1; this is ensured if:

- `User` does not block the execution of `Dispenser`: whenever `Dispenser` executes an action in ω, `User` can perform that action and so the parallel composition can perform the action `tea`.
- `User` does not execute `timeout` before `Dispenser` performs `tea`.

In general, if $P \models \varphi$, the actions relevant to preserve the satisfaction of a $\langle\rangle s\mu$ formula, when P is composed in parallel with another process, are both the communication actions and those ones belonging to the set $\mathcal{R}(\varphi)$ of all actions occurring in the set R of the modal operators $\langle K\rangle_R$ appearing in φ. In order to express the above conditions, we introduce the notion of $\langle\omega,\rho\rangle$-simulation.

Definition 4 ($\langle\omega,\rho\rangle$-simulation) *Let P, Q be two LOTOS processes and $\omega,\rho \subseteq \mathcal{A}$.*

- *A $\langle\omega,\rho\rangle$-simulation, $\mathcal{S}$, is a relation on $\mathcal{P} \times \mathcal{P}$ such that $P\ \mathcal{S}\ Q$ implies:*

$$P \stackrel{a}{\Longrightarrow}_{\omega} P' \text{ implies } Q \stackrel{a}{\Longrightarrow}_{\omega\cup\rho} Q' \text{ for some } Q' \text{ with } P'\ \mathcal{S}\ Q'.$$

- *We say that Q $\langle\omega,\rho\rangle$-simulates P (write $P \leq^{\omega}_{\rho} Q$) iff there exists a $\langle\omega,\rho\rangle$-simulation $\mathcal{S}$ containing the pair (P,Q).*

In other words, Q $\langle\omega,\rho\rangle$-simulates P means that, if P becomes the process P', after performing a sequence of actions which are not communication actions, followed by an action α in ω, then Q performs α, without executing an action in ρ, and becomes a process which $\langle\omega,\rho\rangle$-simulates P'. Note that Q can have more computations than P.

Let φ be a $\langle\rangle s\mu$ formula, we give the following compositional rule:

$$\langle\rangle\textbf{rule}\quad \frac{P \models \varphi, \quad P \leq^{\omega}_{\mathcal{R}(\varphi)} Q}{P\ |[\omega]|\ Q \models \varphi}$$

Recall the coffee and tea dispenser described in Example 3; it holds that:

- $\texttt{Dispenser} \models \varphi_1$: $\langle\texttt{tea}\rangle_{\{\texttt{timeout}\}}\,\texttt{tt}$;
- $\texttt{Dispenser} \leq^{\omega}_{\{\texttt{timeout}\}} \texttt{User}$ where $\omega = \{\texttt{coin}, \texttt{coffee}, \texttt{tea}\}$.

Thus we can deduce that $\texttt{Dispenser}\ |[\omega]|\ \texttt{User} \models \varphi_1$.

With this approach we can check the formula only on $\mathcal{S}$(`Dispenser`). We remark that $\mathcal{S}$(`Dispenser`) has only 5 states, while $\mathcal{S}$(`System`) has 13 states. On the other hand, we have to determine whether it exists a $\langle\omega, \{\mathtt{timeout}\}\rangle$-simulation between `Dispenser` and `User`: we discuss this problem in Section 5. Note that also `User` satisfies φ_1, but `User` $\not\leq_\rho^\omega$ `Dispenser`, and so the $\langle\rangle$**rule** cannot be applied in this direction. Since the $\langle\omega, \rho\rangle$-simulation is not symmetric, if the application of the rule in one direction fails, we can apply the rule in the other direction. The choice of the direction is made arbitrarily, but some remarks concerning the sorts of the processes can be useful. For example, if we want to check whether $P\ |[\omega]|\ Q$ satisfies a formula φ, containing the modal operator $\langle\alpha\rangle_R$, and $\alpha \notin \mathcal{L}(P)$, while $\alpha \in \mathcal{L}(Q)$, then we clearly choose Q to check φ on it and then we verify that P does not alter the satisfaction of φ (i.e. $Q \leq_{\mathcal{R}(\varphi)}^\omega P$).

Finally, note that the above compositional rule is not complete; in fact it is possible that a process P satisfies a $\langle\rangle$sμ formula φ, Q does not $\langle\omega, \mathcal{R}(\varphi)\rangle$-simulates P, but $P\ |[\omega]|\ Q$ satisfies φ.

Let us now consider selective-μ-calculus formulae containing only $[K]_R$ modal operators, in the following referred to as $[]$sμ formulae. Recall the coffee and tea dispenser of Example 3. Let us suppose that we want to check whether `System` satisfies the $[]$sμ formula:

$$\varphi_2 = [\mathtt{coffee}]_{\{\mathtt{coin}\}}\ \mathtt{ff}.$$

It is easy to show that the process `Dispenser` satisfies φ_2. If we want to deduce that also

$$\mathtt{Dispenser}\ |[\{\mathtt{coin}, \mathtt{coffee}, \mathtt{tea}\}]|\ \mathtt{User} \models \varphi_2$$

we must check that `User` does not alter the satisfaction of φ_2. A sufficient condition to ensure this fact is that `User` and `Dispenser` behave in the same way with respect to the action `coffee`.

In general, if $P \models \varphi$ and φ is a $[]$sμ formula, the actions relevant to preserve the satisfaction of φ, when P is composed in parallel with another process, are only those ones belonging to the set $\mathcal{K}(\varphi)$ of all actions occurring in the set K of the modal operators $[K]_R$ appearing in φ. In order to express the above condition, we introduce the notion of σ-bisimulation.

Definition 5 (σ-bisimulation) *Let P, Q be two LOTOS processes and $\sigma \subseteq \mathcal{A}$.*

- *A σ-bisimulation, $\mathcal{B}$, is a relation on $\mathcal{P} \times \mathcal{P}$ such that $P\ \mathcal{B}\ Q$ implies:*

- $P \stackrel{\alpha}{\Longrightarrow}_\sigma P'$ *implies* $Q \stackrel{\alpha}{\Longrightarrow}_\sigma Q'$ *for some* Q' *and* *if* $Q \stackrel{\alpha}{\Longrightarrow}_\sigma Q'$ *then* $P' \, \mathcal{B} \, Q'$*; and*
- $Q \stackrel{\alpha}{\Longrightarrow}_\sigma Q'$ *implies* $P \stackrel{\alpha}{\Longrightarrow}_\sigma P'$ *for some* P' *and* *if* $P \stackrel{\alpha}{\Longrightarrow}_\sigma P'$ *then* $P' \, \mathcal{B} \, Q'$.

- *We say that* P *and* Q *are* σ*-bisimilar (write* $P \sim_\sigma Q$*) iff there exists a* σ*-bisimulation* $\mathcal{B}$ *containing the pair* (P, Q).

$P \sim_\sigma Q$ means that, if P performs an action in σ and becomes the process P', then also Q can perform the same action and becomes a process which is σ-bisimilar to P'. Moreover, each α-derivative of P is σ-bisimilar to all α-derivatives of Q. Similarly with P and Q interchanged.

Suppose that P satisfies a $[]s\mu$ formula φ. Note that the actions belonging to the set R of the modal operators $[K]_R$ occurring in φ are not relevant to preserve the satisfaction of φ, when P is composed in parallel with another process Q. In fact a computation containing an action in K but preceded by an action in R does not alter the truth value of a formula. Further, also the communication actions are not relevant to preserve the satisfaction of φ: if P performs an action α in K and Q blocks the execution of such action, then the parallel composition cannot perform α and consequently the formula is vacuously true.

Let φ be a $[]s\mu$ formula, we give the following compositional rule:

$$[]\textbf{rule} \quad \frac{P \models \varphi, \quad Q \sim_{K(\varphi)} P}{P \; |[\omega]| \; Q \models \varphi}$$

Recall the coffee and tea dispenser described in Example 3; it holds that:

- $\texttt{Dispenser} \models \varphi_2 = [\texttt{coffee}]_{\{\texttt{coin}\}}\, \texttt{ff}$;
- $\texttt{User} \sim_{\{\texttt{coffee}\}} \texttt{Dispenser}$.

Thus we can deduce that $\texttt{Dispenser} \; |[\{\texttt{coin}, \texttt{coffee}, \texttt{tea}\}]| \; \texttt{User} \models \varphi_2$.

Now we consider the full selective-μ-calculus formulae. Again, if we want to check that $P \; |[\omega]| \; Q \models \varphi$, we check φ on P and we check whether P and Q are related by a $\langle \omega, \varphi \rangle$-relation, which, in this case, is obtained combining $\langle \omega, \rho \rangle$-simulation with σ-bisimulation in the following way:

Definition 6 ($\langle \omega, \rho, \sigma \rangle$-relation) *Let* P, Q *be two LOTOS processes and* $\omega, \rho, \sigma \subseteq \mathcal{A}$.

- *A $\langle\omega,\rho,\sigma\rangle$-relation, $\mathcal{R}$, is a relation on $\mathcal{P}\times\mathcal{P}$ such that $P\ \mathcal{R}\ Q$ implies:*
  - $P \stackrel{\alpha}{\Longrightarrow}_{\omega} P'$ *implies* $Q \stackrel{\alpha}{\Longrightarrow}_{\omega\cup\rho} Q'$ *for some* Q' *with* $P'\ \mathcal{R}\ Q'$; *and*
  - $P \stackrel{\alpha}{\Longrightarrow}_{\sigma} P'$ *implies* $Q \stackrel{\alpha}{\Longrightarrow}_{\sigma} Q'$ *for some* Q' *and if* $Q \stackrel{\alpha}{\Longrightarrow}_{\sigma} Q'$ *then* $P'\ \mathcal{R}\ Q'$; *and*
  - $Q \stackrel{\alpha}{\Longrightarrow}_{\sigma} Q'$ *implies* $P \stackrel{\alpha}{\Longrightarrow}_{\sigma} P'$ *for some* P' *and if* $P \stackrel{\alpha}{\Longrightarrow}_{\sigma} P'$ *then* $P'\ \mathcal{R}\ Q'$.
- *We say that P $\langle\omega,\rho,\sigma\rangle$-relates Q iff there exists a $\langle\omega,\rho,\sigma\rangle$-relation $\mathcal{R}$ containing the pair (P,Q).*

Let φ be a selective-μ-calculus formula and $\omega \subseteq \mathcal{A}$, in the following we say that two processes P and Q are related by a $\langle\omega,\varphi\rangle$-relation (write $P \bowtie^{\omega}_{\varphi} Q$) if and only if a $\langle\omega,\rho,\sigma\rangle$-relation exists containing the pair (P,Q), where $\mathcal{R}(\varphi)=\rho$ and $\mathcal{K}(\varphi)=\sigma$.

We can now give the complete compositional rule:

$$\textbf{compositional\_rule}\quad \frac{P\models\varphi,\quad P\bowtie^{\omega}_{\varphi} Q}{P\ |[\omega]|\ Q\models\varphi}$$

Recall the coffee and tea dispenser described of Example 3. Let us suppose that we want to check whether `System` satisfies the selective-μ-calculus formula:

$$\varphi_3 = \langle\texttt{coin}\rangle_{\{\texttt{coin}\}}\Big(\langle\texttt{coffee}\rangle_{\emptyset}\,\texttt{tt}\wedge[\texttt{tea}]_{\{\texttt{coin}\}}\,\texttt{ff}\Big).$$

Applying the compositional rule, we can deduce that the property holds on the complete system by checking such property only on `Dispenser` and by showing that `Dispenser` $\bowtie^{\omega}_{\varphi_3}$ `User`.

We now present the main result of the paper, stating the soundness of the compositional rule.

Theorem 1 (main) *Let P, Q be two LOTOS processes and φ a formula of the selective-μ-calculus.*

$$P\models\varphi \ \textit{ and }\ P\bowtie^{\omega}_{\varphi} Q\ \textit{ implies }\ P\ |[\omega]|\ Q\models\varphi$$

Proof. By induction on the structure of the formula.

The soundness of $\langle\rangle$**rule** and $[]$**rule** follows by Theorem 1.

5. DISCUSSION

The compositional rule described in the previous section can significantly reduce the state explosion problem arising in the direct model checking method, since we can check a formula φ only on a parallel component. However, to complete the verification, we have to check whether there exists a $\langle\omega, \varphi\rangle$-relation between two processes P and Q composed in parallel over the set of actions ω. Thus, a crucial point is to find an efficient algorithm to determine whether $P \bowtie_{\varphi}^{\omega} Q$. In literature, efficient algorithms exist for computing bisimulation equivalences [19], among them there is the algorithm of Paige-Tarjan [20]. We can adapt this algorithm to compute the $\langle\omega, \varphi\rangle$-relation. In the worst case, we obtain an algorithm whose complexity, in space, is $\mathcal{O}(n+m)$, where n is the size of the transition system for P and m is the size of the transition system for Q. As a consequence, we obtain a reduction of the state space since, in the worst case, the size of the parallel composition of P and Q may be equal to the product of the sizes of the two components.

It is important to note that the compositional approach presented in this paper can be applied to standard μ-calculus too. As proved in [3], it holds that: $\langle K\rangle\, \varphi = \langle K\rangle_{\mathcal{A}}\, \varphi$ and $[K]\, \varphi = [K]_{\mathcal{A}}\, \varphi$. As a consequence, we can apply our compositional rule, taking into account that, given a μ-calculus formula φ, $\mathcal{R}(\varphi)$ coincides with the whole set $\mathcal{A}$ if it exists a $\langle\rangle$ operator in φ. Moreover, our approach can handle also infinite processes and we are investigating how the method could work for systems composed of more than two processes.

We are actually developing a tool to check whether there exists a $\langle\omega, \varphi\rangle$-relation between two processes P and Q composed in parallel over the set of actions ω. The tool can be included, for example, in the the Concurrency Workbench environment [10], which is a verification system for process algebra description languages.

6. RELATED WORK

We now compare our method with some related works based on compositional reasoning. In [8] a technique based on the use of *interface processes* is proposed. This technique attempts to minimise the global transition system by focusing on the communication among the component processes. The method considers the set of actions used in the interface between two components and minimises the system by eliminating events that are not related to communication actions. More precisely, assume that P_1 and P_2 communicate using a set of actions ω. Then P_1 can only observe the behaviour of P_2 through ω. Thus P_2 can be replaced by an equivalent process A_2 (called the *interface process for* P_2) which is indistinguishable from P_2 with respect to ω. Our approach differs from the above one since for several reasons, among them: suppose that ω is almost the whole set $\mathcal{L}(P_2)$, the method proposed in [8] does not improve the efficiency of model checking, since the size of A_2 is not significantly less than the size of P_2. Instead our approach can be applied also in this situation, reducing in many cases the complexity of model checking. Moreover, in [8], after finding the interface process, the model checking is always applied to a parallel composition of two processes (although less than the original one) which is the main cause of state explosion; this is avoided in the solution proposed in this paper. On the other hand, the method proposed in [8] can achieve better results than ours, when the set of communication actions is small.

Assume-guarantee reasoning [21] is a semi-automatic technique that verifies each component separately. The behaviour of each component depends on the behaviour of the rest of the system, i.e. its environment. Because of this, the user must specify properties that the environment has to satisfy in order to guarantee the correctness of the component. These properties are *assumed.* If these assumptions are satisfied, the component will satisfy other properties, called *guarantees.* By combining the set of assume/guarantee properties in an appropriate way, it is possible to demonstrate the correctness of the entire system without constructing the global transition system. In order to automate this type of reasoning, in [13] a simulation preorder has been defined which preserves the satisfaction of formulae of a subset of CTL [11]. Our approach differs from the above one since that one uses a subset of a general logic (CTL) and cannot be applied to infinite state systems. Moreover, in the assume-guarantee approach the state explosion problem is replaced with another problem: the *assumption explosion problem*, i.e. assume/guarantee properties are expressed by complex formulae. However, for some kind of formulae, the method described in [13] is more efficient than ours.

In [15] it is proved that, for a given property φ that a system composed of two concurrent processes must satisfy, there exist properties φ_1, φ_2 that the processes in the composition must satisfy. Nevertheless, no method is given to deduce such properties; this is full responsibility of the human verifier of the system. Our approach differs from the above one since it can be completely automated.

In [1] the parallel composition operator was eliminated basically by encoding one of the component of the parallel composition into the formula, which is checked on the other components. In the worst case this results in an exponential blow-up in the size of the formula, and the total complexity remains the same as for non-compositional model checking.

We remark that, unlike all other approaches discussed above, the effectiveness of our method depends on the efficiency of the algorithm to decide whether two parallel processes are related by a $\langle \omega, \varphi \rangle$-relation.

Acknowledgments

The author wish to thank Nicoletta De Francesco and Gigliola Vaglini for their useful comments and suggestions.

References

[1] H.R. Andersen, G. Winskel. Compositional Checking of Satisfaction. *Formal Methods in System Design, 1(4)*, 1992.

[2] R. Barbuti, N. De Francesco, A. Santone, G. Vaglini. Selective Mu-Calculus: New Modal Operators for Proving Properties on Reduced Transition Systems. In *Proceedings of FORTE X/PSTV XVII '97.* Chapman & Hall, 1997. 519-534

[3] R. Barbuti, N. De Francesco, A. Santone, G. Vaglini. Selective Mu-Calculus and Formula-Based Abstractions of Transition Systems. *Journal of Computer and System Sciences, 59(3)*, 1999. 537-556.

[4] T. Bolognesi, E. Brinksma. Introduction to ISO Specification Language LOTOS. *Comp. Networks and ISDN Systems, 14*, 1987. 25-59.

[5] A. Bouali, S. Gnesi, S. Larosa. The integration Project for the JACK Environment. *Bulletin of the EATCS, 54*, 1994. 207-223.

[6] J. Burch, E. Clarke, K. McMillan, D. Dill, L. Hwang. Symbolic Model Checking: 10^{20} States and Beyond. In *Proceedings of Fifth Annual IEEE Symposium on Logic in Computer Science*, 1990. 428-439.

[7] E.M. Clarke, O.Grumberg, D.E. Long. Model Checking and Abstraction. *ACM Transactions on Programming Languages and Systems, 16(5)*, 1994. 1512-1542.

[8] E.M. Clarke, D.E. Long, K.L. McMillan. Compositional Model Checking. In *Proceedings of the Fourth Annual IEEE Symposium on Logic in Computer Science*, 1989. 353-362.

[9] R. Cleveland. Tableau-based Model Checking in the Propositional Mu-Calculus. *Acta Informatica, 27*, 1990. 725–747.

[10] R. Cleaveland, S. Sims. The NCSU Concurrency Workbench. In *Proceedings of the Eighth International Conference on Computer-*

Aided Verification (CAV'96), Lecture Notes in Computer Science 1102, 1996. 394–397.

[11] E.A. Emerson, J.Y. Halpern. "Sometimes" and "Not Never" Revisited: On Branching Time Versus Linear Time. *Journal of ACM, 33(1)*, 1986. 151–178.

[12] J.C. Fernandez, L. Mounier. Verifying Bisimulation on the fly. In *Proceedings of the Third International Conference on Formal Description Techniques, FORTE'90*, 1990.

[13] O. Grumberg, D.E. Long. Model Checking and Modular Verification. *ACM Transactions on Programming Languages and Systems, 16(3)*, 1994. 843-871.

[14] C. Jard, T. Jéron. Bounded-memory Algorithms for Verification on-the-fly. In *Proceedings of the Third International Conference on Computer-Aided Verification (CAV'91)*, Lecture Notes in Computer Science 575, 1991. 192-201.

[15] R. Kaivola. Compositional Model Checking for Linear-Time Temporal Logic. In *Proceedings of the Fourth International Conference on Computer-Aided Verification (CAV'92)*, Lecture Notes in Computer Science 663, 1991. 248–259.

[16] D. Kozen. Results on the Propositional Mu-Calculus. *Theoretical Computer Science, 27*, 1983. 333–354.

[17] E. Madelaine, D. Vergamini. Finiteness Conditions and Structural Construction of Automata for all Process Algebras. In *Proceedings of 2nd Workshop on Computer-Aided Verification. DIMACS Technical Report 90-31*, 1990.

[18] K. McMillan. *Symbolic Model Checking.* Boston: Kluwer Academic Publishers, 1993.

[19] R. Milner. *Communication and Concurrency.* Prentice-Hall, 1989.

[20] R. Paige, R. Tarjan. Three Partition Refinement Algorithms. *SIAM Journal of Computing, 16(6)*, 1987. 973–989.

[21] A. Pnueli. In Transition for Global to Modular Temporal Reasoning about Programs. *Logics and Models of Concurrent Systems.* NATO ASI Series. Series F, Computer and System Sciences, 13. Springer-Verlag, 1984.

[22] C. Stirling. An Introduction to Modal and Temporal Logics for CCS. In *Concurrency: Theory, Language, and Architecture*, Lecture Notes in Computer Science 391, 1989.

[23] C. Stirling, D. Walker. Local Model Checking in the Modal Mu-Calculus. *Theoretical Computer Science, 89*, 1991. 161-177.

A MODEL CHECKING METHOD FOR PARTIALLY SYMMETRIC SYSTEMS

Serge Haddad[1], Jean-Michel Ilié[2] and Khalil Ajami[2]
(1) lab. Lamsade, Université Paris Dauphine, France, Serge.Haddad@lamsade.dauphine.fr
(2) lab. LIP6, Université Paris VI, France, Jean-Michel.Ilie@lip6.fr, Khalil.Ajami@lip6.fr

Abstract A new method of model checking is proposed based on the existence of symmetries in system. We show how to fully handle the partial symmetries of both properties and systems. Our method does not depend on a particular formalism and *a priori* can be applied to any one. Well-formed Petri Nets are used as an illustration.

Keywords: Verification and validation, Temporal logic, model-checking, symmetries, partial symmetries, Büchi automata, well-formed Petri nets.

1. INTRODUCTION

Model checking of temporal logic formulas over finite state systems is now a widely used method of verification of protocols and distributed algorithms. Such a technique has led to numerous tools ([11]). However the main drawback of this kind of verification is the complexity factor depending on the size of the space of reachable states. Thus different improvements have been proposed (and implemented in tools). The partial order analysis exploits the independence of events of the system in order to avoid explorations of equivalent paths in the state graph ([9]). Another fruitful research direction is based on the symmetries of the system to be analysed. The main idea is to build a quotient graph where nodes denote set of equivalent reachable states. Then the reduced graph is shown to be equivalent to the original one w.r.t. to some generic properties ([12],[2]).

However the modeller often needs to check particular properties related to the behaviour of its system. It is then necessary to adapt the quotient graph building with the aim of verifying a temporal logic formula. The key point is the characterisation of symmetries of a formula and we now discuss the previous approaches to this problem. We limit

Tommaso Bolognesi and Diego Latella (Eds.), *Formal Methods for Distributed System Development.*

our presentation to the linear time logic (e.g. LTL). So let us briefly recall its usual verification algorithm ([8]):

- the negation of the formula is translated into a Büchi automaton,
- the synchronised product of the automaton and the state graph of the system is built (product of compatible states and transitions),
- a path is searched within the synchronised product, such that it ends by a loop containing a state associated with one of the acceptation states of the Büchi automaton. We call such a path, an invalidating path.

The formula is true if and only if such an invalidating path does not exist. Although the size of the automaton is exponentially larger than the size of formula, it remains low compared to the number of states in the system. This last parameter is the main complexity factor.

In a first approach ([5]), one restricts the atomic propositions of the language to symmetric propositions. For instance, "*In a future state, all processes will be idle*" or "*If some process is waiting for a resource, then some process will get it*" are symmetric formulas. In other words, a formula is symmetric if its atomic propositions are invariant under any process identities permutation. For checking such formulas, one substitutes a quotient state graph to the state graph and one applies the above algorithm without any change. Unfortunately, many usual formulas are not considered as symmetric. For instance, the fairness formula "*If some process is waiting for a resource, then it will get it*" is not considered as a symmetric formula.

The second approach ([7]), defines symmetric Büchi automata. An automaton is symmetric if, given a (accepting) state and a process identities permutation, there is another (accepting) state whose atomic propositions are obtained by the permutation applied on the atomic propositions of the first state and for any successor of one state, there is a successor of the other one for which the property is again fulfilled (and this recursively). Starting from this automaton and the model of the system, one directly builds a quotient synchronised product. It is then shown that the existence of an invalidating path in the original synchronised product and the quotient structure are equivalent. The symmetric formulas of the previous method, and also new temporal properties like the discussed fairness property are considered as symmetric formulas.

Anyway, this technique does not cover the case of partially symmetric formulas which are considered as asymmetrical. Here is a partially symmetric formula: "*If some process is waiting for a resource then it will get it, provided none of the processes with higher identity will require the resource in the future*". Therefore in ([1]), the authors have proposed a more refined definition of the quotient synchronised product than the previous one:

- One computes an equivalence relation between states of the Büchi automaton per item of the "*symmetry group*" of the model in such a way that two states are equivalent whenever they induce the same present and future w.r.t. the action of this item.
- A quotient synchronised graph is built by applying these equivalence relations on the pairs (state of the model, state of the automaton).

In case of a symmetric automaton, the quotient graph is identical to the previous one, but the technique also reduces the complexity of the verification in case of a partially symmetric automaton. In other words, it generalises the previous techniques.

However none of the existing methods cover two important situations. At first, they require the system to be symmetric. Many models do not fulfill this requirement. In most cases, they are partially symmetric, meaning that large parts of the specification is symmetric but some critical modules are asymmetric. In ([10]), a specific algorithm is proposed for partially symmetric high-level Petri nets, but it mainly deals with the reachability problem and it does not seem that it can be extended to the model checking. In ([6]), asymmetric behaviour is allowed in "*symmetric states*" (left invariant by any process identities permutation). Unfortunately, this kind of asymmetry is very restrictive.

The second situation which cannot be exploited by the previous methods concerns the formula. Actually, the two automaton-based techniques assume that some symmetric relations hold on the structure of the automaton. However in automata of partially symmetric formula, most of the states of the automaton are partially symmetric, but the automaton is globally asymmetric (i.e. the equivalence relations between states are almost reduced to the identity).

In the present work, we will show how to overcome these two problems. In the second section, a new method is presented which applies on a symmetric system and any Büchi automaton. Each state of this automaton is associated with the group of symmetries which left its set of atomic propositions globally invariant. No supplementary condition is required on the structure of the automaton. The method does not depend on a specific formalism for the system and can be applied over any model where symmetries can be syntactically checked. So in the third section, we illustrate a possible implementation on well-formed Petri nets for which symmetries have been intensively studied [3]. More importantly, we show how to apply our method on asymmetric systems. The key point is that such systems are defined as synchronised products of a symmetric model and an asymmetric Büchi automaton. In our opinion, many protocols and algorithms can be represented with the help of this formalism. In the conclusion, we discuss about combinations of techniques and experimentations.

2. PRINCIPLES OF OUR METHOD

2.1. VERIFICATION OF TRANSITION SYSTEMS

In order to represent the general aspects of our method, we consider it at a semantic level, in other terms we consider the transition system that is generated from the syntactic model, e.g. a Petri net.

Definition 1 *A (finite) transition system* $S = (Q, Q_0, R, Prop, \Pi)$ *is defined by : Q, a (finite) set of states;* Q_0*, the set of initial states; R, a transition between states* $R(q, q')$*, also denoted* $q \rightarrow q'$*; Prop, the set of atomic propositions;* Π *an* ***injective*** *mapping from Q to* 2^{Prop}.

The requirements that Π is injective ensures that each state is totally characterised by the values of its atomic propositions. Since we focus on the verification of state formulas, there is no label attached to arcs. Our approach could be adapted easily to formulas expressing conditions on state transitions. The final result (see proposition 8) holds for any finite branching transition systems, i.e. the number of initial states is finite and each reachable state has a finite number of successors.

One may express a linear time temporal logic formula with languages like LTL ([15]) or μ-calcul ([13]). During the model checking process, the considered formula can be translated in a Büchi automaton ([16]).

Definition 2 *A Büchi automaton* $A = (B, B_0, R, Prop, \Pi, F)$ *is defined by : B, a finite set of states;* B_0*, the subset of B of the initial states; R, a transition relation between states* $R(b, b')$*, also denoted* $b \rightarrow b'$*; Prop, the set of atomic propositions;* Π *a mapping from B to* 2^{Prop}*; F, the subset of B of the accepting states.*

The verification problem of a formula expressed by the Büchi automaton is usually reduced to the search of an infinite run within the system : such a run $\{q_i\}_{i=0..\infty}$ with $q_0 \in Q_0$ must correspond to an infinite path $\{b_i\}_{i=0..\infty}$ with $b_0 \in B_0$ within the automaton. More precisely, the atomic propositions labelling every state b_i must also be present in state q_i : $\Pi(b_i) \subset \Pi(q_i)$. In addition, an accepting state must occur infinitely often in the considered path. This leads to the definition of the following key concept, called the synchronised product.

Definition 3 *Let S be a transition system and A a Büchi automaton. The synchronised product between S and A is a graph $Gr(S, A) = (V, V_0, R)$, defined by : $V = \{(q, b)$ s.t. $\Pi(b) \subset \Pi(q)\}$, the set of nodes; $V_0 = \{(q, b) \in V$ s.t. $q \in Q_0 \wedge b \in B_0\}$, the set of initial states; R, the transition relation between nodes, denoted $\rightarrow$, and such that $(q, b) \rightarrow (q', b')$ iff $q \rightarrow q'$ and $b \rightarrow b'$.*

One standard model checking principle consists of a translation of the negation of the formula in a Büchi automaton, the building of the synchronised product, and the search of an "invalidating" path within the synchronised product.

Definition 4 *Let S be a transition system and A a Büchi automaton expressing the negation of formula f. Formula f is valid iff there is no invalid path in $Gr(S, A)$. An invalidating path of formula f is an infinite path within the synchronised product $Gr(S, A)$, starting from one initial node and including infinitely often the subset of nodes $\{(q, b)$ s.t. $b \in F\}$*

If the system is finite, the verification procedure is reduced to the search of an elementary path ending by a state yet visited, s.t. between the two occurrences, a state referring an accepting condition is met. Many improvements to this method can be found in the literature, among them there are the on-the-fly techniques which search for an invalidating path simultaneously to the building of $Gr(S, A)$ ([8]). All these techniques can be applied in the context of the quotient synchronised product we develop in section 2.3.

2.2. SYMMETRIC TRANSITION SYSTEMS

In order to highlight all the developed concepts, we consider in this section a model of a protocol where four processes, denoted by $C = \{u, v, w, x\}$, execute the same program. For sake of simplicity, we assume that communications are instantaneous, so that the global system is entirely defined by the local states of the processes. In the following, the local state of any process, e.g. x, is symbolically represented by a variable, e.g. $l(x)$. The definition domain for $l(x)$ is {idle, transmit, wait, access}.

Because "symmetries" are usually handled by group theory, we now recall some elementary notions of group ([14]).

Definition 5 *Let G be a group, containing a neutral element id, which operation is denoted by $(\bullet)$.*

- *Let E be a set, an action of G over E is a mapping from $G \times E$ to E s.t. the image of (g, e), denoted by $g.e$, fulfills : $\forall e \in E\ id.e = e\ \ \forall g, g' \in G\ (g \bullet g').e = g.(g'.e)$*
- *The isotropy (sub)group G_e of an element e is defined by : $G_e = \{g \text{ s.t. } g.e = e\}$*
- *Let H be a subgroup of G, the orbit $H.e$ of e under H is defined by : $H.e = \{g.e \text{ s.t. } g \in H\}$*
- *This action can be straightforwardly extended to the powerset of E by : $g.E' = \{g.e \text{ s.t. } e \in E'\}$*

Let us assume that E is the set of the atomic propositions of our model. For instance, proposition $[l(u) = idle]$ means that the local state of process u is idle. If G is the group of permutations of C and if g is the permutation which exchanges processes u and v, then $g.[l(u) = idle] = [l(v) = idle]$. Let $\{[l(u) = idle], [l(v) = idle]\}$ be a set of atomic propositions, then the isotropy group of this subset is the subgroup of permutations which let the subset $\{u, v\}$ globally invariant.

We are now able to characterise what is a symmetric transition system.

Definition 6 *Let S be a transition system and G be a group acting on Prop.*
S is said to be symmetric (w.r.t. G) iff :

- *Every state has a " symmetric " state w.r.t. any element of G : $\forall q \in Q, \forall g \in G,\ \exists q' \in Q,\ \Pi(q') = g.\Pi(q)$ The action of the group on the states is extended, by denoting $g.q$, the unique q' of the former formula.*
- *The set of initial states is invariant under the action of G : $G.Q_0 = Q_0$.*
- *The action of G is congruent w.r.t. the transition relation : $\forall q, q' \in Q,\ \forall g \in G,\ \ q \to q' \Leftrightarrow g.q \to g.q'$*

Observe that all these conditions hold regarding our model of protocol, by defining G as the group of process permutations.

2.3. MODEL CHECKING OF SYMMETRIC TRANSITION SYSTEM

As a state is associated with a subset of *Prop* in both the Büchi automaton and the model, we will use the action of G on the powerset of *Prop*. Let b be a state of Büchi automaton A, $G_{\Pi(b)}$ will simply be denoted G_b (although one cannot define an action of G on B). Observe that G_b is strongly related to the symmetry degree of b independently of the structure of the Büchi automaton.

Let us consider different cases of G_b in the context of our example. If this subgroup equals the group G, then the state is totally symmetric, e.g. this is the case for proposition set $\{[l(u) = idle], [l(v) = idle], [l(w) = idle], [l(x) = idle]\}$. In contrast, whenever the isotropy group is reduced to the identity (denoted $\{id\}$), the state is totally asymmetric, e.g. $\{[l(u) = idle], [l(v) = transmit], [l(w) = wait], [l(x) = access]\}$. In most cases, the isotropy group is not trivial (i.e. it differs from id and G). Let us note that in the context of our model, the isotropy group of processes implicitly corresponds to a partition of processes, namely the group of permutations which left each set of the partition invariant. For $\{[l(u) = idle], [l(v) = idle]\}$, the partition of C is $\{\{u, v\}, \{w, x\}\}$. According to a state of the Büchi automaton, we call such a partition *the local partition of the state.*

Our aim is to build a quotient of the synchronised product over which it is possible to search an invalidating path directly. In this " quotient" structure, each node is characterised by a triple (H, O, b), where H is a subgroup of G, $O \subset Q$, $b \in B$. Moreover, the following two points must hold:

(C1) $\forall q \in O, \Pi(b) \subset \Pi(q)$

(C2) $H.O = O$

The intuitive idea is that the node (H, O, b) is an aggregation of a set of nodes of the synchronised product, more precisely $\{(q, b)\}_{q \in O}$. As we will see now, H is used in the definition in order to give a sound definition of the successor relation.

The building of the quotient structure starts from the set of initial nodes, defined as follows : $(G_{b0}, G_{b0}.q_0, b_0)$ with $q_0 \in Q_0$, $b_0 \in B_0$ and $\Pi(b_0) \subset \Pi(q_0)$. Observe that condition $C2$ is fulfilled trivially because G_{b0} is a group, moreover condition $C1$ is deduced from the definition of G_{b0}.

Then, there remains to define the successor relation. (H_2, O_2, b_2) is a successor of (H_1, O_1, b_1), denoted $(H_1, O_1, b_1) \rightarrow (H_2, O_2, b_2)$, **iff** $b_1 \rightarrow b_2$

and $\exists q_1 \in O_1$, $\exists q_2 \in Q$ s.t. $q_1 \rightarrow q_2$ and $\Pi(b_2) \subset \Pi(q_2)$. **Then** $O_2 = (H_1 \cap G_{b_2}).q_2$ and $H_2 \subset G_{O_2}$.

Observe that both conditions $C1$ and $C2$ are fulfilled by the new node. Effectively, from $O_2 \subset G_{b_2}.q_2$, one deduces that: $\forall q_2' \in O_2, \exists g \in G_{b_2}$ *s.t.* $q_2' = g.q_2$. So, $\Pi(b_2) = g.\Pi(b_2) \subset g.\Pi(q_2) = \Pi(q_2')$. Moreover, as $H_2 \subset G_{O_2}$ then $H_2.O_2 = O_2$.

In order to make the successor relation operational, one must specify how the subgroup H_2 is chosen. With regard to the successor relation, it is interesting to maximise H_2, i.e. to choose $H_2 = G_{O_2}$, but this choice might not be possible w.r.t. the syntactical aspects of the model. In any case, H_2 can be chosen as the subgroup $(H_1 \cap G_{b_2})$. The following lemma and proposition 1 show the validity of our construction.

Lemma 1 *Let S be a symmetric transition system and A be a Büchi automaton. Let $Gr(S, A)$ be the corresponding synchronised product and $GRQ(S, A)$ the corresponding quotient structure.*

- *Let (H_1, O_1, b_1) and (H_2, O_2, b_2) be two nodes of GRQ(S,A).* $(H_1, O_1, b_1) \rightarrow (H_2, O_2, b_2) \Rightarrow \forall q_2 \in O_2,\ \exists q_1 \in O_1,\ (q_1, b_1) \rightarrow (q_2, b_2)$

- *Let $(q_0, b_0) \rightarrow (q_1, b_1) \rightarrow ... \rightarrow (q_n, b_n) \rightarrow ...$ be a (infinite) path in $Gr(S, A)$, then $\exists (H_0, O_0, b_0) \rightarrow (H_1, O_1, b_1) \rightarrow ... \rightarrow (H_n, O_n, b_n) \rightarrow ...$ a (infinite) path in $GRQ(S, A)$ s.t. $q_i \in O_i$ $(i \in 1..n)$.*

It is worth noting that the first point becomes false in the case where one exchanges its consequence by the following one: " $\forall q_1 \in O_1, \exists q_2 \in O_2,\ (q_1, b_1) \rightarrow (q_2, b_2)$".

Proposition 1 *For a symmetric finite branching transition system, there is an invalidating path in the synchronised product if and only if there is an invalidating path in the quotient structure.*

3. APPLICATION TO THE WELL-FORMED PETRI NETS

The well-formed Petri net model offers a global solution for the design, verification and performance evaluation of distributed systems graphs. A well-formed Petri net is a high level Petri net model. Its specific syntax has made possible to define methods which take profit from the symmetry relations existing in systems. One of the associated methods is the building of the symbolic reachability graph which is a highly compact structure used to represent the state space. It can be used to

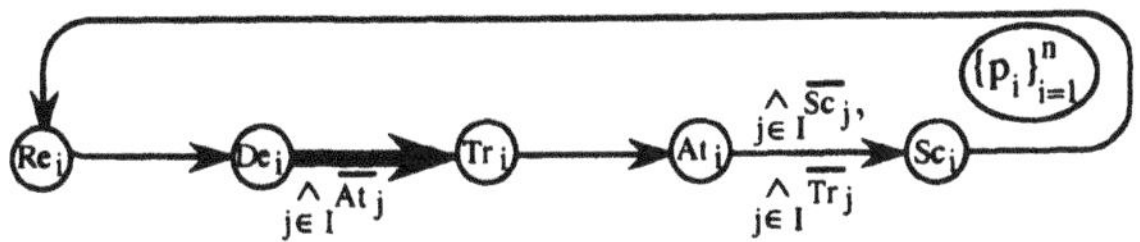

Figure 1 Automaton of one process

analyse some standard properties of the Petri nets as well as to derive an aggregated Markov chain in order to compute performance indices ([3]). We are interested in adapting this construction to our method, but we do not fully describe neither the well-formed Petri net definition nor the implementation of the symbolic reachability graph building (see [2] for more details). Actually, we will focus on the representation of nodes and on the building of the successor relation.

3.1. PRESENTATION OF THE WELL-FORMED PETRI NET MODEL

Description of the system and properties to verify. The system which is described here consists of n processes $\{P_i\}_{i=1..n}$ accessing to a critical section. Like in the former model, we consider that the processes run the same critical section access protocol. They are now synchronised by shared variables.

This protocol is described for a process P_i in Figure 1. Initially, the process is idle (local state Re) then it may ask for an access to the critical section (local state De). Before accessing this section (local state Sc) the process follows a sequel of two local states : Transmitting (Tr) and Waiting (At). We now detail the state transitions : (1) A requesting process can be in state Tr if there is no process in waiting state. The scheduler considers as privileged all the processes in state De. This is represented in the figure by a bold arc. (2) A process changes from state Tr to At without specific condition. (3) To enter the critical section, there must be no process in this section, but also no process in state Tr.

We aim at demonstrating that the system is (weakly) fair : Every process which asks for the critical section will obtain it in a finite time. Intuitively, this property holds due to the conditions which ensures that requests are treated *wave after wave*. A wave starts from the moment when there is a requesting process and is completed as soon as one process enters the waiting state.

Hence, a wave contains all the processes in state Tr and At, and these processes will enter the critical section without specific order. Due to the privilege of the change from De to Tr, a requesting process belongs

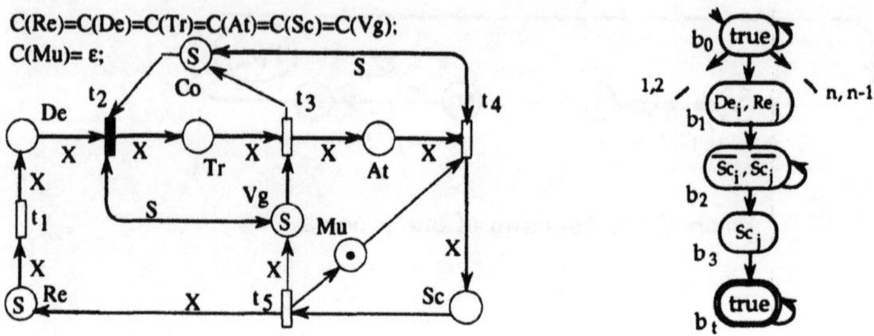

Figure 2 Modelling of the symmetrical behaviour of the system

Figure 3 Büchi automaton of $\neg f$

either to the current wave or to the following one. This ensures the fairness property.

In a second stage, the system is modified so that within a wave, the executions of the critical section are ordered according to a priority relation, induced by the identity number of processes. Actually, the first system is symmetric, while the second is not.

With respect to this new specification, the two expected properties are the followings :

- Every requesting process will finally obtain the access
- The processes in state Tr or At will be served according to the order induced by the specified priority relation.

In contrast, the strong fairness property "*A requesting process will be served before every process in state idle*" is false.

Formal modelling for the System and its properties.

In order to represent our system, we proceed in two stages : a symmetric system is specified first in terms of the well-formed Petri nets of Figure 2, then the runs which do not preserve the priority relation are prohibited by means of the additional automaton of Figure 4, namely control automaton. Actually, the system formally corresponds to the synchronised product of the well-formed Petri net and the control automaton.

In a well-formed Petri net, a colour domain is attached to each place and transition. Each token in a place is typed by a colour of the place colour domain. Each colour of a transition indicates a different way to fire the transition, hence, the labels of the arcs are no more integers but colour functions which denote for each firing the quantities of tokens to

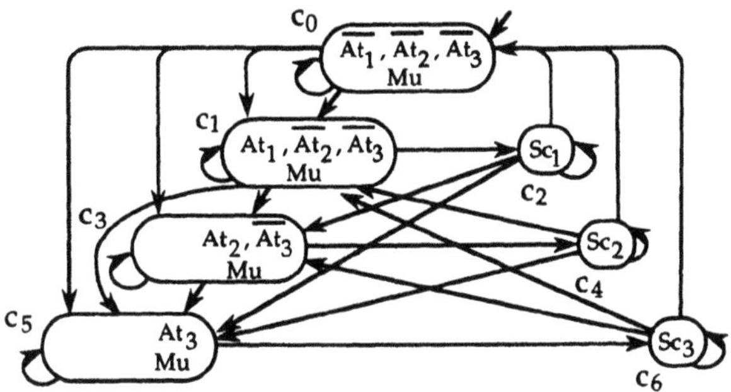

Figure 4 Control automaton w.r.t. a priority relation

be consumed and produced per each colour. In well-formed Petri nets, the colour functions are written under a specific syntax which allows one to exploit the system symmetries. Function S represents a constant function, called a diffusion. Function S is a sum of tokens s.t. there is one token per colour of the domain, e.g. one for each process. Function X represents the colour which is used to instantiate the firing of a transition in the net.

In addition, a priority number is associated with each transition of the well-formed Petri nets. Thus, among the transitions that are enabled from the current marking of the net, only the ones which have the highest priorities can be fired.

In the well-formed Petri net of Figure 2, the local states of the processes are represented by means of the corresponding places. The set $C = \{\langle 1\rangle), \langle 2\rangle), \langle 3\rangle\}$ is the colour domain of processes. It is the domain of all the places (except Mu) and the domain of all transitions of the net. The marking of place Re by the (constant) diffusion function corresponds to the fact that, initially, all processes are idle. Three more places allow one to implement the protocol mechanism : (1) Place Mu is used to specify the control of accesses to the critical section; initially, it is set to an (uncoloured) token and prohibits the firing of transition t_4, whenever place Sc is marked. (2-3) Places Vg and Co (initially set to S) are used to specify the wave mechanism. Vg prohibits the firing of transition t_2, whenever one of the At or Sc places is marked. Place Co prohibits the firing of transition t_4 whenever place Tr is not empty. At last, t_2 is a transition the firing of which is privileged with respect to the others transition, so that the expected behaviour of the scheduler is obtained by the following priority relation : $\forall t \neq t_2$, $Pri(t_2) > Pri(t)$.

The specification of the well-formed Petri net is completed by the control automaton A_c given in Figure 4. The first state of A_c specifies that there is no process currently in the critical section, and that there is no request for that. The three states at the left side specify the different cases where a process must be privileged, provided the critical section is empty. Moreover, the three processes in the right side determine which processes is in the critical section. The arcs from left to right ensure the priority mechanism. The others follow the behaviour of the net.

The semantic of the system is defined by the synchronised product between the control automaton and the reachability graph of the net.

In order to illustrate our method, we wish evaluate the following strong fairness property : "*A requesting process will be served before every idle process*". The corresponding LTL formula is the following ([15]) : $f = \wedge_{i,j \in I} G[(De_i \wedge Re_j) \Rightarrow X(\overline{Sc_j} U Sc_i)]$

The Büchi automaton $A_{\neg f}$ partially represented by Figure 3 models the possible runs of the system for which the negation of property f holds. For sake of simplicity, only one branch (i, j) is represented. In this elementary case, the structure of the automaton is immediately deduced from the formula to invalidate. With the help of the first state of the automaton, one waits to the moment when $(De_i \wedge Re_j)$ holds, which will cause the invalidation of the property. Then, the next states of the automaton ensures that process j will enter the critical section before i. The last state represents the remaining part of the infinite run, without meaning according to the considered property.

We highlight now (for this kind of specification) that *the checking of a formula on an asymmetric system can be reduced to the checking of an asymmetric formula on a symmetric system.* The search of a path must be computed through the synchronised product of the three structures : the reachability graph of the well-formed Petri net, the control automaton and the automaton of the negation of the formula. However, the synchronised product is a commutative and associative operation. It is thus possible to perform the synchronised product of the two automata, A_c and $A_{\neg f}$, in order to obtain a new one, then to apply our construction on this new automaton and the well-formed Petri net (a symmetric model).

On can check that the following path in the synchronised product invalidates the desired formula :

$$(m_0, (c_0, b_0)) \xrightarrow{t_1(\langle 1 \rangle)} (m_1, (c_0, b_1)) \xrightarrow{t_2(\langle 1 \rangle)} (m_2, (c_0, b_2)) \xrightarrow{t_1(\langle 2 \rangle)} (m_3, (c_0, b_2)) \xrightarrow{t_2(\langle 2 \rangle)}$$

$$(m_4, (c_0, b_2)) \xrightarrow{t_3(\langle 2 \rangle)} (m_5, (c_3, b_2)) \xrightarrow{t_3(\langle 1 \rangle)} (m_6, (c_3, b_2)) \xrightarrow{t_4(\langle 2 \rangle)} (m_7, (c_4, b_3))$$

Each node (m,(c,b)) of such a sequence corresponds respectively to one marking of the net and one pair obtained by a state of the Büchi automaton and a state of the control automaton. Observe that, the label of an arc represents some firing in the net. For instance, $\xrightarrow{t_1(\langle 1 \rangle)}$ corresponds to the firing of transition t_1 with respect to process $\langle 1 \rangle$.

3.2. BUILDING OF THE SYMBOLIC SYNCHRONISED PRODUCT

The symbolic synchronised product is a representation of the *quotient synchronised product.* We call them symbolic due to the analogy with the symbolic reachability graph representation ([2]). Its major peculiarity is that objects are never more represented explicitly, but are viewed under a symbolic form. For instance, one can say symbolically that two processes require the same resource, instead of saying that processes 1 and 2 require resource 1. From such a symbolic representation, a symbolic firing rule can be defined which allows one to compute the symbolic reachability graph, automatically and directly from the object specification within the well-formed net.

Computation of local partitions of colours. Before computing the quotient synchronised product, one must compute and represent the isotropy group of each state of the automaton. The most easy way to proceed consists in *detecting* the colours which marks the same propositions in the states. These colours are gathered in subsets and these subsets form a partition, named local partition. The isotropy group is implicitly defined as the group of permutations which left each set of the partition globally invariant. Let us give three examples : (1) For a state where the atomic propositions are $\{\overline{At_1}, \overline{At_2}, \overline{At_3}\}$, the partition is composed by a unique singleton $\{\langle 1 \rangle, \langle 2 \rangle, \langle 3 \rangle\}$ (i.e. all the permutations are admissible). (2) For a state where the atomic propositions are $\{\overline{At_2}\}$, the partition is composed of a singleton $\{\langle 2 \rangle\}$ and a set composed of the remaining colours, here $\{\langle 1 \rangle, \langle 3 \rangle\}$ (the permutations must left $\langle 2 \rangle$ invariant). (3) For a state where the atomic propositions are $\{At_2, \overline{At_3}\}$, the partition is composed of three elements, $\{\langle 1 \rangle\}, \{\langle 2 \rangle\}, \{\langle 3 \rangle\}$. The isotropy group is reduced to the identity permutation.

States of the symbolic synchronised product. Let us recall that we want to represent triples (H,O,(b,c)), such that O is subset of markings which is unchanged under the action of the subgroup H ($O = H.o$, $\forall o \in O$). The subgroup H will be implicitly represented by a local partition, which may differ from the partition attached to the state of the automaton (H may be different from G_b). The subset of markings will be represented in a symbolic way : the colour domain is partitioned in abstract subsets, called *dynamic subclasses* (i.e. the composition is not known). However for each subset, the cardinality and the reference to a local partition is preserved. The markings of the net are now symbolic since their tokens are no more colour but dynamic subclasses. *Symbolic markings* are denoted like $\widehat{m}$.

In our example, there is one initial symbolic node, such that : $(\widehat{m}_0, (b_0, c_0))$ where $\widehat{m}_0$ is a symbolic marking of the net such that :

- $\widehat{m}_0$ is defined from a unique local partition $\{\langle 1\rangle, \langle 2\rangle, \langle 3\rangle\}$ composed of a unique dynamic subclass (denoted Z_1 s.t. $\mid Z_1 \mid= 3$),
- $\widehat{m}_0 = (Z_1.Re + Z_1.Co + Z_1.Vg) + Mu$, meaning that there is one token of each process in places Re, Co and Vg and that there is one uncoloured token in place Mu.

The atomic propositions of b_0 and c_0 $(\{\overline{At_i}\}_{i=1..3}, Mu)$ are satisfied by $\widehat{m}_0$. Observe that $\widehat{m}_0$ is a symbolic representation for both subgroup H of permutations and subset O of markings.

Symbolic firing in the quotient synchronised product. In order to build a successor of a node $(H_0, O_0, (b_0, c_0))$ within the symbolic synchronised product, a successor of the current state in the automaton is selected. For instance from (b_0, c_0), one can choose state (b_1, c_0) as a possible successor. Then a new local partition is computed, by intersecting the local partition of the state currently considered within the symbolic synchronised product and the local partition of the state newly considered within the automaton. This is equivalent to consider the group $H_0 \cap G_{(b_1, c_0)}$.

For state $(\widehat{m}_0, (b_0, c_0))$, the local partition in the symbolic synchronised product is given by $\{\langle 1\rangle, \langle 2\rangle, \langle 3\rangle\}$ and the one of the state within the automaton is given by $\{\langle 1\rangle\}, \{\langle 2\rangle\}, \{\langle 3\rangle\}$. Hence, a decomposition identical to colours is obtained. Let $\widehat{m'}_0$ be the symbolic representation $\widehat{m}_0$ after the above decomposition:

- $\forall i \in \{1, 2, 3\}$ Z_i is the unique dynamic subclass of $\{\langle i\rangle\}$
- $\widehat{m'}_0 = (Z_1.Re + Z_2.Re + Z_3.Re) + (Z_1.Co + Z_2.Co + Z_3.Co) + (Z_1.Vg + Z_2.Vg + Z_3.Vg) + Mu$

Then, a symbolic firing is performed in the net which is similar to the ordinary one (see [2] for more details). For instance, the firing of $t_1(Z_1)$ leads to symbolic marking $\widehat{m'}_1 = (Z_1.De + Z_2.Re + Z_3.Re) + (Z_1.Co + Z_2.Co + Z_3.Co) + (Z_1.Vg + Z_2.Vg + Z_3.Vg) + Mu$.
Additionally, one must verify that $\widehat{m'}_1$ satisfies the atomic properties of $(b_1, c_1) : (\{\overline{At_i}\}_{i=1..3}, Mu, De_1, Re_2)$.

The last stage consists in grouping some local partitions which contain a single dynamic subclass, provided the distributions of the concerned dynamic subclasses over the places are the same. Actually, this does not modifies the set of ordinary markings associated with the symbolic marking and this means that the group of admissible permutations $H_0 \cap G_{(b_1,c_0)}$ is extended. For instance in $\widehat{m'}_1$, sets $\{\langle 2 \rangle\}$ and $\{\langle 3 \rangle\}$ can be grouped yielding symbolic marking $\widehat{m}_1$: Z_1 is the unique dynamic subclass of $\{\langle 1 \rangle\}$ and Z_2 is the unique dynamic subclass of $\{\langle 2 \rangle, \langle 3 \rangle\}$; moreover, $\widehat{m}_1 = (Z_1.De + Z_2.Re) + (Z_1.Co + Z_2.Co) + (Z_1.Vg + Z_2.Vg) + Mu$.

4. CONCLUSION

Complexity reduction of the verification by use of the symmetries is now well established. However most of the proposed techniques do not handle (or in a limited way) the partially symmetric systems. In this work we have proposed an efficient way to deal with asymmetric models and/or formulas. Our method does not depend on a particular formalism and *a priori* can be applied to any one. Here we have illustrated the method on well-formed Petri nets in order to clarify the principles based on actions of groups on transition systems.

This work needs to be completed in two ways. At first, it must be integrated in a tool. So the implementation of the presented work is under progress. We have chosen to develop our method in the GreatSPN tool ([4]) as symmetries are already exploited for well-formed Petri nets. To this implementation, must succeed a stage of evaluation in order to characterise systems for which our algorithm is efficient.

Moreover, we plan to combine our method with the global analysis of symmetries of the automaton as done in ([1]). At last, we investigate the way to extend this method to performance evaluation of stochastic well-formed Petri nets.

References

[1] K. Ajami, S. Haddad, and J.-M. Ilié. Exploiting Symmetry in Linear Temporal Model Checking : One Step Beyond. In *Proc. of Tools and Algorithms for the Construction and Analysis of Systems TACAS'98, part of Theory and practice of Software ETAPS'98*, volume 1384 of *LNCS*, pages 52–67, Lisbon - Portugal, April 1998. Springer Verlag.

[2] G. Chiola, C. Dutheillet, G. Franceschinis, and S. Haddad. On Well-Formed Coloured Nets and Their Symbolic Reachability Graph. In K. Jensen and G. Rozenberg, editors, *High-Level Petri Nets. Theory and Application*, pages 373–396. Springer Verlag, 1991.

[3] G. Chiola, C. Dutheillet, G. Franceschinis, and S. Haddad. Stochastic Well-Formed Colored Nets and Symmetric Modeling Applications. *IEEE Transactions on Computers*, 42(11):1343–1360, 1993.

[4] G. Chiola, G. Franceshinis, R. Gaeta, and M. Ribaudo. GreatSPN1.7: GRaphical Editor and Analyzer for Timed and Stochastic Petri Nets. *Performance Evaluation, North Holland Journal*, 24, 1997.

[5] E.M. Clarke, R. Enders, T. Filkorn, and S. Jha. Exploiting Symmetry in Temporal Logic Model Chacking. *Formal Methods and System Design*, 9:77–104, 1996.

[6] E. A. Emerson and R. J. Trefler. From Asymmetry to Full Symmetry: New Techniques For Symmetry Reduction in Model Checking. In *Proc of CHARME'99*, Lecture Notes in Computer Science, pages 142–156, Bad Herrenalb - Germany, September 1999. Springer Verlag.

[7] E.A. Emerson and A. Prasad Sistla. Symmetry and Model Checking. *Formal Methods and System Design*, 9:307–309, 1996.

[8] R. Gerth, D. Peled, M. Vardi, and P. Wolper. Simple On-the-fly Automatic Verification of Linear Temporal Logic. In *Proc. Int Conf. on Protocol Specification Testing and Verification*, 1993.

[9] P. Godefroid and P. Wolper. A Partial Approach to Model Checking. *Information and Computation*, 110(2):305–326, May 1994.

[10] S. Haddad, J.M. Ilié, M. Taghelit, and B. Zouari. Symbolic Reachability Graph and Partial Symmetries. In *Proc. of the* 16th *Intern. Conference on Application and Theory of Petri Nets*, volume 935 of *LNCS*, pages 238–257, Turin, Italy, June 1995. Springer Verlag.

[11] G. J. Holzmann. The Spin Model Checker. *IEEE Transaction on Software Engineering*, 23(5):279–295, May 1997.

[12] K. Jensen. Coloured Petri Nets. In W. Brauer, W. Reisig, and G. Rozenberg, editors, *Petri Nets: Central Model and their Properties, Advances in Petri Nets, Part 1*, volume 254 of *Lecture Notes in Computer Science*, pages 248–299, Bad Honnef, Germany, September 1986. Springer Verlag.

[13] D. Kozen. Results on the propositional mu-calculus. *Theoretical Computer Science*, 27:333–354, 1983.

[14] S. Lang. *Algebra.* 7th printing. Addison Wesley, 1977.

[15] A. Pnueli. The temporal logic of programs. In *Proceedings of the 18th IEEE Symposium on Foundations of Computer Science*, pages 46–57, 1977.

[16] M. Vardi. An Automata-theoretic Approach to Linear Temporal Logic. *Lecture Notes in Computer Science*, 1043:238–266, 1996.

Part Four

MODEL CHECKING – APPLICATIONS

SPECIFICATION AND VERIFICATION OF MESSAGE SEQUENCE CHARTS

Doron Peled

Bell Labs

Murray Hill, NJ 07974

doron@research.bell-labs.com

Abstract The use of message sequence charts (MSCs) is popular in designing and documenting communication protocols. A recent surge of interest in MSCs has led to various algorithms for their automatic analysis, e.g., finding race conditions. In this paper we adopt a causality based temporal logic to specify properties of MSCs. This alleviates some problems that arise when specifying properties of MSCs using the traditional interleaving-based linear temporal logic: systems of MSCs are not necessarily finite state systems, leading to undecidability of LTL model checking. Even when dealing with finite state MSC systems, the set of linearizations can easily generate an exponential state space explosion. We provide an efficient model checking algorithm for systems of MSCs. Our construction models the FIFO MSC systems using a restricted version of ω-automata with *two* successor relations. We implemented a model checking environment for MSCs as an extension to the SPIN model checking system.

Keywords: Message sequence charts, Model checking, Specification, Verification.

1. INTRODUCTION

Software verification is a very challenging task. One of the reasons is that software lacks some of the regularity that is typically found in hardware circuits. Some of the successful attempts to apply verification technology to software focused on communication protocols, abstract versions of algorithms, and finite state systems. One potential target of verification algorithms is the early design of software systems. There, the cost-performance of finding bugs is better than in later development stages, and the description is typically already an abstract version of the desired system.

Tommaso Bolognesi and Diego Latella (Eds.), *Formal Methods for Distributed System Development.*

Message Sequence Charts (MSCs) is a standard ITU notation (see [7]), describing the executions of communication systems. It focuses on the description of messages passed between the participating concurrent processes, and abstracts away most other details, e.g., program variables and their values. Like other ISO (International Standardization Organization) notations, MSCs can be specified both using a visual and a textual form. This simple standard is in particular appealing for specifying communication protocols. A similar notation of 'use cases' is used in object-oriented design. Different collections of *executions* or *scenarios* are described using MSCs, providing a clear representation of different features of systems. Grouping together different scenarios is facilitated using high-level MSCs (HMSCs). These are graphs that include in each of the nodes a single MSC. The graph structure allows describing different alternative executions, according to the different branches of the graph.

In this paper we suggest a generic model checking paradigm that allows the automatic verification of systems of message sequence charts. This paradigm is based on choosing a temporal specification that is based on *partial order semantics*, instead of the traditional *interleaving semantics*. Such a logic cannot distinguish between different linearizations of the same partial order execution, and hence is most suited for causality based models such as MSCs. We provide a concrete suggestion of such a logic, and present a simple implementation of the model checking algorithm, by a translation of the MSC notation and the temporal property to the model checker SPIN [8].

Because HMSCs do not impose any constraint on the number of messages that were sent and were not yet received, HMSCs are not necessarily limited to *finite state systems*. This makes the problem of automatic verification of MSCs tricky. Standard model checking algorithms, which are usually applied to finite state systems [5] fail. Specifically, it was shown that checking linear temporal logic (LTL) properties of HMSCs is undecidable [2](e.g., by a reduction from PCP – post correspondence problem).

Checking LTL properties of MSCs is based on the linearizations of HMSCs, i.e., completing the partial order between events that is imposed by an HMSC scenario into a total order. This approach asserts that *all* the linearizations of an MSC execution satisfy the given LTL specification. To alleviate this problem, one may impose a bound on the number of messages in any communication channel. This allows translating the HMSCs to finite state models that are sometimes exponentially bigger than the original HMSC description. Then it is possible to apply standard model checking algorithms. This approach has the unintuitive property that different linearizations of the same MSC execution may not agree on satisfying the given LTL specification. Another constraint that makes HMSCs decidable is described in [2, 9]. Besides restricting the systems that can be checked, this approach also requires deciding whether a given system satisfies this constraint, a task which is in the complexity class co-NP complete.

A different approach that is based directly on the partial order semantics of HMSCs is to specify properties of HMSCs using HMSCs. Intersecting two HMSCs is analogous to checking the emptiness of intersecting a system automaton and a property automaton in the automata approach to model checking [11]. The property automaton specifies the disallowed executions. Unfortunately, the problem of emptiness of HMSC intersection turns to be undecidable as well [10] (again by a reduction from PCP).

In the current work, we show how a temporal logic TLC$^-$, a subset of TLC [1], can be used to assert properties of HMSCs. We provide a model checking algorithm for this logic, which is linear in the size of the checked HMSC, and exponential in the size of the formula and the number of processes. The main observation is that the logic TLC, and its subset TLC$^-$ are interpreted over the partial order structure of the HMSCs. Hence they do not distinguish between the different linearizations of each partial order execution. Since these formulas can also be interpreted over linearizations of the partial order, a single linearization suffices for each execution. The subset TLC$^-$ of TLC was carefully chosen. It is capable of asserting interesting properties of communication systems. Moreover, it allows a simple translation from TLC$^-$ into automata. Yet, one can extend the framework suggested here to other partial order semantics based temporal logics.

2. PRELIMINARIES

Each MSC describes a scenario where some processes communicate with each other. Such a scenario includes a description of the messages sent, messages received, the local events, and the ordering between them. In the visual description of MSCs, each process is represented by a vertical line, while a message is represented by a horizontal or slanted arrow from the sending process to the receiving one, as in Figure 1.

Definition 1 *An MSC M is a sextuple $\langle V, <, L, T, m, \mathcal{P}\rangle$ where V is a set of events, $< \subseteq V \times V$, $\mathcal{P}$ is a set of processes, $L : V \rightarrow \mathcal{P}$ is a mapping that associates each event with a process, $T : V \rightarrow \{s, r, l\}$ is a mapping that describes the type of each event (send, receive or local, respectively), and $m \subset V \times V$ is a relation between send and receive events. If $(e, f) \in$*

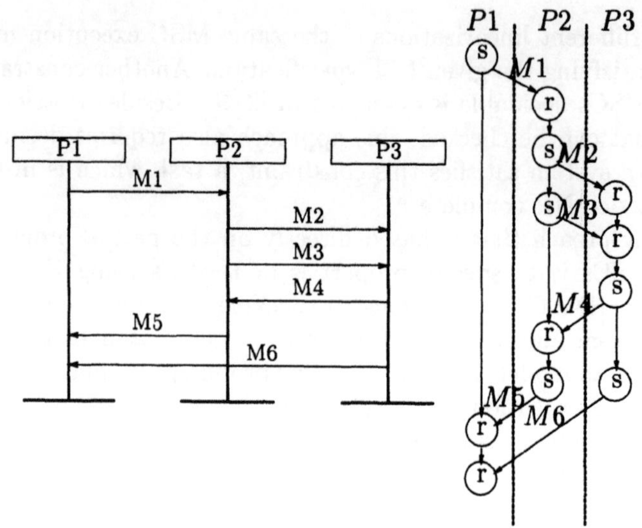

Figure 1 An MSC and its partial order semantics

m then $T(e) = s$, *and* $T(f) = r$. *In this case, we call* e *and* f *a* matching pair. *Each send and receive event is paired up using* m *with exactly one other event. For two events* e *and* f, *we have* $e < f$ *when one of the following holds: (1)* e *and* f *are a matching pair, or (2)* e *and* f *belong to the same process* P, *with* e *appearing before* f *on the process line.*

We assume FIFO *(first in first out) message passing, i.e.,* $(T(e_1) = T(e_2) = s \land T(f_1) = T(f_2) = r \land m(e_1, f_1) \land m(e_2, f_2) \land L(e_1) = L(e_2) \land L(f_1) = L(f_2) \land e_1 < e_2) \to f_1 < f_2$.

Denote by $e \longrightarrow f$ when $e < f$ and either e and f are the *send* and *receive* of the same message, or when they belong to the same process and there is no event between e and f on some process line. That is, e *immediately precedes* f. The relation '$<$' is called the *visual ordering* of events and it is obtained from the syntactical representation of the chart (e.g. represented according to the standard syntax ITU-Z120). The transitive closure '$<^*$' of the relation '$<$' is a partial order between the events V.

The partial order between the send and receive events of the MSC in Figure 1 is shown in the right side of Figure 1. There, only the 'immediately precedes' order '$\longrightarrow$' is shown. This MSC describes an interaction between three processes, $P1$, $P2$ and $P3$. Process $P1$ sends the message $M1$ to $P2$. After receiving that message, process $P2$ sends two messages, $M2$ and $M3$ to $P3$. After receiving $M3$, process $P3$ sends the message $M4$ back to $P2$ and later also sends the message $M6$ to $P1$.

Process $P2$, after receiving $M4$, sends $M5$ to $P1$. The message $M5$ is received by process $P1$ before the message $M6$. The *send* events of the two messages, $M5$ and $M6$, are unordered.

Some papers (e.g., [3]) distinguish between the above visual order and a more relaxed order among the events that allows for example some receive events of the same process to be unordered. This is because some receive events of messages that are sent from different processes cannot be guaranteed to arrive in a particular order. This relaxed order is used in the MSC tool [3], in conjunction with the visual order, to detect race conditions, where messages may arrive in an order that is not prescribed by the MSC. For example, in the MSC of Figure 1, the message $M6$ may arrive at $P1$ before $M5$, as both are independently sent from different processes. In this paper, we will commit to the simple (visual) order defined above. Next, we define how to sequentially compose a pair of MSCs.

Definition 2 *The* concatenation *of MSCs* $M_1 = \langle V_1, <_1, L_1, T_1, m_1, \mathcal{P}\rangle$ *and* $M_2 = \langle V_2, <_2, L_2, T_2, m_2, \mathcal{P}\rangle$ *over the same set of processes* $\mathcal{P}$ *and disjoint sets of events* $V_1 \cap V_2 = \emptyset$*, denoted* $M_1\, M_2$*, is* $\langle V_1 \cup V_2, <, L_1 \cup L_2, T_1 \cup T_2, m_1 \cup m_2, \mathcal{P}\rangle$*, where* $< = <_1 \cup <_2 \cup \{(p,q) | L(p) = L(q) \wedge p \in V_1 \wedge q \in V_2\}$

That is, the events of M_1 precede the events of M_2 for each process, respectively. Notice that there is no synchronization of the different processes when moving from one node to the other. Hence, it is possible that one process is still involved in some actions of a previous node, while another process has advanced to a subsequent node.

Since a communication system usually includes many different scenarios, a high level description is needed for combining them together. This description is represented as a graph, where each node contains one MSC as in Figure 2. Each maximal path in this graph (i.e., a path that is either infinite or ends with a node without outgoing edges), starting from a designated initial state, corresponds to a single *execution* or *scenario*.

Definition 3 *An HMSC N is a triple* $\langle \mathcal{M}, \tau, M_0, \rangle$ *where* $\mathcal{M}$ *is a set of MSCs, all with the same set of processes, and with sets of events disjoint from one another.* $\tau \subseteq \mathcal{M} \times \mathcal{M}$ *is the* edge relation *and the* initial MSC *is* $M_0 \in \mathcal{M}$*. An* execution *of N is a path* $\xi = M_0\, M_1\, M_2\, M_3 \ldots$ *of N that starts with the initial MSC* M_0 *and either ends with a state without outgoing edges, or is infinite.*

Figure 2 shows an example of an MSC graph, where the state in the upper left corner is the starting state. Note that the executions of this

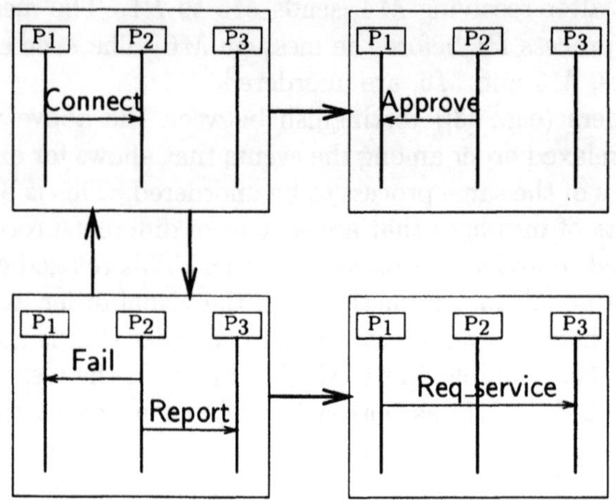

Figure 2 An HMSC graph

system are either finite or infinite. In Figure 2, process $P2$ may send its Report message after process $P1$ has progressed into the next node (on the left) and has sent its Req_service message.

2.1. THE LOGIC TLC$^-$

The logic TLC$^-$ is a subset of the logic TLC [1]. A model of the logic is a finite or infinite partial order $\zeta = (V, <, \longrightarrow)$, where $<\subset V \times V$ is a partial order relation, and $\longrightarrow\subset<$ is the 'immediately precedes' relation. The set of formulas $\mathcal{L}$ of TLC$^-$ over a set of atomic formulas AP is as follows: *true*, *false* $\in \mathcal{L}$, if $p \in AP$, then $p \in \mathcal{L}$, and if φ, ψ are in $\mathcal{L}$ then $\varphi \wedge \psi$, $\varphi \vee \psi$, $\neg\varphi$, $\exists\bigcirc\varphi$, $\forall\bigcirc\varphi$, $\varphi U \psi$, $\varphi R \psi \in \mathcal{L}$.

An *interpretation function* $I : V \mapsto 2^{AP}$ assigns to each event of V a set of propositions from AP. Each proposition in AP represents some property (e.g., of an event, or the local state before or after the event, when the events are taken from some system execution). Then, $I(v)$ returns the set of atomic propositions that hold for v. The semantics of the logic is defined as follows.

$(\zeta, v) \models true.$

$(\zeta, v) \models p$ if $p \in I(v)$

$(\zeta, v) \models \varphi \wedge \psi$ if $(\zeta, v) \models \varphi$ and $(\zeta, v) \models \psi$.

$(\zeta, v) \models \neg\varphi$ if it is not the case that $(\zeta, v) \models \varphi$.

$(\zeta, v) \models \exists\bigcirc\varphi$ if for some w such that $v \longrightarrow w$, it holds that $(\zeta, w) \models \varphi$.

$(\zeta, v) \models \varphi U \psi$ if there is a path $v = v_0 \longrightarrow v_1 \longrightarrow \ldots \longrightarrow v_n$, such that $(\zeta, v_n) \models \psi$, and for $0 \leq i < n$, $(\zeta, v_i) \models \varphi$.

We define $false \equiv \neg true$, $\varphi \vee \psi \equiv \neg(\neg\varphi \wedge \neg\psi)$, $\varphi R \psi \equiv \neg(\neg\varphi U \neg\psi)$, $\forall\bigcirc\varphi \equiv \neg\exists\bigcirc\neg\varphi$. Two additional modalities, F and G, can be defined in terms of the previous ones: $F\varphi \equiv trueU\varphi$, and $G\varphi \equiv falseR\varphi$. For TLC$^-$ we have selected an *existential until* 'U' operator, hence its dual *release* 'R' operator is universal. The full logic TLC contains also a universal *until*, an existential *release*, and a *concurrent with* operator '$||$'. The modalities U and R satisfy the following equations: $\varphi U \psi \equiv \psi \vee (\varphi \wedge \exists \bigcirc \varphi U \psi)$, $\varphi R \psi \equiv \psi \wedge (\varphi \vee \forall \bigcirc \varphi R \psi)$. An HMSC satisfies a TLC$^-$ specification exactly when each execution of it satisfies the specification.

Some examples for TLC$^-$ Specification are as follows:

$G\,(req \rightarrow F\,ack)$ Every request is causally followed by an acknowledgement.

$G\,(recA \rightarrow \exists\bigcirc sendB)$ A message B is sent immediately after receiving a message A.

$\neg F\,(tranA \wedge F\,(tranB \wedge F\,tranA))$ Transaction B cannot interfere with the events of transaction A.

$G\,(beginA \rightarrow \exists\bigcirc(tranA\,U\,finishA))$ The execution of transaction A is not interrupted by any other event.

3. LINEARIZATIONS OF MSCS

In order to simplify model checking, and to reuse code that was written for verifying concurrent protocols, we will interpret TLC$^-$ formulas over linearizations of MSCs. Each linearization is the completion of an MSC execution into a total order. All the linearizations of a single MSC execution must agree with their origin MSC upon the interpretation of each TLC$^-$ formula.

Consider the following linearization of the execution in Figure 1: (s, P_1) (r, P_2) (s, P_2) (r, P_3) (s, P_2) (r, P_3) (s, P_3) (r, P_2) (s, P_2) (r, P_1) (s, P_3) (r, P_1). Each event is denoted as a pair, including the type of event (*send* or *receive*), and the corresponding process.

We also keep with each linearization *forward edges* from each *send* node to its matching *receive*. In the above linearization example, the forward edge from each *send* event always points to the immediate successor on the sequence. However, this is not necessarily always the case. For example (this example was found by Anca Muscholl), consider the case where $P1$ sends a message to $P2$. Before this message is received by $P2$, $P1$ sends a message to $P3$, and then $P3$ sends another message to $P2$, which arrives before the message that was sent by $P1$. Such cases, which are rare in practical use of MSCs, complicate the verification algorithm that is presented in the sequel.

When the current event is a *send*, there are one or two immediate causal successors according to the order '$\longrightarrow$': an essential successor is the matching *receive* and, unless the current event is the last event for the current process, there is another event on the same process Q. Any other type of event (*receive* or *local*) has at most the second kind of immediate successor. The immediate successor of an event in a linearization is not necessarily also a successor under the original partial order relation. For example, the fourth event in the linearization, which is (r, P_3), is immediately followed by (s, P_2). But the latter event is not the immediate successor of the former in the partial order.

The interpretation of a TLC$^-$ formula φ over a *linearization* ξ of a partial order $\zeta = (V, <, \longrightarrow)$ follows the interpretation of φ in ζ. That is, $(\xi, i) \models \varphi$ if for the ith event v of ξ, we have that $(\zeta, v) \models \varphi$, according to the above partial order semantics. We add the special event-type propositions *send*, *receive* and *local* with the obvious interpretation, and in addition, a special proposition for each process: we denote the proposition that corresponds to a process $P \in \mathcal{P}$ also by P. We also add the special atomic proposition start, which marks nodes that start the linearization of each new HMSC node.

In order to facilitate the translation of TLC$^-$ formulas into automata over sequences, we extend the syntax of TLC$^-$ with three additional types of formulas:

- $\langle P \rangle \varphi$, where P is a process, and φ is a TLC$^-$ formula. $(\xi, i) \models \langle P \rangle \varphi$ holds when $(\xi, j) \models \varphi$ for some $j \geq i$, and for each k such that $j > k \geq i$, the kth event does not belong to process P.

- $\{match\}\varphi$, where φ is a TLC$^-$ formula. $(\xi, i) \models \{match\}\varphi$ holds when $(\xi, j) \models \varphi$ for some $j > i$, where the jth event of ξ is the matching *receive* for the ith event, which is a *send*.

- NEVER P, where P is a process. $(\xi, i) \models$ NEVER P holds exactly when for each $j \geq i$, the jth event of ξ does not belong to P.

For convenience, we assume that for each property φ there is some initiating process Q. That is, a processes whose first event is the one from which we want to interpret φ. The first event of process Q cannot be a *receive*. Over linearizations, we can then write $\langle Q \rangle \varphi$. Of course, we can may vary Q over all processes.

4. AUTOMATA THEORETIC MODEL CHECKING

A *Büchi automaton* is a sextuple $(S, S_0, \Sigma, L, \delta, F)$, where S is a finite set of states, $S_0 \subseteq S$ are the initial states, Σ is the finite alphabet, $L : S \to \Sigma$ is a mapping from the states to the alphabet, $\delta \subseteq S \times S$ is the transition relation (for simplicity of presentation, we deviate from the traditional edge-based labeling and use a state based labeling), and $F \subseteq S$ are the accepting states. A run ξ over a word $a_1 a_2 \ldots \in \Sigma^\omega$ is an infinite sequence of states $s_0 s_1 s_2 \ldots$, with $s_0 \in S_0$, such that for each $i > 0$, $(s_i, \alpha_i, s_{i+1}) \in \delta$. A run is *accepting* if at least one state of F occurs in ξ infinitely many times. A word is accepted by a Büchi automaton exactly when there exists a run accepting it. The *language* $\mathcal{L}a(A)$ of a Büchi automaton A is the set of words that it accepts.

A *specification* automaton $A = (S^A, S_0^A, \Sigma, L^A, \delta^A, F^A)$ is a Büchi automaton that represents a set of *disallowed* executions. Thus, if η is some temporal specification of an HMSC system, then the corresponding specification automaton accepts exactly the sequences that do not satisfy η, i.e., those that satisfy $\neg\eta$. An *implementation* automaton $B = (S^B, S_0^B, \Sigma, L^B, \delta^B, S^B)$ is a restricted version of a Büchi automaton representing the executions of the checked HMSC; it requires all the states S^B to be accepting. Thus, the accepting runs of B are the infinite paths that start from an initial state in S_0^B. An implementation automaton represents the modeled system (finite paths are completed to infinite paths for convenience).

The *intersection* (or *product*) $B \times A$ is $(S', S'_0, \Sigma, \delta', F')$, where $S' = \{(s,t) | s \in S^A \wedge t \in S^B \wedge L^A(s) = L^B(t)\}$, $S'_0 = S' \cap (S_0^A \times S_0^B)$, $\delta' = \{((s,s'),(t,t')) | (s,s') \in S' \wedge (t,t') \in S' \wedge (s,t) \in \delta^A \wedge (s',t') \in \delta^B\}$, $F' = S' \cap (F^A \times S^B)$. We have that $\mathcal{L}a(A \times B) = \mathcal{L}a(A) \cap \mathcal{L}a(B)$.

The intersection of a specification automaton and an implementation automaton gives all the executions of the implementation that are not allowed by the specification. Thus, the intersection is empty exactly if the implementation does not satisfy the specification. Any sequence in the intersection can be used as a counterexample. Checking the emptiness of a Büchi automaton, and in particular the automaton F' that results from the above intersection can be done by applying the DFS algorithm to find reachable maximal strongly connected components of the automaton graph. If there exists such a component that includes at least one state from each accepting subset $f \in F'$, then the automaton is not empty. In particular, in this case there is an *ultimately periodic* sequence of the form $v\,u^\omega$ accepted by F'.

5. MODEL CHECKING FOR HMSCS

We suggest here a generic framework for model checking HMSCs, with TLC$^-$ as a specific instance of a specification formalism that can be used for this purpose. We first describe a translation from any TLC$^-$ formula η into a corresponding Büchi automaton A, which accepts the linearizations that satisfy η. This is a Büchi automaton over linearizations of MSC executions, represented by sequences of events that contain forward edges. The forward edges will introduce another constraint on acceptance, as will be shown later. The construction presented here is in particular appropriate for systems with FIFO asynchronous message passing. There is no fixed dependence relation between the different events according to the process to which they belong, as was the case in previous work, e.g., [1].

We first represent formulas and subformulas in *negation normal form*; in this form, the negation operator can be applied only to atomic propositions. Each TLC$^-$ formula can be converted into an equivalent formula in normal form by repeated use of the DeMorgan laws and the dualities between 'U' and 'R', presented above. For example, we can make the following transformations: $\neg\exists \bigcirc (pR(q \vee r)) \equiv \forall \bigcirc \neg(pR(q \vee r)) \equiv \forall \bigcirc (\neg pU \neg(q \vee r)) \equiv \forall \bigcirc (\neg pU(\neg q \wedge \neg r))$.

Due to the existence of forward edges in the linearizations, we will deal with automata with *two* transition relations $(S, S_0, \Sigma, L, \delta, C, F)$, where the additional component C is a relation between pairs of nodes. Thus, the intersection of a specification automaton $(S^A, S_0^A, \Sigma, L^A, \delta^A, C^A, F^A)$ with an implementation automaton $(S^B, S_0^B, \Sigma, L^B, \delta^B, C^B, F^B)$ has an additional component C', which is $C^A \cap C^B \cap (S' \times S')$.

In the construction of the specification automaton, each node in S will contain a *safety component* and a *liveness component*. The safety component is responsible for guaranteeing the consistency between each node and its successors. It contains the following:

- Exactly one of the event-type propositions: *send*, *receive* or *local*, for the type of the current event. In addition, it may include the proposition start.
- A maximal non-contradicting subset of the sub-formulas of the checked property and their negations (in negation normal form). That is, for each subformula φ, each node will contain either φ or $\neg\varphi$, but not both.
- If the node does not contain *never* P, then it contains a maximal noncontradicting subset of the formulas of the form $\langle P\rangle\varphi$, where P is a process and φ is a subformula of the checked property. If a node contains *never* P, it cannot contain formulas of that type.

- If the node is a *send* event, then it contains a maximal noncontradicting subset of the formulas of the form $\{match\}\varphi$, where φ is a subformula of the checked property. Otherwise, it cannot contain formulas of that type.
- Exactly one process predicate P, for one of the processes of the checked HMSC, except for the first node, which contains all the process predicates.
- If a node contains the process predicate P, then for each formula of type $\langle P\rangle\varphi$ that it contains, it also contains φ.

The liveness component is responsible, together with an appropriate acceptance condition that will described later, to guarantee the following: in every accepted run of the automaton, for all the nodes where a subformula of the form $\varphi U\psi$ hold, ψ will hold in the current or a subsequent node. It contain a subset of the formulas of the form $\varphi U\psi$ and $\langle P\rangle\varphi U\psi$ that were mentioned above.

We have a set of directed edges between pairs of nodes such that the following constraints are satisfied. The constraints are described according to the type of subformulas that appears in the safety component of a given node. They apply to each node that contains such a formula and to its immediate successor according to the relation δ^A. When an enumerated list of constraints is given for a certain type of subformula, then at least *one* of them must apply. The condition on the safety part appears in roman fonts, while the condition on the liveness component is given in *italics*. In the construction, we denote the process predicate in the current node by Q.

- $\exists\bigcirc\varphi$. (1) $\langle Q\rangle\varphi$ holds for each successor, or (2) the current node has both the proposition *send*, and $\{match\}\varphi$.
- $\forall\bigcirc\varphi$. (1) The current node has *receive* or *local*. Each successor has $\langle Q\rangle\varphi$, or (2) the current node has *send* and $\{match\}\varphi$. Each successor has $\langle Q\rangle\varphi$, or (3) the current node has *send* and $\{match\}\varphi$. Each successor has NEVER Q.
- $\varphi U\psi$. (1) The current node has ψ, or (2) the current node has φ, *send* and $\{match\}\varphi U\psi$, or (3) the current node has φ. Each successor has $\langle Q\rangle\varphi U\psi$. *If the liveness component of the current node contains $\varphi U\psi$, then the liveness component of each successor must have $\langle Q\rangle\varphi U\psi$.*
- $\varphi R\psi$. (1) The current node has both φ and ψ, or (2) the current node has ψ, *send* and $\{match\}\varphi R\psi$, each successor has $\langle Q\rangle\varphi R\psi$, or (3) the current node has ψ, *send* and $\{match\}\varphi R\psi$, each successor has NEVER Q, or (4) the current node has ψ and either *receive* or *local*. Each successor has $\langle Q\rangle\varphi R\psi$.

- $\varphi \vee \psi$. (1) The current node has φ, or (2) the current node has ψ.
- $\varphi \wedge \psi$. The current node has both φ and ψ.
- $\langle P \rangle \varphi$. (1) The current node has $Q = P$, and φ. *If φ is of the form $\delta U \psi$ and the liveness component has $\langle P \rangle \delta U \psi$, then the liveness component of the current node must also have $\delta U \psi$*, or (2) the current node has $Q \neq P$. Each successor has $\langle P \rangle \varphi$, *If φ is of the form $\delta U \psi$ and the liveness component has $\langle P \rangle \delta U \psi$, then the liveness component of each successor node must have $\langle P \rangle \delta U \psi$.*
- NEVER P. The current node has $Q \neq P$. Each successor has NEVER P.
- start. *If the liveness component of the current node is empty, then the liveness component of each successor consists of the formulas of the type $\varphi U \psi$ and $\langle P \rangle \varphi U \psi$ that are included in the safety component of the successor.*

The last case takes care of the situation where all the previous liveness constraints were satisfied, and new ones are loaded from the safety component. Synchronizing with nodes that start a linearization of an MSC, marked with the proposition start. guarantees that there are no forward edges that propagate liveness constraints 'over' the current node. A state s is *forward compatible* with a state t if the following conditions hold: (1) s has a *send* proposition, (2) t has a *receive* proposition, (3) for each formula of type $\{match\}\varphi$ in s, we have the formula φ in t, and (4) *If the safety component of s contains a formula of the form $\{match\}\delta U \psi$, then the liveness component of t must contain $\delta U \psi$.*

Forward compatibility means here that we *allow* a forward edge between the nodes, rather than *enforce* it. We can define the nodes of S^A of the constructed specification automaton by spliting the nodes defined above as follows: for a compatible pair of nodes s and t, we split s into s_1 and s_2, such that both copies have the same incoming edges as the original s. Furthermore, we set $C^A(s_1, t)$ but not $C^A(s_2, t)$. However, this may cause an explosion in the number of states. We take an alternative approach: we set up $C^A(s, t)$, but redefine the relation C' of the intersection automaton as follows:

$$\{((s, s'), (t, t')) | (s, s') \in S' \wedge (t, t') \in S' \wedge (s, t) \in C^B \rightarrow (s', t') \in C^A\}$$

The asymmetric definition of the intersection is due to the fact that the set S^A is a compact representation of the actual states of the property automaton A, as described above.

We also have to take care that formulas that need to be satisfied further down the linearization will eventually be satisfied, rather than be postponed forever. This is done using the Büchi acceptance set: it contains all the states that include the empty liveness component. This guarantees that we satisfy all the constraints that were loaded at the previous node with start. New such constraints will be loaded at the next such node. Notice that unlike the LTL construction in [11], the liveness constraints here may change, e.g., from $\langle P\rangle \eta U \mu$ into $\eta U \mu$.

The automaton B recognizes a single linearization for each execution of a given HMSC $N = (\mathcal{M}, \tau, M_0)$. Take any MSC $M \in \mathcal{M}$. It is simple to construct an automaton A_M that recognizes a single linearization of M. The ith node along the single execution of this automaton is labeled with the propositions of the ith event in the linearization. The linearization can be done using a topological sort in linear time in the number of events of M. The first node in each such linearization is marked with start.

Denote by $first(A_M)$ the initial state of A_M and by $last(A_M)$ the final state. We would like to combine the different automata A_M for $M \in \mathcal{M}$. Let M^1, M^2, ..., M^k be the successors of M under the relation τ. That is $M \tau M^i$ for $1 \leq i \leq k$. To that, we add the set of transitions $(last(A_M), \alpha, init(A_{M_i}))$ for $1 \leq i \leq k$. That is, after executing A_M, there is a nondeterministic choice to any of the initial states of the automata representing the successor MSCs of M. Label these edges with the proposition start. The implementation automaton is linear in the size of the checked HMSC. Such an automaton for the HMSC that appears in Figure 2 is shown on Figure 3. Forward edges are denoted with dotted arrows. Nodes that are marked with an asterisk '*' are labeled with start.

5.1. COMBINING THE TWO AUTOMATA

Emptiness of automata with two successor relations is in general undecidable [6]. However, we obtain here decidability, and in fact, a decision procedure that is similar to the one for traditional Büchi automata through of the following restrictions: One of the relations (the forward edges) is embedded in the transitive closure of the other relation (representing the linearization successors). There is a finite bound on the number of successor edges between the source and the target of each forward edge. This is due to the fact that communication is limited to occur within a single HMSC node. Furthermore, there are no forward edges (i.e., the second relation) that 'overtake' the accepting states, i.e., start before an accepting node and ending after it.

Due to the the forward edges, the intersection of the two automata has an additional constraint over the construction in Section 4. During

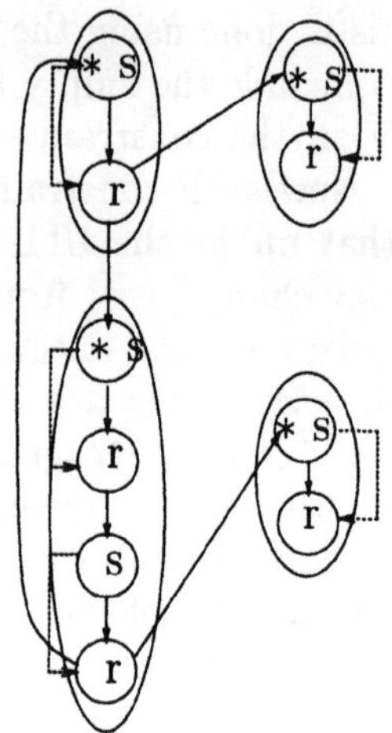

Figure 3 An implementation automaton for the HMSC in Figure 2

the DFS performed to check the emptiness of the intersection, when we reach a *receive* node, we trace back the corresponding *send*. It must be already on the DFS stack. In fact, we can consult the implementation automaton B, which fixes the distance of the *send* event corresponding to the current *receive* event. In the intersection DFS stack, the *send* node would be down the stack the same distance from the *receive* node as it is in B. We then check the components of the property automaton A on the DFS stack for forward compatibility. If there is no compatibility, we backtrack. Due to the above mentioned constraints on the forward edges relation, checking for emptiness of the intersection is done by simply looking for a reachable cycle through an accepting state of the intersection. The above property translation gives a Büchi automaton A whose size may be exponential in the size of the checked formula and the number of processes. The implementation automaton B is linear in the size of the checked HMSC.

6. DISCUSSION

Selecting a specification formalism often involves tradeoffs. A more concise specification often involves a higher complexity decision procedure. Adding expressiveness can also increase the complexity of even result in undecidability. In the case of HMSCs, the unboundedness of the message queues renders traditional specification formalisms such as LTL undecidable. One way to fight this problem is by imposing constraint on the HMSCs, as done e.g., in [2, 9].

In this paper, an alternative solution was suggested. The specification for HMSCs is based on partial order semantics rather than on interleaving semantics. The logic TLC^- is only one candidate for such a specification formalism. Other logics such as $ISTL^F$ [4] can also be used.

Executions of HMSCs are partial order structures that do not have a fixed dependency relation between events according to the processes in which they participate. For example, if we have two sends from process $P1$ to process $P2$, we can have the linearizations $(s, P1)$ $(r, P2)$ $(s, P1)$ $(r, P2)$ and $(s, P1)$ $(s, P1)$ $(r, P2)$ $(r, P2)$, but cannot commute between the *send* and *receive* events on these sequences further. Thus, we *cannot* freely commute *send* and *receive* events of different processes. A previous construction for the full logic TLC [1] relied on a fixed dependency relation between events. The construction in the current paper, with forward edges and forward compatibility checks, is appropriate for FIFO asynchronous message passing, as in the visual semantics of HMSCs.

Since MSCs are used in very early stages of the design, verifying them can help in achieving early fault detection. Several logics for partial order executions were suggested in the last decade [4]. However, the linear semantics approach is still by far the more common way for specifying concurrent systems. For systems of MSCs, the partial order based model checking approach presented here seems natural and efficient. This may promote further research into such specification formalisms, their expressiveness and complexity.

A prototype implementation of HMSC model checking was done in SML/NJ. The HMSC was translated from an HMSC notation where each individual MSC is described using the ITU Z120 standard. Code that was used for HMSC pattern matching [10] was reused for implementing the translation of the HMSCs and the linearizations. The translation results in PROMELA code, which is the input language for the SPIN model checking system [8]. After translating an HMSC and a TLC^- formula into PROMELA code, the SPIN model checking tool can be used to make the automatic verification. If the HMSC does not satisfy the specification, a counterexample is found. Simulating the counterexample using SPIN prints out a sequence of the HMSCs nodes that participate in the counterexample. Then, one can use the POGA tool to display the MSC execution.

References

[1] Alur R., Peled D., and W. Penczek (1995), Model-Checking of Causality Properties, LICS'95, 10th Symposium on Logic in Com-

puter Science, IEEE, 1995, 90–100,

[2] Alur R., and Yannakakis M. (1999), Model Checking of Message Sequence Charts, CONCUR'99, Concurrency Theory, LNCS 1664, Eindhoven, Holland, 1999.

[3] Alur R., Holzmann G.J., and Peled D. (1996), An Analyzer for Message Sequence Charts, Software Concepts and Tools, Vol. 17, No. 2, pp. 70–77, 1996.

[4] Alur R. McMillan K., and Peled D. (1998), Deciding global partial-order properties, ICALP'98, Aalborg, Denmark, July, 1998, LNCS 1443, Springer, 41-52.

[5] Clarke E.M., and Emerson E.A. (1981), Design and synthesis of synchronization skeletons using branching time temporal logic. Workshop on Logic of Programs, Yorktown Heights, NY, 1981, LNCS 131, Springer-Verlag.

[6] Giammarresi D., Restivo A., Seibert S., and Thomas W. (1993), Monadic second-order logic over rectangular pictures and recognizability by tiling systems, Information and Computation 125 (1996), 32–45.

[7] ITU (1993), ITU-T Recommendation Z.120, Message Sequence Chart (MSC), March 1993.

[8] Holzmann G.J. (1992), *Design and Validation of Computer Protocols*, Prentice Hall Software Series, 1992.

[9] Muscholl A., and Peled D. (1999), Message Sequence Graphs and Decision Problems on Mazurkiewicz Traces, Mathematical Foundations of Computer Science, Szklarska Poreba, Poland, September 1999, LNCS 1672, Springer, 81–91.

[10] Muscholl A., Peled D., and Su Z. (1998), Deciding Properties for Message Sequence Charts, FoSSaCS 1998, Foundations of Software Science and Computation Structures, Lisbon, Portugal, LNCS 1378, 226-242.

[11] Vardi M.Y., Wolper P. (1986), An automata-theoretic approach to automatic program verification. Proc. 1st Annual Symposium on Logic in Computer Science IEEE, 1986, 332-344.

A STATE-EXPLORATION TECHNIQUE FOR SPI-CALCULUS TESTING-EQUIVALENCE VERIFICATION

Luca Durante, Riccardo Sisto
Politecnico di Torino
Dipartimento di Automatica e Informatica
corso Duca degli Abruzzi 24
I-10129 Torino
Italy
{luca.durante, riccardo.sisto}@polito.it

Adriano Valenzano
Istituto di Ricerca sull'Ingegneria
delle Telecomunicazioni e dell'Informazione
Politecnico di Torino
corso Duca degli Abruzzi 24
I-10129 Torino
Italy
adriano.valenzano@polito.it

Abstract Several verification techniques based on theorem proving have been developed for the verification of security properties of cryptographic protocols specified by means of the spi calculus. However, to be used successfully, such powerful techniques require skilled users. Here we introduce a different technique which can overcome this drawback by allowing users to carry out the verification task in a completely automatic way. It is based on the definition of an extended labeled transition system, where transitions are labeled by means of the new knowledge acquired by the external environment as the result of the related events. By means of bounding the replication of parallel processes to a finite number, and by using an *abstract* representation of all explicitly allowed values in interactions between the spi process and the environment, the number of states and transitions remains finite and tractable, thus enabling the use of state-space exploration techniques for performing verification automatically.

Keywords: Spi Calculus, Cryptographic Protocols, Testing Equivalence.

Tommaso Bolognesi and Diego Latella (Eds.), *Formal Methods for Distributed System Development.*

1. INTRODUCTION

Due to the increasing importance of secure distributed applications such as electronic commerce, formal verification of cryptographic protocols is being extensively studied by several researchers. Some of them have investigated proof techniques, based on various proof systems and description formalisms [10, 11]. Although partial automation of proofs is possible using theorem provers, this approach is generally highly time consuming and requires a lot of expertise. An alternative simpler and quicker approach is to use state exploration methods, such as model checking [4, 5, 6, 7, 8]. This requires modeling protocol behaviors as reasonably sized finite state systems, which generally entails introducing simplifying assumptions that can reduce the accuracy of the analysis. Nevertheless, this kind of verification has the invaluable advantage of being fully automatic.

Both theorem proving and state exploration have been used with several description formalisms. In this paper attention is focused on spi calculus [1], a process algebra derived from the π-calculus [9] with some simplifications and the addition of cryptographic operations. The main strength of spi calculus with respect to other similar formalisms stands in its simplicity in describing cryptographic protocols and their security requirements. In particular, in [1] it is shown how security properties such as authentication and secrecy can easily be expressed by means of a testing equivalence notion. For example, if $P(M)$ is the description of a cryptographic protocol to exchange a secret message M, secrecy can be simply expressed saying that for any M', $P(M)$ and $P(M')$ must be testing equivalent, i.e. any tester process must be unable to distinguish their behaviors. Instead, when other specification formalisms are used, it is generally required that both the protocol and the attacker behavior be specified. The attacker specification is not only extra work, but also a potential weak point, because it is somewhat arbitrary and might not include some possible attacks on the protocol. Expressing properties in terms of testing equivalence means implicitly considering any attacker that can be specified in spi calculus. This is because the tester process actually represents the protocol environment, i.e. the attacker.

The main problem that remains with the spi calculus approach is the checking of testing equivalence in an efficient and easy way. This is difficult because of universal quantification over testers: checking equivalence means checking that two processes are indistinguishable for *any* tester process, and there are infinitely many such processes. This problem has been addressed in [2] and [3], where tractable proof methods aimed at checking the testing equivalence of spi calculus processes are introduced. In [2], the proof method is based on a bisimulation relation that is a sufficient but not necessary condition for testing equivalence. In [3], a more accurate method is proposed to check may-testing, an equivalence very similar to the testing equivalence defined in [1]. The approach presented in [3] starts from the definition of a contextual labeled transition system which represents the protocol behavior constrained by the knowledge which the environment has about names and keys. The proof method exploits the fact that the trace equivalence defined on this model is a necessary and sufficient condition for may-testing equivalence.

Differently from the above two approaches, this paper investigates the possibility of checking the spi calculus may-testing equivalence using state exploration instead of theorem proving. The quantification over contexts problem is solved in a way similar to the one reported in [3], i.e. by defining a labeled transition system such that trace equivalence is necessary and sufficient for may-testing equivalence. The other new problem that has to be solved in order to make state exploration possible and effective is to keep the size of the trace sets to be explored within finite and reasonable bounds. This objective is achieved mainly in two different ways. First of all, to keep the state set finite, only spi calculus processes having a finite number of parallel instances are dealt with. In practice, our approach is to substitute any replication expression of the form $!P$, which is interpreted in spi calculus as an infinite number of copies of P running in parallel, with a finite number n of parallel copies of P. Since the replication operator is generally used to represent parallel sessions of a cryptographic protocol, this restriction is equivalent to considering up to n parallel runs of the protocol. Consequently, attacks that are possible only with more than n parallel sessions cannot be detected in this way. A similar restriction is adopted in the literature whenever state exploration methods are used and is generally considered fairly acceptable, since bugs tend to show up with few numbers of parallel sessions.

A second way to reduce the size of the trace sets consists in avoiding the explicit representation of transitions corresponding to inputs from the environment. Since the environment can in principle send at any time any data that is part of its knowledge as well as any fresh name or integer, the number of such transitions can be huge if not infinite. Our approach is to represent a set of such transitions as a single transition with an abstract label called generic term or generic value. This kind of reduction does not imply any loss in accuracy and is a key means to make traces enumerable and to limit their number.

The paper is organized as follows: section 2 briefly introduces the language and our conventions about symbols, section 3 deals with the

σ, ρ θ, η ::= terms		P, Q, R ::=	processes
m	name	$\overline{\sigma}\langle\rho\rangle.P$	output
(σ, ρ)	pair	$\sigma(x).P$	input
0	zero	$P \mid Q$	composition
$suc(\sigma)$	successor	$(\nu b)\ P$	restriction
x	variable	$!P$	replication
$\{\sigma\}_\rho$	shared-key encryption	0	nil
		$[\sigma\ is\ \rho]\ P$	match
		$let\ (x, y) = \sigma\ in\ P$	pair splitting
		$case\ \sigma\ of\ 0 : P\ suc(x) : Q$	integer case
		$case\ \eta\ of\ \{x\}_\rho\ in\ P$	shared-key decryption

Table 1 Syntax of spi calculus

definition of our labeled transition system and section 4 explains the derivation rules for it and gives an example. In section 5 the concept of *trace equivalence* for our labeled transition system is defined while section 6 contains some final remarks.

2. THE LANGUAGE

The spi-calculus is a process algebraic language defined in [1] as an extension of the π-calculus, specifically designed for the specification of cryptographic protocols. No explicit types exist in spi calculus and spi entities are divided into processes and terms only. However, since spi calculus terms are extended here with generic terms, we prefer to use the following partially different naming convention:

- m ranges over names;
- n ranges over natural numbers;
- x and y range over variables;
- b is a fresh name, and $\tilde{b}$ is a tuple of fresh names;
- P, Q and R are spi calculus processes;
- σ, ρ, η, θ and ψ denote terms in the most general sense (i.e. generic terms also);
- Σ is a set of terms;
- Γ is the set containing all generic terms;
- γ ranges over Γ;
- λ is a substitution list i.e. $\lambda = \sigma_1/\rho_1, \sigma_2/\rho_2, \cdots \sigma_n/\rho_n$. $\theta[\lambda]$, $P[\lambda]$ and $\Sigma[\lambda]$ mean that each term ρ_i must be replaced by σ_i in θ, P and Σ respectively.

Tab.1 shows the language syntax. Informally, the meaning of each construct of the language is the following:

- Term $\{\sigma\}_\rho$ is the encrypted message obtained by encrypting σ under key ρ using a shared-key cryptosystem.

- $\overline{\sigma}\langle\rho\rangle.P$ is an *output process*, ready to output ρ on channel σ when a synchronization occurs. After the synchronization, P starts. σ should be a name.
- $\sigma(x).P$ is an *input process*, ready to perform an input from channel σ when a synchronization occurs. After the synchronization, $P[\eta/x]$ starts, where η is the input message. σ should be a name.
- $P \mid Q$ is a *parallel composition* where P and Q run in parallel. They may either synchronize between themselves or with the external environment separately.
- $(\nu b)P$ is a *restriction* where a fresh, private name b is made and then P starts.
- $!P$ is a *replication* where an unbounded number of instances of P run in parallel.
- $[\sigma\ is\ \rho]P$ is a *match*. It means that P starts when σ and ρ are the same; it stucks otherwise.
- 0 is the *nil* process: it does nothing.
- $let\ (x,\ y)\ =\ \sigma\ in\ P$ is a *pair splitting* and it behaves as $P[\rho/x, \theta/y]$, provided that term σ is a pair $(\rho,\ \theta)$; it stucks otherwise.
- $case\ \sigma\ of\ 0 : P\ suc(x) : Q$ is an *integer case*. If σ is 0, it behaves as P; if σ is $suc(\rho)$, $Q[\rho/x]$ starts. It stucks otherwise.
- $case\ \eta\ of\ \{x\}_\rho\ in\ P$ is a *shared-key decryption* and behaves as $P[\sigma/x]$ if η is a cyphertext as $\{\sigma\}_\rho$. Otherwise it stucks.

About encryption the following implicit assumptions hold:
- an encrypted message can be decrypted by means of the corresponding key only;
- the encryption key cannot be detected from the encrypted message;
- an encrypted message is sufficiently redundant so that the decryption algorithm can detect whether its task succeeded in.

All the above assumptions are formally defined at the end of section 4.

The spi calculus operational semantics is defined in [1]. For lack of space, they are not recalled here. In this paper we adhere to such definitions, with the exception of the replication operator $!P$, which is interpreted here as a syntactical shortcut for the parallel composition of n parallel copies of P (i.e. $!P = \underbrace{P|P|\cdots|P}_{n}$), where n is a predetermined finite natural number.

3. THE ENVIRONMENT-SENSITIVE LTS

Our first objective is the definition of an *environment-sensitive labeled transition system* (ES-LTS) describing all the possible interactions of a given spi calculus process with its environment. Although the concept is borrowed from [3], our specific ES-LTS differs substantially from the one described in [3].

The environment that is considered in building the ES-LTS is the most powerful one, power being measured by the knowledge the environment acquires by interacting with the spi process. For this reason, in analogy with [3], each state of our ES-LTS is made up of a spi calculus process P and an environment's knowledge Σ, and is denoted $\Sigma \rhd P$. Σ can be informally defined as the *minimal* set of spi calculus terms that allows the environment to build everything it has learned during the preceding interactions with P.

Transitions are described by the following syntactical form, already introduced in [3]:

$$\Sigma \rhd P \quad \overset{\mu}{\underset{\epsilon}{\longmapsto}} \quad \Sigma' \rhd P' \tag{1}$$

where μ is the action performed by process P, and ϵ is a complementary action used to represent additional information that must be recorded in the ES-LTS traces. While in [3] such complementary actions represent environment actions, in our model they are also used for other purposes.

Transitions are categorized into 4 different types, according to their meaning:

$$\Sigma \rhd P \quad \overset{\tau}{\underset{-}{\longmapsto}} \quad \Sigma \rhd P' \quad (2) \qquad \Sigma \rhd P \quad \overset{\sigma}{\underset{\gamma}{\longmapsto}} \quad \Sigma \rhd P' \quad (4)$$

$$\Sigma \rhd P \quad \overset{\overline{\sigma}}{\underset{\delta_\Sigma}{\longmapsto}} \quad \Sigma' \rhd P' \quad (3) \qquad \Sigma \rhd P \quad \overset{-}{\underset{\lambda}{\longmapsto}} \quad \Sigma' \rhd P' \quad (5)$$

Transitions taking form (2) are related to synchronization events occurring inside the spi process and for this reason they leave the environment unaware of what has happened, and do not involve any *knowledge migration* (Σ does not change).

Transitions taking the second and third form instead are related to synchronization events between the spi process and the environment.

Form (3) represents an output on channel σ, and implies a data transfer from the process to the environment. In this case there is also a complementary action representing an increment in the environment knowledge. This is denoted as δ_Σ and represents the set of new terms added to Σ to yield the new knowledge Σ'.

Form (4) represents an input on channel σ, and implies a data transfer from the environment to the process. Thus, no modification in the environment knowledge takes place. In this case, the complementary action represents the data sent by the environment to the process. In order to limit the number of transitions, such data is not represented explicitly, but it is represented in an abstract way by means of a generic value denoted γ, symbolizing each value the environment is able to build at that time. Formally, γ is a pair $\langle m_\gamma, \Sigma_\gamma \rangle$ where m_γ is a fresh name that identifies the term and Σ_γ is the set of terms representing the knowledge of the environment at the time the data represented by the term was generated (i.e. Σ). The behavior that follows the input of a *generic value* is indeed the abstract representation of a set of behaviours, each one corresponding to a different actual value of the generic term. Such behaviors are *indistinguishable* from one another (in the sense they are not affected by the particular value of the input) until the generic value that has been received is *filtered* in some way by means of a *let*, *case* or *is* operator. Whenever this happens, the contents of the environment's knowledge at the time the *generic value* had been generated is analyzed, in order to find the satisfying value(s), i.e. the actual values of γ which enable the behavior to continue after the filtering operation.

To represent the fact that only some of the behaviors continue after a filtering operation, we use another kind of transition, which takes the form (5). This transition is labeled by a substitution list λ representing the substitution of a generic term with a corresponding set of satisfying values. In this case, P' and Σ' are obtained actualizing P and Σ by λ ($P' = P[\lambda]$, $\Sigma' = \Sigma[\lambda]$).

4. DERIVATION RULES FOR THE ES-LTS

The spi calculus semantics is defined in [1] by means of the reaction relation $\rightarrow$, where $P \rightarrow P'$ means that process P can evolve into P' by performing an (internal) action. To formally define our environment sensitive LTS we use a set of derivation rules based upon the derivation system defined in [1] for the reaction relation. In practice, the reaction relation definition rules are assumed implicitly. Although generic terms are new with respect to the original spi calculus, we assume that derivation rules as defined in [1] apply to generic terms as to any other term. This does not introduce any technical problem, because all terms are treated in the same way in [1].

In order to minimize the number of derivation rules needed, we exploit the following property:

$$\forall R \begin{cases} \exists\, P, Q, \tilde{b} \mid R \equiv (\nu \tilde{b})(\overline{\sigma}\langle\rho\rangle.P \mid Q) \\ \qquad\qquad \vee \\ \exists\, P, Q, \tilde{b} \mid R \equiv (\nu \tilde{b})(\sigma(x).P \mid Q) \\ \qquad\qquad \vee \\ \qquad R \equiv 0 \end{cases} \tag{6}$$

where symbol $\equiv$ means *structural equivalence* as defined in [1]. Informally this property means that each spi expression R can always be reduced to a structural equivalent expression taking one of the three syntactical forms shown in (6), i.e. input and output action prefix and inaction. Consequently, it is enough only to define derivation rules for such syntactical forms (indeed, no rule is needed for inaction because it represents a stuck process).

The following rules specify the conditions for transitions of types (2), (3) and (4):

$$\frac{(\nu \tilde{b})\,(\overline{\sigma}\langle\rho\rangle.P \mid \sigma(x).Q) \longrightarrow (\nu \tilde{b})\,(P \mid Q[\rho/x])}{\Sigma \rhd (\nu \tilde{b})\,(\overline{\sigma}\langle\rho\rangle.P \mid \sigma(x).Q) \;\underset{-}{\overset{\tau}{\longmapsto}}\; \Sigma \rhd (\nu \tilde{b})\,(P \mid Q[\rho/x])} \tag{7}$$

$$\frac{\Sigma \vdash \sigma}{\Sigma \rhd (\nu \tilde{b})\,(\overline{\sigma}\langle\rho\rangle.P \mid Q) \;\underset{\overline{\Sigma \cup \{\rho\}} \setminus (\overline{\Sigma \cup \{\rho\}} \cap \Sigma)}{\overset{\overline{\sigma}}{\longmapsto}}\; \overline{\Sigma \cup \{\rho\}} \rhd (\nu \tilde{b})\,(P \mid Q)} \tag{8}$$

$$\frac{\Sigma \vdash \sigma}{\Sigma \rhd (\nu \tilde{b})\,(\sigma(x).P \mid Q) \;\underset{\gamma}{\overset{\sigma}{\longmapsto}}\; \Sigma \rhd (\nu \tilde{b})\,(P[\gamma/x] \mid Q)} \tag{9}$$

Rule (7) describes interactions on channel σ, internal to the spi process itself. Each of them corresponds exactly to a reaction step as defined in [1]. The environment is not involved in such an event, thus it is unaware of what has happened, and it's knowledge Σ remains unchanged after the event. Label τ emphasizes the silence of the event.

Rule (8) describes what happens when the spi process is ready to perform an output on a certain channel σ, provided that σ belongs to the environment's knowledge (this precondition is expressed by $\Sigma \vdash \sigma$, formally defined at the end of this section). The transition is labeled with the output channel $\overline{\sigma}$ and with the new knowledge that message ρ brings to the environment's knowledge. Such new knowledge can be computed as $\overline{\Sigma \cup \{\rho\}} \setminus (\overline{\Sigma \cup \{\rho\}} \cap \Sigma)$, where $\overline{\Sigma \cup \{\rho\}}$ represents the final knowledge reached after the addition of ρ to Σ.

It is worth noting that the update of the environment's knowledge is not simply $\Sigma \cup \{\rho\}$, because the addition of ρ to Σ could enable a further decoding of terms already present in Σ. The notation $\overline{\Sigma}$ informally represents the minimal set of terms needed to generate all the elements that can be generated by Σ. In practice, we can imagine that $\overline{\Sigma}$ is the result of two operations: first, Σ is expanded by adding to it all the terms that derive from all possible decoding operations. It is then minimized, considering that, when an element of Σ has been decoded into simpler parts (already added to Σ too), it is no longer necessary to continue to keep it in Σ, since in can be built by means of its parts. A formal definition of this operation is given at the end of this section.

For example, let us consider $P(M, k) ::= \overline{a}\langle\{M\}_k\rangle.\overline{b}\langle k\rangle.Q$ with $\Sigma = \{a, b\}$. By means of rule (8) we have the following behavior:

$$\{a,\, b\} \;\triangleright\; \overline{a}\langle\{M\}_k\rangle.\overline{b}\langle k\rangle.Q \;\underset{\{\{M\}_k\}}{\overset{\overline{a}}{\longmapsto}}\; \{a,\, b,\, \{M\}_k\} \triangleright \overline{b}\langle k\rangle.Q \;\underset{\{M,\, k\}}{\overset{\overline{b}}{\longmapsto}}\; \{a,\, b,\, M,\, k\} \;\triangleright\; Q$$

The example shows clearly how the environment's knowledge is re-arranged after k has been received from b and that $\{M,\, k\}$ is the *new* knowledge added to the environment's knowledge, instead of only $\{k\}$. In particular, looking at event $\underset{\{M,k\}}{\overset{\overline{b}}{\longmapsto}}$, we start with $\Sigma = \{a,\, b,\, \{M\}_k\}$ and, with $\rho = k$, $\Sigma \cup \{\rho\} = \{a,\, b,\, \{M\}_k,\, k\}$. By means of k, now $\{M\}_k$ can be decrypted and replaced by M. Such an operation is performed by operator $\overline{}$, thus $\overline{\Sigma \cup \{\rho\}} = \{a,\, b,\, M,\, k\}$. $\overline{\Sigma \cup \{\rho\}} \cap \Sigma = \{a,\, b\}$ describes the knowledge that is present both in Σ and in $\overline{\Sigma \cup \{\rho\}}$, i.e. knowledge not affected by the reception of ρ, thus $\overline{\Sigma \cup \{\rho\}} \setminus (\overline{\Sigma \cup \{\rho\}} \cap \Sigma)$ represents the knowledge brought, directly and indirectly, by ρ, i.e. the *new* knowledge used to label the transition.

Rule (9) describes what happens when the spi process is ready to perform an input from a certain channel, provided that the channel belongs to the environment's knowledge ($\Sigma \vdash \sigma$). The transition is labeled with the input channel σ and with the generic term γ which describes *everything the environment can build by using elements of its knowledge* Σ. After synchronization, the input variable x is bound to γ. The environment's knowledge Σ does not need to be updated, as it is the action of transferring data from the environment to the spi process.

Rule (9) empasizes the moment when a behavior comprising all possible behaviors for each acceptable value of γ starts. Thus semantic rules must be introduced which are able to emphasize when a certain abstract behavior, representing all possible behaviors for a generic term γ, continues only for some particular values of γ. Such rules are the ones needed to define type (5) transitions and are applied whenever two terms must be compared. It is worth noting that after a comparison operation each generic term is not necessarily substituted with a non-generic value, but it can be substituted with another more specialized term, made up of other generic terms. For example, rule (12) replaces a generic term with a pair of generic terms.

$$\frac{\Omega(\sigma \bullet \rho) \;\wedge\; \sigma \bullet \rho = \langle \psi, \langle \lambda \rangle \rangle \;\wedge\; \lambda \neq \emptyset}{\Sigma \triangleright (\nu\, \tilde{b})\, (\overline{\sigma}\langle\theta\rangle.P \mid \rho(x).Q \mid R) \;\underset{\lambda}{\overset{-}{\longmapsto}}\; \overline{\Sigma[\lambda]} \triangleright (\nu\, \tilde{b})\, \left(\overline{\psi}\langle\theta\rangle.P[\lambda] \mid \psi(x).Q[\lambda] \mid R[\lambda]\right)} \tag{10}$$

Rule (10) must be used whenever the spi process becomes ready on both one input and one output channel, in order to check whether the channels can be the same or not.

$\Omega(\sigma \bullet \rho)$ indicates whether the match exists, and, if it does, ψ is the matching value, and λ is the substitution list which must be applied to turn σ and/or ρ into ψ. In practice ψ and λ make it possible to satisfy the following: $\sigma[\lambda] = \rho[\lambda] = \psi$, and the transition labeled with λ means that future behavior is allowed only when $\sigma[\lambda] = \rho[\lambda] = \psi$. $\sigma \bullet \rho$ and $\Omega(\sigma \bullet \rho)$ are formally defined in [12].

The case $\lambda = \emptyset$ means that the match exists and no substitution list is needed i.e. σ and ρ are the same term; but this case is already handled in [1].

$$\frac{\Omega(\sigma \bullet \rho) \wedge \sigma \bullet \rho = \langle \psi, \langle \lambda \rangle \rangle \wedge \lambda \neq \emptyset}{\Sigma \rhd ([\sigma \; is \; \rho] \; P) \mid Q \; \underset{\lambda}{\overset{-}{\longmapsto}} \; \overline{\Sigma[\lambda]} \rhd P[\lambda] \mid Q[\lambda]} \tag{11}$$

Rule (11) tries to equate terms σ and ρ. In case of success, the behavior expressions P and Q are updated with the substitution list λ and then enabled. The environment's knowledge Σ is also updated with λ. The matching value ψ is not explicitly used, but λ contains all the information needed to turn σ and/or ρ into ψ in P and Q.

$$\Sigma \rhd (let \; (x,y) = \gamma \; in \; P) \mid Q \; \underset{\frac{(\gamma',\gamma'')}{\gamma}}{\overset{-}{\longmapsto}} \; \overline{\Sigma\left[\frac{(\gamma',\gamma'')}{\gamma}\right]} \rhd P\left[\frac{(\gamma',\gamma'')}{\gamma}, \frac{\gamma'}{x}, \frac{\gamma''}{y}\right] \mid Q\left[\frac{(\gamma',\gamma'')}{\gamma}\right] \tag{12}$$

where $\gamma' = \langle m_{\gamma'}, \Sigma_\gamma \rangle$ and $\gamma'' = \langle m_{\gamma''}, \Sigma_\gamma \rangle$; $m_{\gamma'}$ and $m_{\gamma''}$ are fresh names.

Rule (12) equates the pair of variables (x, y) with a generic term which is turned into a pair of generic terms whose names are fresh and whose knowledge is inherited from γ. As shown by the substitution list, this new pair replaces γ in P and its components are respectively assigned to x and y.

$$\Sigma \rhd (case \; \gamma \; of \; 0 \; : \; P \; suc(x) \; : \; Q) \mid R \; \underset{0/\gamma}{\overset{-}{\longmapsto}} \; \overline{\Sigma[0/\gamma]} \rhd P[0/\gamma] \mid R[0/\gamma] \tag{13}$$

$$\Sigma \rhd (case \; \gamma \; of \; 0 \; : \; P \; suc(x) \; : \; Q) \mid R \; \underset{\gamma'/\gamma}{\overset{-}{\longmapsto}} \; \overline{\Sigma[\gamma'/\gamma]} \rhd Q[\gamma'/\gamma, \gamma'/x] \mid R[\gamma'/\gamma] \tag{14}$$

where $\gamma' = \langle m_{\gamma'}, \emptyset \rangle$ and $m_{\gamma'}$ is a fresh name. Such a restriction allows to constrain the generic data to be an integer number only (due to rule (17), an empty Σ can always produce integer numbers).

When a generic term is involved in such a construction, the behaviour forks. In fact, if the value 0 is assumed, then behavior P is enabled, otherwise Q. In the former case, the generic term γ has been replaced by 0, while in the latter one all integer values are still allowed, thus the domain of the new generic term γ' has been restricted to the set of natural numbers (17).

$$\frac{\Omega(\eta \circ \rho) \wedge \eta \circ \rho = \langle \psi, \langle \lambda \rangle \rangle \wedge \lambda \neq \emptyset}{\Sigma \rhd (case \; \eta \; of \; \{x\}_\rho \; in \; P) \mid Q \; \underset{\lambda}{\overset{-}{\longmapsto}} \; \overline{\Sigma[\lambda]} \rhd P[\lambda, \psi/x] \mid Q[\lambda]} \tag{15}$$

Behavior P is enabled when term η can be considered as *something encrypted by means of* ρ. In this case $\Omega(\eta \circ \rho)$ returns true, ψ is the value of the encrypted message, and λ is the substitution list that gives the equivalence $\eta[\lambda] = \{\psi\}_{\rho[\lambda]}$. The environment's knowledge Σ is accordingly updated. $\eta \circ \rho$ and $\Omega(\eta \circ \rho)$ are formally defined in [12].

The following rules allow us to verify whether a certain term can be generated by means of the contents of Σ:

$\Sigma_\gamma \sqsubseteq \Sigma \Leftrightarrow \Sigma \vdash \gamma$ (16)	$\Sigma \vdash \rho' \wedge \Sigma \vdash \rho'' \Leftrightarrow \Sigma \vdash (\rho', \rho'')$ (20)
$\Sigma \vdash n$ (17)	$\Sigma \vdash \{\rho\}_\sigma \wedge \Sigma \vdash \sigma \Rightarrow \Sigma \vdash \rho$ (21)
$\rho \in \Sigma \Rightarrow \Sigma \vdash \rho$ (18)	$\Sigma \vdash \rho \wedge \Sigma \vdash \sigma \Rightarrow \Sigma \vdash \{\rho\}_\sigma$ (22)
$\Sigma \vdash suc(\rho) \Leftrightarrow \Sigma \vdash \rho$ (19)	

where

$$\Sigma' \sqsubseteq \Sigma'' \triangleq \Sigma'' \vdash \sigma' \;\; \forall \sigma' \mid \Sigma' \vdash \sigma' \tag{23}$$

Definition (23) does not introduce circularity in (16) since $\gamma \notin \Sigma_\gamma$.

In the following, we introduce a set of rules that should allow to compute $\overline{\Sigma}$ from Σ:

$\dfrac{\Sigma \vdash \gamma}{\Sigma \twoheadrightarrow \Sigma \setminus \{\gamma\}}$ (24)	$\dfrac{(\rho', \rho'') \in \Sigma}{\Sigma \twoheadrightarrow \Sigma \setminus \{(\rho', \rho'')\} \cup \{\rho'\} \cup \{\rho''\}}$ (27)
$\dfrac{\rho \in \Sigma \wedge \Sigma \setminus \{\rho\} \vdash \rho}{\Sigma \twoheadrightarrow \Sigma \setminus \{\rho\}}$ (25)	$\dfrac{\{\rho\}_\sigma \in \Sigma \wedge \Sigma \vdash \sigma}{\Sigma \twoheadrightarrow \Sigma \setminus \{\{\rho\}_\sigma\} \cup \{\rho\}}$ (28)
$\dfrac{suc(\rho) \in \Sigma}{\Sigma \twoheadrightarrow \Sigma \setminus \{suc(\rho)\} \cup \{\rho\}}$ (26)	

It can be proven [12] that the number of all allowed evolutions, starting from Σ, is finite and each one has a finite number of steps. In particular the last set of each evolution is $\overline{\Sigma}$. The proof can be made by underlining that each rule tends to replace a composite term with its components (i.e. no rule does insert in Σ any element previously extracted by another one) and, given two or more rules, their application (evaluation) order does not affect the obtained set.

4.1. AN EXAMPLE OF ES-LTS

Fig.1 shows the ES-LTS of a simple spi process. We have two processes: $(\nu k)(\bar{c}\langle k\rangle.c(y).[C(y, \{z\}_k)]\; P(z))$ and $(\nu M)(c(x).\bar{c}\langle\{M\}_x\rangle.Q)$. The former generates a fresh key which is sent to the latter which encrypts M with the received key and sends back the encrypted message to the former. The former checks the received message ($[C(y, \{z\}_k)]$ stands for *case* y *of* $\{z\}_k$ *in*) and enters P. In fig. 1 each behavior is bounded by a box, and the related environment's knowledge is enclosed in braces

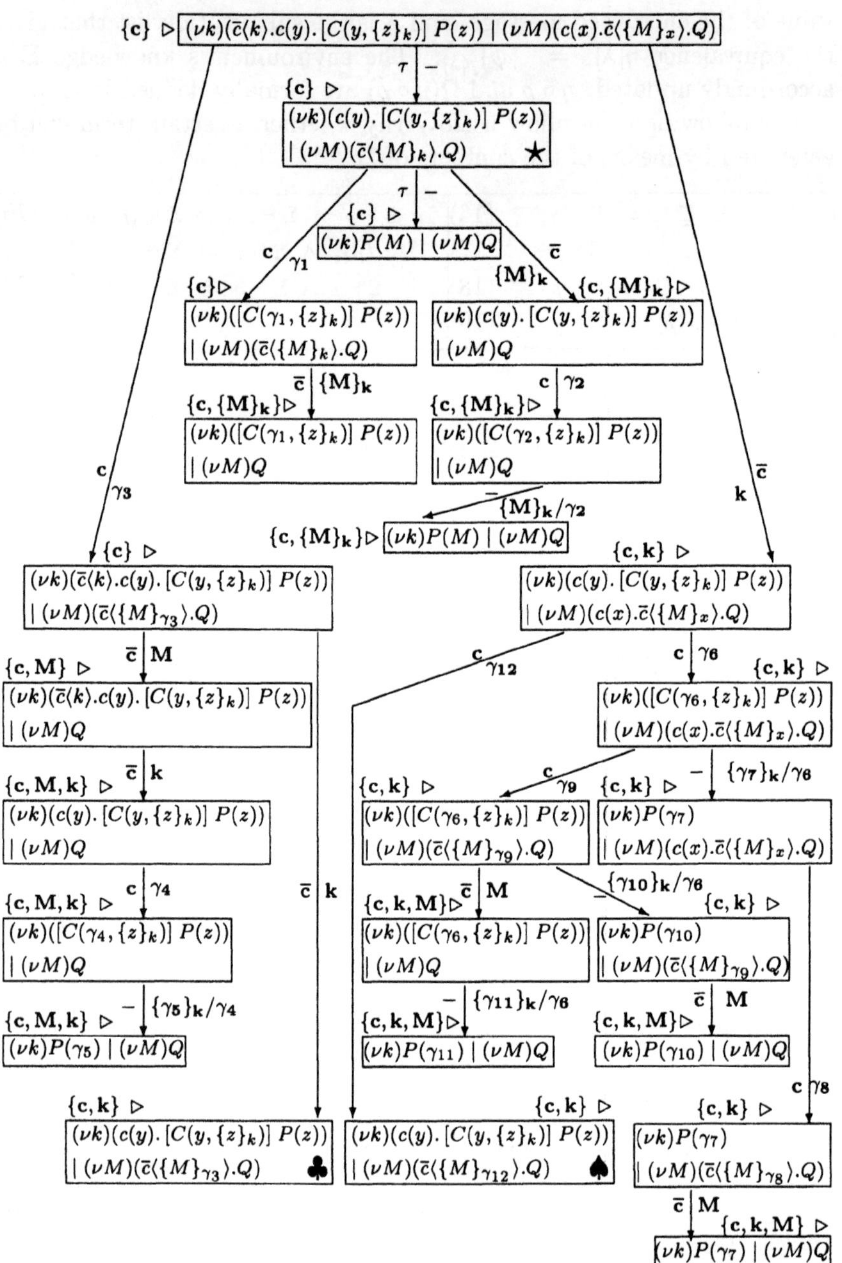

Figure 1 Example of ES-LTS

near to the box itself. The initial environment's knowledge contains only the unrestricted name c. The initial specification can evolve along the three following paths:

a) an internal synchronization occurs ($\overset{\tau}{\underset{-}{\longmapsto}}$). The behavior expression is reduced accordingly.

   If a second internal synchronization occurs, a final state (P and Q are not specified) is reached without any interaction with the environment. The two alternative paths concern interactions with the environment; in particular they are the two interleavings of the allowed input and output interaction with the environment. In case of the input interaction first ($\overset{c}{\underset{\gamma_1}{\longmapsto}}$), the environment knows only c, thus in the next state it is not possible to find a specializing value for γ_1 that matches the constraint $([C(\gamma_1, \{z\}_k)])$. By contrast, when the process output precedes the input, the environment receives $\{M\}_k$ (even if it cannot decrypt it) before generating γ_2, which can be turned into $\{M\}_k$ ($\overset{-}{\underset{\{M\}_k/\gamma_2}{\longmapsto}}$)in the next state.

b) an input interaction occurs ($\overset{c}{\underset{\gamma_3}{\longmapsto}}$): in this case a path leads to a final state, moreover a second path leads to a behavior labeled with ♣. Due to lack of space, this behavior has not been expanded further; however it is similar to the one labeled with $\star$, where γ_3 replaces k and the environment also knows k. Thus the set of allowed evolutions is richer than the one obtained from $\star$.

c) an output interaction occurs ($\overset{\bar{c}}{\underset{k}{\longmapsto}}$): in this case, a path leads to a behavior labeled with ♠, similar to the one with ♣. Another path leads to a little more complex path, which takes into account all allowed interleavings between actions.

Now let us analyze the leftmost path in more detail: the environment knows c, thus it sends a generic term γ_3 to the spi process over channel c. γ_3 represents each message the environment is able to build at this point. The spi process uses γ_3 to encrypt M; when it outputs $\{M\}_{\gamma_3}$ on channel c, the environment is able to capture M since it knows c and M has been encrypted by means of *something* it knows. Thus the transition has been labeled with M, which has also been added to the environment's knowledge. The spi process is now ready to output k on channel c; k is captured by the environment (which enlarges its knowledge) and it is used to label the transition. At the next input event, the environment gives a generic γ_4 to the spi process. The last transition means that the last behavior is reached only when γ_4 can be seen as *something encrypted with k*. It is allowed since Σ_{γ_4} includes k.

The meaning of the whole path is the following: the path to the last but one state includes all paths that can be built using each allowed message the environment can build in correspondence to the two input events. The last state is reachable only when the second generic message (γ_4) has been made as *something encrypted with k*. The *something* cannot only be M, but also (M, M), $\{\{M, c\}_k\}_k$ (even if meaningless) and so on.

5. TRACE SEMANTICS

As seen above, transitions take the forms (2), (3), (4) and (5). By introducing $\Longmapsto$ as the reflexive and transitive closure of $\overset{\tau}{\underset{-}{\longmapsto}}$, we inductively define $\overset{a}{\underset{b}{\Longmapsto}}$ as $\Longmapsto \overset{\alpha}{\underset{\beta}{\longmapsto}} \overset{a'}{\underset{b'}{\Longmapsto}}$ where $a = \alpha \cdot a'$ and $b = \beta \cdot b'$ and the pair (α, β) can assume the forms $(\overline{\sigma}, \delta_\Sigma)$, (σ, γ) and $(-, \lambda)$ in the cases of (3), (4) and (5) respectively.

In order to check whether two processes are trace equivalent, for a process P, we define:

$$traces(P) = \{(a, b) \mid \exists P' \mid P \overset{a}{\underset{b}{\Longmapsto}} P'\} \tag{29}$$

where P is a spi process and, inductively, the trace equivalence:

$$(\alpha'.a', \beta'.b') \simeq (\alpha''.a'', \beta''.b'') \Leftrightarrow (\alpha', \beta') \sim (\alpha'', \beta'') \wedge (a', b') \simeq (a'', b'') \tag{30}$$

where $(\alpha', \beta') \sim (\alpha'', \beta'')$ is defined as it follows:

$$(c', \gamma') \sim (c'', \gamma'') \iff c' \equiv c'' \wedge m_{\gamma'} \equiv m_{\gamma''} \tag{31}$$

$$(\overline{c'}, \delta\Sigma') \sim (\overline{c''}, \delta\Sigma'') \iff \overline{c'} \equiv \overline{c''} \wedge R(\delta\Sigma', \delta\Sigma'') \tag{32}$$

$$(-, \lambda') \sim (-, \lambda'') \iff S(\lambda', \lambda'') \tag{33}$$

$R(\delta\Sigma', \delta\Sigma'')$ holds true if a bijective mapping exists between $\delta\Sigma'$ and $\delta\Sigma''$ so that each pair of the mapping (σ', σ'') satisfies the following:

$$\sigma' \equiv \sigma'' \vee \left(\sigma' \equiv \{\eta'\}_{\theta'} \wedge \sigma'' \equiv \{\eta''\}_{\theta''}\right) \tag{34}$$

$S(\lambda', \lambda'')$ holds true if a bijective mapping exists between λ' and λ'' so that each pair of the mapping $(\sigma'/\gamma', \sigma''/\gamma'')$ satisfies the following:

$$m_{\gamma'} \equiv m_{\gamma''} \wedge \left(\sigma' \equiv \sigma'' \vee \left(\begin{array}{l} \sigma' \equiv \{\eta'\}_{\theta'} \\ \wedge \\ \sigma'' \equiv \{\eta''\}_{\theta''} \\ \wedge \\ \Sigma_{\gamma'} \setminus \{\sigma'\} \not\vdash \sigma' \\ \wedge \\ \Sigma_{\gamma''} \setminus \{\sigma''\} \not\vdash \sigma'' \end{array} \right) \right) \tag{35}$$

In practice, in the case of input transitions (31), we require that the names of the two channels are the same, and that the two generic terms are also the same; in the case of output transitions (32), we require that the new data added to the environment's knowledge are the same or are indistinguishable from the environment's point of view (34). It must be remembered that when an encrypted term is present in $\delta\Sigma$, it means that the environment cannot decrypt it, thus it cannot distinguish among such kinds of terms.

In the case of *specializing* transitions (33), two corresponding substitution items must replace the same generic term ($m_{\gamma'} \equiv m_{\gamma''}$) and the replacing terms must be the same, or must be indistinguishable, in the sense of the environment's knowledge. In fact two different replacing terms can be considered equivalent when they are encrypted terms that the environment's knowledge cannot decrypt (last sentence in (35)).

Everything works well since the following assumptions holds: the same criteria must be used in generating the name of generic terms in both processes (rules (31) and (35)), and the initial environment's knowledge must be the same for both processes. In particular, it is assumed that the initial environment's knowledge of a process explicitly contains each unrestricted name of the process specification.

The fact that two trace equivalent processes in our definition are also testing equivalent, and vice versa, will be proven in a future work.

6. CONCLUSIONS

A tractable method to check the spi calculus may-testing equivalence by state space exploration has been presented. It deals with the spi calculus as defined in [1], with the only exception of the replication operator. As far as we know, this is the first attempt in this direction. Previous related work, reported in [2] and [3], investigated only the theorem proving approach.

Further research efforts are oriented to formally prove soundness and completeness of the method and to develop a tool that can implement the verification task automatically.

References

[1] M. Abadi, and A. D. Gordon, "A Calculus for Cryptographic Protocols The Spi Calculus", *Digital Research Report*, vol. 149, January 1998, pp. 1-110.

[2] M. Abadi, and A. D. Gordon, "A bisimulation method for cryptographic protocols", *Nordic Journal of Computing*, Vol. 5, pp. 267-303, 1998.

[3] M. Boreale, R. De Nicola, and R. Pugliese, "Proof Techniques for Cryptographic Processes", *Proc. of the 14th IEEE Symposium Logic In Computer Science (LICS'99)*, IEEE Computer Society Press, pp. 157-166, 1999.

[4] G. Lowe, "Breaking and fixing the Needham-Schroeder public-key protocol using FDR", *Proc. of TACAS'97*, Springer LNCS 1055, 1996.

[5] G.Lowe, "Casper: a compiler for the analysis of security protocols", *Proc. of 1996 IEEE Computer Security Foundations Workshop*, IEEE Computer Society Press, 1996.

[6] G. Lowe, B. Roscoe, "Using CSP to Detect Errors in the TMN Protocol", *IEEE Transactions on Software Engineering*, Vol. SE-23, No. 10, pp. 659-669, October 1997.

[7] J. K. Millen, S. C. Clark, and S. B. Freedman, "The Interrogator: Protocol Security Analysis", *IEEE Transactions on Software Engineering*, Vol. SE-13, No. 2, pp. 274-288, February 1987.

[8] G. Leduc, O. Bonaventure, L. Léonard, E. Koerner, and C. Pecheur, "Model-Based Verification of a Security Protocol for Conditional Access to Services", *Formal Methods in System Design*, Vol. 14, No. 2, pp. 171-191, March 1999.

[9] R. Milner, J. Parrow, and D. Walker, "A Calculus of mobile processes, parts I and II", *Information and Computation*, pages 1-40 and 41-77, September 1992.

[10] L. C. Paulson, "The inductive approach to verifying cryptographic protocols", *Journal of Computer Security*, Vol. 6, pp. 85-128, 1998.

[11] S. Schneider, "Verifying Authentication Protocols in CSP", *IEEE Transactions on Software Engineering*, Vol. SE-24, No. 9, pp. 741-758, September 1998.

[12] L. Durante, R. Sisto, and A. Valenzano, "A state-exploration technique for spi-calculus testing equivalence verification", *Technical Report DAI/ARC 1-00*, Politecnico di Torino, Italy, 2000.

VERIFICATION OF CONSISTENCY PROTOCOLS VIA INFINITE-STATE SYMBOLIC MODEL CHECKING

A Case Study: the IEEE Futurebus+ Protocol

Giorgio Delzanno
DISI - University of Genova
via Dodecaneso 35, 16146 Italy
giorgio@disi.unige.it

Abstract We apply infinite-state model checking to verify safety properties of a *parameterized* formulation of the IEEE Futurebus+ coherence protocol modeled at the behavior level in a system with split transaction. This case-study shows that verification techniques previously applied to hybrid and real-time systems can be used as tools for validating parameterized protocols. This technology transfer is achieved by combining abstraction techniques, symbolic representation via constraints, efficient operations based on real arithmetics, and reachability algorithms. To our knowledge this is the first time that safety properties for a parameterized version of the Futurebus+ protocol has been automatically verified.

Keywords: Parameterized Consistency Protocols, Symbolic Model Checking

1. INTRODUCTION

In a multiprocessor system with local caches, coherence protocols ensure the consistency of the data stored in main memory and in the caches during the execution of a program [17]. For a fixed number of caches, these protocols can be described as finite-state systems obtained by composing the specifications of individual caches [23]. The state of a cache depends on the hardware (usually supporting a *valid* and a *dirty* bit) and denotes (shared or exclusive) ownership of (clean or modified) data. The state changes in accord with write and read commands issued by the corresponding CPU or coming from the system bus. Automatic methods based on finite-state symbolic model checking with BDDs have been successfully applied to the verification (and debugging) of consis-

Tommaso Bolognesi and Diego Latella (Eds.), *Formal Methods for Distributed System Development.*

tency protocols for systems with a fixed number of processors [6, 22]. Unfortunately, proving the correctness of the protocol for few special cases does not guarantee its correctness for *any* number of processors.

In recent years new techniques that extend the applicability of symbolic model checking to infinite-state systems have been used to attack the problem of verification of *parameterized* formulations of concurrent systems [1], and communication protocols [15, 12, 13, 11]. Based on these ideas, in this paper we apply infinite-state symbolic model checking methods working over real arithmetics to verify a parameterized (where the parameter=number of processors) version of the IEEE Futurebus+ cache coherence protocol specified at the *behavior* level for a single-bus multiprocessor system with split-transaction.

Our work is based on the description of the Futurebus+ coherence protocol given by Jim Handy in *The Cache Book* [17]. Following [23], we consider caches with a single line, and we abstract from the data read from and written into memories and buffers. Despite these limitations, we still model the special features of the protocol like the conditions based on the *transaction flag* used to give special meaning to read operations. The behavior level of the caches will be described via the four states *invalid*, *shared unmodified*, *exclusive modified*, and *exclusive modified*, and we will model split-transaction via non-atomic synchronization actions. For this purpose, we need to introduce auxiliary states that represent the behavior of caches when pending read and write commands occur.

Our verification methodology is based on the use of *abstraction*, *constraints*, and *backward* reachability analysis. Let us discuss all these aspects more in detail. The state of a cache ranges over a *finite* (and usually small) set of possible local states. Then, we can apply the following *abstraction* from global states (global state=collection of local cache states) to tuples of natural numbers: we only keep track of the *number* of caches in every possible local state. This abstraction allows us to cluster together all *symmetric* states. The resulting abstract protocol can be represented as a transition system with data variables ranging over positive integers formally described as an Extended Finite State Machine (EFSM) [5]. EFSM-transitions are linear transformations where global and local conditions are expressed as guards (=linear constraints). Note that in the abstract protocol the only fixed dimension is the number of local states (not the number of processors), i.e., the EFSM-model gives us a parameterized formulation of the original protocol. Parameterized initial and unsafe configurations can be represented as linear constraints as well. This way, we reduce the problem of checking safety properties for parameterized protocols to a *reachability* problem for a system with integer data variables. To solve the reachability problem, we adopt the backward reachability algorithm of [2]. Starting from the set of (abstract) *unsafe* states we compute the closure of the *predecessor* operator associated to the abstract protocol (an EFSM). To make this algorithm effective, we use a symbolic representation of potentially infinite sets of EFSM-states (i.e. tuples of natural numbers) via linear constraints. This way, we represent compactly sets of global states of the original (family of) protocols. As a last ingredient of our recipe, in order to obtain efficient constraint operations we apply a relaxation from integers to reals when testing satisfiability, checking entailment, and eliminating existentially quantified variables (the operations used in the symbolic reachability algorithm). This technique is widely used in program analysis and verification [8, 19]. The relaxation introduces a further abstraction that, however, is applied only during the execution of the backward reachability analysis. The combination of all of these ingredients leads us to the following *technology transfer*: tools devised for the automatic verification of hybrid, concurrent and real-time systems like HyTech [18], Polka [19], and DMC [10], can now be applied to check *temporal properties* of consistency protocols. In order to enhance the underlying verification algorithms based on symbolic fixpoint computations, some of these tools implement further approximation techniques like widening and acceleration operators [19, 10].

As practical results, we have automatically verified that several safety properties for the Futurebus+ protocol hold for any number of processors. An example of safety property we have considered is *different caches can not be simultaneously in exclusive states.* In our experiments we have used the infinite-state symbolic model checkers DMC [10] (based on a constraint solver over reals) and HyTech [18] (based on Halbwachs' polyhedra library). To our knowledge, this is the first time that a parameterized version of (an abstraction of) the Futurebus+ protocol has been automatically verified. Furthermore, the use of infinite-state symbolic model checking tools working over real arithmetics seems to be a novel approach to attack the verification of consistency protocols with split-transaction bus structure. We have used the same methodology to validate many other cache coherence protocols taken from the literature [9].

Related Works. Finite-state symbolic model checking has been successfully applied to the verification of consistency protocols for multiprocessor systems with a relatively small number of processors. In [6], Clarke et al. found several design errors in the IEEE specification of the Futurebus+ for a hierarchical bus system with up to 8 processors.

As in our approach, they consider single cache lines, one action as an abstraction of a complete transaction, non-determinism to simplify the management of read and write requests, no exception handling, and no *write-* and *read-invalid* commands. However, being in the finite-state case they can also distinguish processes. It would be interesting to investigate in the combination of BDDs and constraint-based symbolic model checking in order to handle more detailed parameterized models. In [23], Pong and Dubois apply a technique called Symbolic State Model to the verification of several coherence protocols formulated at the behavior level. However, they do not consider the Futurebus+ protocol. In [20], Norris-Ip and Dill have incorporated the abstractions (based on repetition operators) used in [23] in Murφ.

In [9], we have applied our methodology to analyze other existing protocols taken from the literature (see e.g. [17]). In none of the examples in [9], we have considered a split-transaction bus structure (we considered indeed only atomic synchronization actions). Our method is inspired to [15, 12, 13, 11] where the authors present decidability results for proper subclasses of the EFSMs we consider in this paper and relate them to verification problems of parameterized systems. Semi-automatic methods for parameterized system have been investigated in [4, 7, 22], whereas abstraction techniques for parameterized systems have been investigated in [16, 20, 21, 23]. In [2, 3, 10], constraints are used as symbolic representation of (potentially infinite) sets of states. Finally, relaxation techniques for handling linear constraints in program analysis and automatic verification have been investigated in [8, 10, 21, 19].

Plan of the paper. In Section 2, we give a formal model of the behavior level of the IEEE Futurebus+ Protocol in terms of the finite-state machines of Pong and Dubois [23]. In Section 3, we describe the abstract protocol obtained applying the abstraction discussed before. In Section 4, we apply our verification algorithm to abstract protocol. Finally, in Section 5 we address future directions of research.

2. THE IEEE FUTUREBUS+ PROTOCOL

In this paper we focus our attention on the open standard IEEE Futurebus+ coherence protocol as described by Jim Handy in *The Cache Book* [17]. Following [23], we make the following assumptions: we consider a single-bus multiprocessor system where each CPU has a local cache connected on the system bus, we consider caches with a single line (i.e. cache state=state of the cache line), and we abstract from the data read from and written into (cache) memory. Furthermore, following [6] we hide all handshaking necessary to issue a command, i.e., we model a

Current State	Message	To CPU	To BUS	Cond.	New State
invalid	READ		Read Shared	tf*	shared-U
				not tf*	exclusive-U
	WRITE		Read Modified		exclusive-U
sharedU exclusiveU exclusiveM	READ	Data			(same state)
sharedU	WRITE		Invalidate		exclusive-M
exclusiveU exclusiveM	WRITE				exclusive-M

Figure 1 CPU Cycles for Futurebus+

command as an atomic action and we use non-determinism to simplify the bus-protocol introducing arbitrary delays in response to read and write commands. Futurebus+ uses a *split transaction* synchronous bus with *reflection.* In a split-transaction, a processor issues a command and then removes itself from the bus until the memory or a I/O device issues a response. In the sequel of the paper, we will introduce non-atomic actions to take into account split-transactions. When using reflection, memory is allowed to snoop cache-to-cache data transfers on the system bus. This way, it always keeps an updated copy of data. Furthermore, Futurebus+ specifies a *copy-back* coherency protocol that supports *direct data intervention*, *write-allocation* and *write-snarf. Direct data intervention* indicates that on a read miss data can be read from another cache. *Write-allocation* indicates that in a write cycle a valid copy of data is first read and then merged with the (part of) line to be written. *Write-snarf* indicates that when a processor is attempting to read data and the data is not available the cache is capable of listening on other transactions in order to grab copies of data. Snoop hit acknowledgments are broadcast from the snooped cached onto the bus via a signal called tf* (transaction flag), a bus signal whose definition depends on the type of transaction which is occurring on the bus. By monitoring a tf* signal, the requesting cache has the option of taking a line immediately into an *exclusive* state, if that line is copied in no other cache. The behavior level of the protocol is illustrated in the tables of Fig. 1 (CPU cycles) and Fig. 2 (Bus cycles) taken from [17]. Following [6], we don't consider here the *read-* and *write-invalid* commands used in DMA transfers to and from memory. A cache can be in one of the following states: *invalid*, the cache has no valid data; *exclusive unmodified* (exclusiveU), the cache has an exclusive copy of the data; *shared unmodified* (sharedU), the cache has a potentially shared copy of the data; *exclusive modified* (exclusiveM),

Current State	Message	To CPU	To BUS	New State
invalid	*any*			invalid
sharedU exclusiveU	Read Shared		assert tf*	shared-U
exclusiveM		Data	assert tf*	shared-U
sharedU exclusiveU	Read Modified			invalid
exclusiveM		Data		invalid
sharedU exclusiveU exclusiveM	Invalidate			invalid

Figure 2 Bus Cycles.

the cache has a modified copy of the data. Let us describe the protocol more in detail starting from the *read cycles.* On a CPU read miss (the cache line is in the *invalid* state), a Read Shared command is placed on the bus. If the snooped cache is in state *shared unmodified* or *exclusive unmodified* then it asserts the signal tf* and its state changes to *shared unmodified*; the data are supplied by main memory. If the snooped cache is in state *exclusive modified* then main memory is disabled, the modified (dirty) line is put on the bus, the snooped cache asserts tf* and its state changes to *shared unmodified.* Reflection ensures that main memory has a consistent copy of data at the end of each Read Shared cycle. If tf* has been asserted, the requesting cache changes its state to *shared unmodified*; otherwise, its state changes to *exclusive unmodified.* When the cache is in one of *shared*, *exclusive unmodified* or *exclusive modified*, read commands do not require any bus activity. On a CPU write miss, according to the write-allocation policy, a valid copy of the data must be loaded first into the cache and then merged with the new data. Modified lines are first copied back to main memory. The Read Modified command is then placed on the bus; the owner supplies the data and goes to *invalid*, while the copy of all other caches are invalidated. The requesting cache goes to *exclusive modified.* Write hits on *exclusive modified* or *unmodified* line do not require bus interaction: the data are written into the cache line, and the cache changes its state to *exclusive modified.* On write hits on *shared unmodified* lines, the other copies are invalidated through the Invalidate command. The requesting cache loads the data and changes its state to *exclusive modified.*

Safety properties. In this paper we limit ourselves to consider *safety* properties that are related to the *semantics* of the cache states. Specifically, we would like to prove that the protocol is free from the

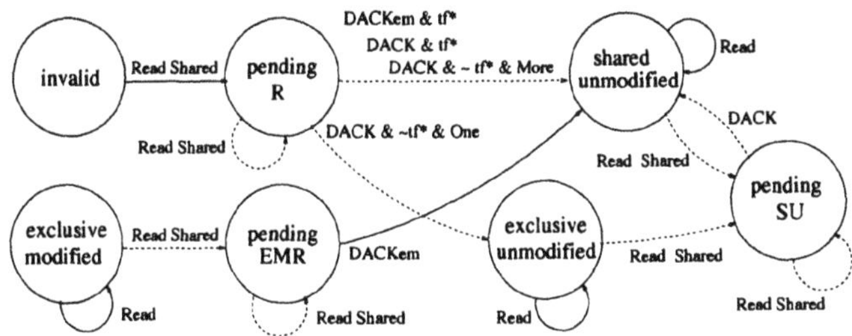

Figure 3 Read Cycles.

following sources of data inconsistency: (a) more than one cache is in an exclusive state; (b) different caches are simultaneously in an exclusive state and in the shared state. In the following section we will show that for a fixed number of caches these properties can be checked using reachability analysis over a finite-state system describing the global behavior of the protocol.

A Finite-state Model with Non-atomic Actions

In this section we use the finite-state model of Pong and Dubois [23] to formally specify the protocol. In the model of [23], the behavior of each cache is specified as a finite-state machine where additional global conditions (i.e. involving the state of other caches) are used as guards for transitions. The behavior of the global system with n caches is obtained composing the individual machines. In the global machine, transitions may require a synchronization of all caches on a given event (e.g. a bus invalidate broadcast message). To simplify the presentation of the formal model for our case-study we discuss separately read and write cycles (plus additional global conditions we will introduce later in this section). The complete model is obtained merging the transitions described at each step.

Read Cycles. In order to describe the read cycles of Futurebus+ taking into account split-transaction, in addition to the basic states *invalid*, *sharedU*, *exclusiveU*, and *exclusiveM*, we introduce the following new states: *pendingR* indicates that the cache has issued a Read Shared command and it is waiting for the data acknowledgement; *pendingEMR* indicates that a cache in *exclusive modified* state is ready to respond with the data to the Read Shared command; *pendingSU* indicates that caches either in *shared* or in *exclusive unmodified* state are ready to assert the tf* signal. The finite-state machine from the perspective of an individual cache is depicted in Fig. 3. Solid edges labeled by an action ℓ from state s_1 to state s_2 indicate that one cache in state s_1 can perform the action ℓ and move to s_2. Dashed edges labeled with ℓ denote the actions of all other caches when the event ℓ occurs. Read hits do not cause state transitions as indicated by the loops in state *sharedU*, *exclusiveU*, and *exclusiveM*. When a cache issues the Read Shared command and goes to *pendingR*, all caches in state *sharedU* or *exclusiveU* go to *pendingSU* and prepare to issue the signal tf*; alternatively, the cache in state *exclusiveM* goes to *pendingEMR* and prepares to send a data acknowledgement (ready signal) (DACK_{em}). If the data acknowledgement (together with tf*) is sent by a cache in state *pendingEMR* (DACK_{em}) or by main memory (DACK) and tf* has been asserted then the caches with pending read go to *sharedU*. if tf* has not been asserted and only one cache has a pending read (this condition is indicated as *One* in Fig.3) than its state changes to *exclusiveU*. If tf* has not been asserted and more than one cache has pending read requests (this condition is indicated as *More* in Fig.3) than all caches move to *sharedU* (note in fact that if one cache first moves to *exclusiveU* then it will immediately move to *sharedU* as soon as the other cache will receive an acknowledgment and move to *sharedU*). We model the test for tf* as the following global conditions:

tf* $\equiv$ there exists a cache in state *pendingEMR* or *pendingSU*.

Write Cycles. In order to model the write miss cycle we introduce other two states. *pendingW* indicates caches that have issued a Read Modified command and wait for a data acknowledgment. *pendingEMW* indicates that a cache in state *exclusiveM* is preparing to issue the data acknowledgement. The Read Modified cycle from the perspective of an individual cache is depicted in Fig. 4. When the cache issues a Read Modified command and goes to *pendingW*, the cache in state *exclusiveM* goes to *pendingEMW*, while all other caches go to *invalid*. When a data acknowledgment is issued than the cache goes to *exclusiveM*.

Let us consider for a moment the *global* machine obtained fixing the number n of caches and merging the read and write miss cycles described above. In the previous model we allow both multiple pending read and write commands. However, it is easy to check that this model gives rise immediately to data inconsistencies where, e.g., different caches are in state *exclusiveM* and *sharedU*. For instance, let $n = 3$ and let $G_0 = \langle invalid, invalid, invalid \rangle$ be the initial state. If the first cache issues a Read Modified before the second one issues a Read Shared the system ends up in the global state $G_1 = \langle pendingW, pendingR, invalid \rangle$. An acknowledgment from memory will force the two caches simultaneously

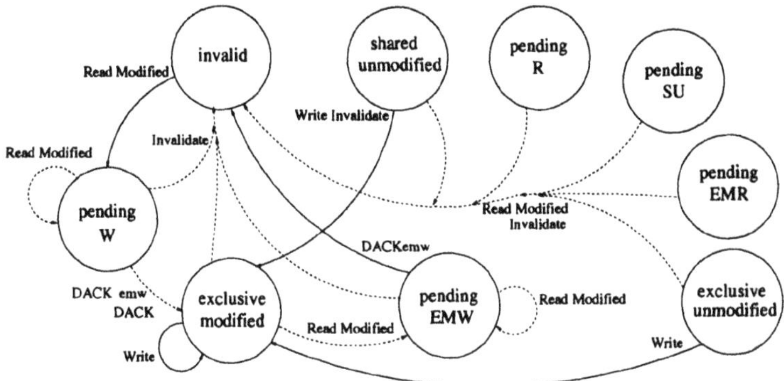

Figure 4 Write cycles.

to the unsafe state $G_2 = \langle exclusiveM, exclusiveU, invalid \rangle$. To avoid this situation we add the following global condition on the transitions Read Shared and Read Modified in state *invalid.*

> A cache can issue a Read Shared or a Read Modified command only if there are no pending writes.

Finally, as shown in Fig. 4 write hits is in state *exclusiveU* and *exclusiveM* need no consistency actions. On the other hand, when the cache is in state *sharedU* all other caches in state *sharedU* are forced to the state *invalid* via the Invalidate signal.

The finite-state model obtained fixing the number of processors can be used to test the safety properties described in Section 2 via finite-state symbolic model checking [6]. However, we would like to check the properties for any number of caches. We attack the problem in the following section.

3. FROM PROTOCOLS TO EFSMs

In order to check parameterized safety properties we apply the following *abstraction.* Let Q be the finite set of states of a cache. Then,

> we keep track only of the *number* of caches in every state $q \in Q$.

Via this abstraction, a global state G with n components (n=number of processes) is mapped to a tuple of *positive integers* with k components where (k=number of cache states). Consider for instance our case-study with three caches. The states $\langle invalid, invalid, exclusiveU \rangle$, $\langle invalid, exclusiveU, invalid \rangle$, and $\langle exclusiveU, invalid, invalid \rangle$ are represented by the abstract state consisting of a tuple with 9 counters; the counter for *invalid* is $= 2$, that for *exclusiveU* is $= 1$ and all the others are = 0. Using this abstraction, we cannot prove properties of individual caches like 'cache i and cache j cannot be simultaneously in a *dirty* state'. However, we can still try to prove global properties like 'two different caches cannot be simultaneously in a *dirty* state'. This is the kind of properties we are interested in to prove that the protocol will not give inconsistent (wrt. the semantics of states) results.

Abstract Protocol. The behavior of an *arbitrary* number of caches can be described *finitely* as a set of linear transformations describing the effect of the actions on the *counters* associated to the states in Q. For this purpose, we (automatically) *compile* the finite-state model into an 'abstract protocol' defined as an *Extended Finite State Machine* (EFSM) [5], i.e., a finite automaton with data variables (ranging over integers) associated to the locations and with guarded linear transformations associated to the transitions. Formally, let $Q = \{s_1, \ldots, s_n\}$. The abstract protocol is an EFSM with one location (omitted for simplicity) and n data variables $\langle x_1, \ldots, x_n \rangle$ (denoted as $\mathbf{x}$) ranging over *positive integers*. An EFSM-state is a tuple of natural numbers $\langle c_1, \ldots, c_n \rangle$ (=$\mathbf{c}$) where c_i denotes the number of caches in state $s_i \in Q$. Transitions are represented via a collection of *guarded linear transformations* defined over the vector of variables $\langle x_1, \ldots, x_n \rangle$ and $\langle x'_1, \ldots, x'_n \rangle$ where x_i and x'_i denote the *number of caches* in state s_i, respectively, before and after the occurrence of an event. More precisely, transitions have the following form $G(\mathbf{x}) \rightarrow T(\mathbf{x}, \mathbf{x}')$, where $G(\mathbf{x})$ is the *guard* and $T(\mathbf{x}, \mathbf{x}')$ is the *transformation.* The guard $G(\mathbf{x})$ is an arbitrary *linear constraint.* A linear constraint is a conjunction (written $\varphi_1, \ldots, \varphi_n$) of atomic predicates φ_i built over the relation symbols $=, <, >, \leq, \geq$ and on linear expressions over $\mathbf{x}$. For instance, the global condition *all caches are in state 'invalid'* can be expressed as the guard $x_1 \geq 1, x_2 = 0, \ldots, x_n = 0$ if x_1 is the counter for *invalid.* The transformation $T(\mathbf{x}, \mathbf{x}')$ is defined as $\mathbf{x}' = M \cdot \mathbf{x} + \mathbf{c}$ where M is an $n \times n$-matrix with unit vectors as columns. This way, we can represent the changes of states of the caches in the system (including the invalidation signals). Since the number of caches is an invariant of the system, we require the transformation to satisfy the condition $x'_1 + \ldots + x'_n = x_1 + \ldots + x_n$. A run of an EFSM is a (possibly infinite) sequence of EFSM-states $\mathbf{c_1}, \ldots, \mathbf{c_i} \ldots$ where $G_r(\mathbf{c_i}) \wedge T_r(\mathbf{c_i}, \mathbf{c_{i+1}}) = true$ for some transition $G_r \rightarrow T_r$; $\mathbf{c} \rightarrow^* \mathbf{c}'$ denotes the existence of a run starting from $\mathbf{c}$ and passing through $\mathbf{c}'$. The *predecessor* operator *pre* defined over a *set* S of EFSM-states as $pre(S) = \{\mathbf{c} \mid \mathbf{c} \rightarrow \mathbf{c}', \ \mathbf{c}' \in S\}$, where $\rightarrow$ indicates a one-step EFSM state transition.

The EFSM for the Futurebus+ Protocol. In this section *invalid, sharedU, exclusiveU, exclusiveM, pendingR, pendingW, pendingEMR, pendingEMW*, and *pendingSU* will denote the counters (integer variables) associated to the cache states. Furthermore, we will omit the equalities of the form $x = x'$ from $T(\mathbf{x}, \mathbf{x}')$ ($T(\mathbf{x}, \mathbf{x}') = \_$ will indicate a transformation where all variables remain unchanged). Finally, we use $x_1 = x_2 = \ldots = x_n = c$ as an abbreviation of the conjunction $x_1 = c, x_2 = c, \ldots, x_n = c$.

The read cycle of Fig. 3 is described by the EFSM-transitions of Fig. 5. Rule r_1 represents read hits (no state change). Rule r_2 occurs when a cache issues a Read Shared command (i.e. a cache moves from *invalid* to *pendingR*) with the proviso that there are no pending writes (according to the global condition we discussed in the previous section). The remaining caches change state simultaneously as indicated, e.g., by $pendingSU' = pendingSU + sharedU + exclusiveU$ (all caches in *exclusive* and *shared unmodified* move to *pendingSU*). Rule r_3 occurs when the data are supplied by another cache (in state *pendingEMR*). Rule r_4 occurs when the data are supplied by main memory and tf* has been asserted: all caches with pending read go to *sharedU*. Rule r_5 and r_6 occur when the data are supplied by main memory and tf* has not been asserted. In case there is only one pending read the cache goes to *exclusiveU* r_6. The write cycles of Fig. 4 are described by the EFSM-transitions of Fig. 6. Rule w_1 occurs when a cache issues a Read Modified command. The Invalidate command is modeled enforcing all caches to *invalid*. As for Read Shared we require that there are no pending writes (alternatively, existing write cycles could be aborted). Rule w_2 occurs when the data are supplied by a cache in state *pendingEMW*. Rule w_3 occurs when the data are supplied by main memory. Rule w_4 occurs when a write is issued on a cache in state *exclusiveM* (no state change). Rule w_5 occurs when a write is issued on a cache in state *exclusiveU*. Rule w_6 occurs when a write is issued on a cache in state *sharedU*.

4. VERIFICATION AS REACHABILITY

For every run of the global machine associated to a protocol there exists a run in the EFSM obtained applying our abstraction. Vice versa, every run of the EFSM corresponds to a set of possible runs of the original protocol. This property (proved in [9]) allows us to reduce a family of reachability problem for a given protocol (indexed on the number of processors) to a reachability problem for the corresponding EFSM. We note that we can use *linear constraints* to represent concisely (possibly

(r_1) $sharedU + exclusiveU + exclusiveM \geq 1 \longrightarrow$ _

(r_2) $invalid \geq 1, pendingW = 0 \rightarrow$
$invalid' = invalid - 1,\ pendingR' = pendingR + 1,$
$pendingEMR' = pendingEMR + exclusiveM,$
$pendingSU' = pendingSU + sharedU + exclusiveU,$
$sharedU' = exclusiveU' = exclusiveM' = 0.$

(r_3) $pendingEMR \geq 1 \longrightarrow$
$pendingEMR' = pendingEMR - 1,\ pendingR' = 0,$
$sharedU' = sharedU + pendingR + 1.$

(r_4) $pendingSU \geq 1 \longrightarrow$
$sharedU' = sharedU + pendingR + pendingSU,$
$pendingR' = pendingSU' = 0.$

(r_5) $pendingR \geq 2, pendingSU = 0,\ pendingEMR = 0 \longrightarrow$
$sharedU' = sharedU + pendingR,\ pendingR' = 0.$

(r_6) $pendingR = 1, pendingSU = 0, pendingEMR = 0 \longrightarrow$
$pendingR' = pendingR - 1,\ exclusiveU' = exclusiveU + 1.$

Figure 5 Read Cycles.

infinite) sets of EFSM-states (hence, indirectly, sets of global states independently from the number of caches in the system). Furthermore, this class of constraints is powerful enough to express *initial* and *target* sets of states for the verification problems we are interested in.

Symbolic Representation via Constraints. In this section we will use the lower-case letters $\varphi, \psi, \ldots$ to denote linear constraints and the upper-case letter $\Psi, \Phi, \ldots$ to denote *sets* (disjunctions) of constraints. The *denotation* of a constraint φ is defined as $[\![\varphi]\!]$ is the set of *solutions* (tuples of positive integers) that satisfy φ. The denotation of a set of constraints is the union of the denotation of its components. Furthermore, we say that a constraint ψ *entails* a constraint φ, written $\varphi \sqsubseteq \psi$, iff $[\![\psi]\!] \subseteq [\![\varphi]\!]$. In order to reduce the complexity of the manipulation of arithmetic constraints, we follow techniques used in program analysis [8]. Specifically, we interpret the *satisfiability* test, *variable elimination*, and *entailment* test needed to implement the symbolic reachability algorithm of [2] over the domain of *reals*. The relaxation allows us to exploit efficient (polynomial) operations over the reals in contrast to (worst-case) exponential operations over the integers. Formally, given a constraint φ, we define $[\![\varphi]\!]_{\mathbb{R}}$ as the set of *real* solutions of φ. The entailment relation over $\mathbb{R}_+$ is defined then as $\varphi \sqsubseteq_{\mathbb{R}} \psi$ if and only if $[\![\psi]\!]_{\mathbb{R}} \subseteq [\![\varphi]\!]_{\mathbb{R}}$.

(w_1) $invalid \geq 1, pendingW = 0 \longrightarrow$
$pendingW' = pendingW + 1,$
$pendingEMW' = pendingEMW + exclusiveM,$
$sharedU' = exclusiveU' = exclusiveM' = 0,$
$pendingR' = pendingEMR' = pendingSU' = 0,$
$invalid' = invalid + exclusiveU + sharedU + pendingSU +$
$+ pendingR + pendingEMR - 1.$

(w_2) $pendingEMW \geq 1 \longrightarrow$
$pendingEMW' = pendingEMW - 1,\ invalid' = invalid + 1,$
$exclusiveM' = exclusiveM + pendingW, pendingW' = 0.$

(w_3) $pendingEMW = 0 \longrightarrow$
$exclusiveM' = exclusiveM + pendingW,\ pendingW' = 0.$

(w_4) $exclusiveM \geq 1 \longrightarrow \_.$

(w_5) $exclusiveU \geq 1 \longrightarrow$
$exclusiveU' = exclusiveU - 1,\ exclusiveM' = exclusiveM + 1.$

(w_6) $sharedU \geq 1 \longrightarrow$
$invalid' = invalid + sharedU - 1,$
$exclusiveM' = exclusiveM + 1,\ sharedU' = 0.$

Figure 6 Write Cycles.

We apply the above relaxation to define a *symbolic predecessor operator*, $\mathbf{sym\_pre}_{\mathbb{R}}$ defined as follow. Let $\varphi(\mathbf{x}')$ be a constraint with variables over $\mathbf{x}'$, $\mathbf{sym\_pre}_{\mathbb{R}}(\varphi(\mathbf{x}')) = \bigvee_{i \in I} \exists \mathbf{x}' \in \mathbb{R}_+ .\ G_i(\mathbf{x}) \wedge T_i(\mathbf{x}, \mathbf{x}') \wedge \varphi(\mathbf{x}')$, where $G_i(\mathbf{x}) \rightarrow T_i(\mathbf{x}, \mathbf{x}')$ is an EFSM transition rule for $i \in I$ (I=index set).

Algorithm for Symbolic Reachability. The symbolic reachability algorithm [2] works as follows. Let Φ_f be a set of constraint representing the *unsafe* states, and let $\Phi_0 = \Phi_f$. At step i, we apply the operator $\mathbf{sym\_pre}_{\mathbb{R}}$ to every constraint in Φ_i. A newly computed constraint ψ is added to Φ_i if and only if there exist no $\varphi \in \Phi_i$ such that $\varphi \sqsubseteq_{\mathbb{R}} \psi$. If we cannot add new constraints, the algorithm terminates. If the algorithm terminates at step k, the solutions of the constraints in Φ_k represent a *superset* of the set of states that are backward reachable from Φ_f. In fact, using the relaxation integer-reals we may compute an over-approximations of the integer solutions of a linear constraint. As a consequence, if the initial set of states is not in $[\![\Phi_k]\!]$, then the original protocol is *safe for any number of caches*. Otherwise, we must return a *don't know* answer. To check the last condition, we simply conjunct the constraint representing the initial states with the constraints in Φ_k.

The previous algorithm is implemented in the backward reachability procedures of existing symbolic model checkers for hybrid and concurrent systems like DMC [10] and HyTech [18]. As for other classes of infinite-state systems (see e.g. [1, 3, 18, 19]), the method is not guaranteed to terminate for all systems and properties. The theoretical termination is guaranteed whenever the abstract protocol is well-structured and the property can be expressed as an upward-closed set of states [11]. Unfortunately, our case study does not satisfy the conditions in [11]. Thus, we must evaluate the method on practical experiments.

Analysis of the Futurebus+ Protocol. The EFSM associated to the Futurebus+ Protocol consists of the set of rules of Fig. 5 and Fig. 6. In accord with the semantics of the states of the protocol, we would like to show that starting from the initial global state where all caches have *invalid* state, it is not possible to reach one of the global states denoting data inconsistency listed at the end of Section 2. The initial state Φ_{init} of the abstract protocol is expressed then as the constraint where $invalid \geq 1$ and all other variables are $= 0$. The unsafe states can be formulated as the set (disjunction) $\Phi_f = \{\varphi_1, \varphi_2\}$, where *more than one cache is in an exclusive state* is expressed as $\varphi_1 \equiv exclusiveU + exclusiveM \geq 2$, and *different caches are simultaneously in exclusive and shared states* is expressed as $\varphi_2 \equiv sharedU \geq 1, exclusiveU + exclusiveM \geq 1$. Using DMC [10], we have automatically checked that none of the unsafe states in Φ_f can be reached from the initial parameterized state Φ_{init}. DMC required 8 steps and 196.1 seconds (on a Pentium) to reach a fixpoint. The fixpoint consists of 47 constraints. This implies that our model of the Futurebus+ is safe (wrt. the data consistency properties we discussed before) for *any number* of processes in the system. We have obtained the same result using HyTech [18]. The description of the EFSM and of the above mentioned analysis together with the experiments described in [9] are available on the web at the address `http://www.disi.unige.it/person/DelzannoG/protocol.html`.

5. CONCLUSIONS

Parameterized protocols for systems with many identical processes can be represented (via suitable abstractions) as infinite-state systems with integer variables [15, 12, 13, 11]. Following this idea, in this paper we have modeled a parameterized version of the Futurebus+ cache coherence protocol at the behavior level taking into consideration a split transaction bus structure. The combination of backward reachability and symbolic representation via constraints is the key technology we used to verify automatically several safety properties (for any number of caches) of an abstraction of the protocol. As future works, it would be interesting to model protocols at a finer level of abstraction (e.g. introducing explicit delays in the actions), and to study other important properties of cache-based systems like *sequential consistency*.

References

[1] P. A. Abdulla, A. Bouajjani, B. Jonsson, and M. Nilsson. Handling Global Conditions in Parameterized System Verification. In *Proc. CAV '99*, LNCS 1633, pages 134–145, 1999.

[2] P. A. Abdulla, K. Cerāns, B. Jonsson and Y.-K. Tsay. General Decidability Theorems for Infinite-State Systems. In *Proc. LICS '96*, pages 313–321, 1996.

[3] T. Bultan, R. Gerber, and W. Pugh. Symbolic Model Checking of Infinite-state Systems using Presburger Arithmetics. In *Proc. CAV '97*, LNCS 1254, pages 400–411, 1997.

[4] M. C. Browne, E. M. Clarke, and O. Grumberg. Reasoning about Networks with Many Identical Finite State Processes. Information and Computation 81(1): 13–31, 1989.

[5] K.-T. Cheng and A. S. Krishnakumar. Automatic Generation of Functional Vectors Using the Extended Finite State Machine Model. ACM Transactions on Design Automation of Electronic Systems 1(1):57–79, 1996.

[6] E.M. Clarke, O. Grumberg, H. Hiraishi, S. Jha, D.E. Long, K.L. McMillan, and L.A. Ness. Verification of the Futurebus+ Cache Coherence Protocol. In *Proc. 11th Int. Symp. on Computer Hardware Description Languages and their Applications*, 1993.

[7] E. Clarke, O. Grumberg, and S. Jha. Verifying Parameterized Networks. TOPLAS 19(5): 726–750 (1997).

[8] P. Cousot and N. Halbwachs. Automatic Discovery of Linear Restraints among Variables of a Program. In *Proc. POPL 78*, pages 84–96, 1978.

[9] G. Delzanno. Automated Verification of Parameterized Cache Coherence Protocols. To appear in *Proc. CAV '00*, July 2000.

[10] G. Delzanno and A. Podelski, Constraint-based Deductive Model Checking in CLP. To appear in *Software Tools for Technology Transfer*.

[11] J. Esparza, A. Finkel, and R. Mayr. On the Verification of Broadcast Protocols. In *Proc. LICS '99*, pages. 352–359, 1999.

[12] E. A. Emerson and K. S. Namjoshi. Automatic Verification of Parameterized Synchronous Systems. In *Proc. CAV '96*, LNCS 1102, pages 87–98, 1996.

[13] E. A. Emerson and K. S. Namjoshi. On Model Checking for Nondeterministic Infinite-state Systems. In *Proc. LICS '98*, pages 70–80, 1998.

[14] A. Finkel and P. Schnoebelen. Well-structured transition systems everywhere! Technical Report LSV-98-4, Laboratoire Spécification et Vérification, Ecole Normale Supérieure de Cachan, April 1998.

[15] S. M. German and A. P. Sistla. Reasoning about Systems with Many Processes. JACM 39(3): 675–735 (1992)

[16] S. Graf and H. Saïdi. Construction of Abstract State Graphs with PVS. In *Proc. CAV '97*, LNCS 1254, pages 72-83, 1997.

[17] J. Handy. The Cache Memory Book. Academic Press, 1993.

[18] T. A. Henzinger, P.-H. Ho, and H. Wong-Toi. HyTech: a Model Checker for Hybrid Systems. In *Proc. CAV '97*, LNCS 1254, pages 460–463, 1997.

[19] N. Halbwachs, Y-E. Proy, and P. Roumanoff. Verification of Real-time Systems using Linear Relation Analysis. *Formal Methods in System Design*, 11(2):157–185, 1997.

[20] C. Norris-Ip and D. L. Dill. Verifying Systems with Replicated Components in Murphi. *Formal Methods in System Design*, 14(3): 273–310, 1999.

[21] D. Lesens, N. Halbwachs, and P. Raymond. Automatic Verification of Parameterized Linear Networks of Processes. In *Proc. POPL '97*, 1997.

[22] K. L. McMillan and J. Schwalbe. Formal Verification of the Gigamax Cache Consistency Protocol. In *Proc. Int. Symp. on Shared Memory Multiprocessors*, pp. 242-51, 1991.

[23] F. Pong and M. Dubois. A New Approach for the Verification of Cache Coherence Protocols. IEEE Transactions on Parallel and Distributed Systems 6(8), August 1995.

Part Five

MULTICAST PROTOCOL ANALYSIS AND SIMULATION

SYSTEMATIC PERFORMANCE EVALUATION OF MULTIPOINT PROTOCOLS

Ahmed Helmy, Sandeep Gupta, Deborah Estrin, A. Cerpa, Y. Yu
University of Southern California
helmy@ceng.usc.edu, sandeep@poisson.usc.edu, {estrin,cerpa,yanyu}@catarina.usc.edu

Abstract

The advent of multipoint (multicast-based) applications and the growth and complexity of the Internet has complicated network protocol design and evaluation. In this paper, we present a method for automatic synthesis of worst and best case scenarios for multipoint protocol performance evaluation. Our method uses a fault-oriented test generation (FOTG) algorithm for searching the protocol and system state space to synthesize these scenarios. The algorithm is based on a global finite state machine (FSM) model. We extend the algorithm with timing semantics to handle end-to-end delays and address performance criteria. We introduce the notion of a virtual LAN to represent delays of the underlying multicast distribution tree.

As a case study, we use our method to evaluate variants of the timer suppression mechanism, used in various multipoint protocols, with respect to two performance criteria: overhead of response messages and response time. Simulation results for reliable multicast protocols show that our method provides a scalable way for synthesizing worst-case scenarios automatically. We expect our method to serve as a model for applying systematic scenario generation to other multipoint protocols.

1. INTRODUCTION

The longevity and power of Internet technologies derives from its ability to operate under a wide range of operating conditions. Perhaps more than any other technology, the range of operating conditions is enormous. It is this enormous set of conditions that motivates us to develop techniques for systematic testing of protocol behavior. We do not expect that complex adaptive protocols will be automatically verifiable under their full range of conditions. Rather, we are proposing a framework in which a protocol designer can follow a set of systematic steps, assisted

Tommaso Bolognesi and Diego Latella (Eds.), *Formal Methods for Distributed System Development.*

by automation where possible, to cover a specific part of the design and operating space.

In our proposed framework, a protocol designer needs to create the initial mechanisms, describe it in the form of a finite state machine, and identify the performance criteria or correctness conditions that need to be investigated. Our automated method will pick up at that point, providing algorithms that eventually result in scenarios or test suites that stress the protocol with respect to the identified criteria. This paper demonstrates our progress in realizing this vision as we present our method and apply it to the performance evaluation of multipoint protocols.

1.1. PROTOCOL DESIGN ISSUES

Through our proposed methodology for test synthesis, we hope to address the following key issues of protocol design and evaluation. *Scenario dependent evaluation, and the use of validation test suites:* In many evaluation studies of multipoint protocols, the results are dependent upon several factors, such as membership distribution and network topology. Hence, conclusions drawn from these studies depend heavily upon the evaluation scenarios. This brings about the need for validation test suites. Constructing these test suites can be an onerous and error-prone task if performed manually. In this paper, we propose a method for synthesizing test scenarios automatically for multipoint protocol evaluation. *Worst-case analysis of protocols:* Identifying breaking points that exhibit worst-case performance behaviors may give insight to protocol designers and help in evaluating design trade-offs. The method presented in this paper automates the generation of scenarios in which multipoint protocols exhibit worst and best case behaviors. *Performance benchmarking:* Newly proposed protocols are usually evaluated using scenarios that show performance improvement. However, without systematic evaluation, these studies often (though unintentionally) overlook other scenarios that may be relevant. To alleviate such a problem we propose to integrate stress test scenarios that provide an objective benchmark for performance evaluation. Using our scenario synthesis methodology we hope to contribute to the understanding of better performance benchmarking and the design of more robust protocols.

1.2. BACKGROUND

The design of multipoint protocols has introduced new challenges and problems. Some of the problems are common to a wide range of protocols and applications. One such problem is the *multi-responder* problem, where multiple members of a group may respond (almost) simultaneously to an event, which may cause a flood of messages throughout the network, and in turn may lead to synchronized responses, causing additional overhead and leading to performance degradation (e.g., the well-known *Ack implosion* problem).

One common technique to alleviate the above problem is the *multicast damping* technique, which employs a *timer suppression* mechanism (TSM). In TSM, a member of a multicast group that has detected loss of a data packet multicasts a request for recovery. Other members of the group, that receive this request and that have previously received the data packet, schedule transmission of a response. In general, randomized timers are used in scheduling the response. While a response timer is running at one node, if a response is received from another node then the response timer is suppressed to reduce the number of responses triggered. Consequently, the response time may be delayed to allow for more suppression. TSM is employed in several multipoint protocols. IP-multicast protocols, e.g., PIM [1] [2], IGMP [3], use TSM on LANs to reduce Join/Prune control overhead. Reliable multicast schemes, e.g., SRM [4], RRM [5], use this mechanism to alleviate *Ack implosion*. Multicast address allocation schemes, e.g., AAP [6], use TSM to avoid an implosion of responses during the collision detection phase. Active services [7] use multicast damping to launch one service agent 'servent' from a pool of servers. TSM is also used in self-organizing hierarchies (SCAN [8]).

We believe TSM is a good building block to analyze as our first end-to-end case study, since it is rich in multicast and timing semantics, and can be evaluated using standard performance criteria. In this study, we examine its worst and best case behaviors in a systematic fashion [1].

Two performance criteria commonly used to evaluate TSM are overhead of response messages and time to recover from packet loss. Depending on the relative delays between group members and the timer settings, the mechanism may exhibit different performance. In this study, our method attempts to obtain scenarios of best case and worst case performance according to the above criteria.

The rest of the paper is organized as follows. Section 2 introduces the protocol and topology models. Section 3 outlines the main algorithm, and Section 4 presents the model for TSM. Sections 5 and 6 present performance analyses for protocol overhead and response time, and Sec-

[1]Such behavior is not protocol specific, and if a protocol is composed of previously checked building blocks, these parts of the protocol need not be revalidated in full. However, interaction between the building blocks still needs to be validated.

tion 7 presents simulation results. Related work is given in Section 8. We present concluding remarks in Section 9.

2. THE MODEL

The model is a processable representation of the system under study that enables automation of our method. Our overall model consists of:

The Protocol Model. We represent the protocol by a finite state machine (FSM) and the overall system by a global FSM (GFSM).

I. FSM model: Every instance of the protocol, running on a single end-system, is modeled by a deterministic FSM consisting of: (i) a set of states, (ii) a set of stimuli causing state transitions, and (iii) a state transition function (or table) describing the state transition rules. A protocol running on an end-system i is represented by the machine $\mathcal{M}_i = (\mathcal{S}_i, \tau_i, \delta_i)$, where $\mathcal{S}_i$ is a finite set of state symbols, τ_i is the set of stimuli, and δ_i is the state transition function $\mathcal{S}_i \times \tau_i \rightarrow \mathcal{S}_i$.

II. Global FSM model: The global state is defined as the composition of individual end-system states. The behavior of a system with n end-systems may be described by $\mathcal{M}_{\mathcal{G}} = (\mathcal{S}_{\mathcal{G}}, \tau_{\mathcal{G}}, \delta_{\mathcal{G}})$, where $\mathcal{S}_{\mathcal{G}}$: $\mathcal{S}_1 \times \mathcal{S}_2 \times \cdots \times \mathcal{S}_n$ is the global state space, $\tau_{\mathcal{G}}$: $\bigcup_{i=1}^{n} \tau_i$ is the set of stimuli, and $\delta_{\mathcal{G}}$ is the global state transition function $\mathcal{S}_{\mathcal{G}} \times \tau_{\mathcal{G}} \rightarrow \mathcal{S}_{\mathcal{G}}$.

The Topology Model. The topology cannot be captured simply by one metric. Indeed, its dynamics may be complex to model and sometimes intractable. We model the delays using the delay matrix and loss patterns using the fault model. We use a virtual LAN (VLAN) model to represent the underlying network topology and multicast distribution tree. The VLAN captures delay semantics using a delay matrix D, where $d_{i,j}$ is the delay from system i to system j.

The Fault Model. A *fault* is a low level (e.g., physical layer) anomalous behavior that may affect the protocol under test. Faults may include packet loss, system crashes, or routing loops. For brevity, we only consider selective packet loss in this study. Selective packet loss occurs when a multicast message is received by some group members but not others. The selective loss of a message prevents the transition that this message triggers at the intended recipient.

3. ALGORITHM AND OBJECTIVES

To apply our method, the designer specifies the protocol as a global FSM model. In addition, the evaluation criteria are given as input to the method [2]. The algorithm operates on the protocol model and synthesizes a set of 'test scenarios'; protocol events and relations between topology delays and timer values, that stress the protocol according to the evaluation criteria (e.g., exhibit maximum overhead or delay).

The basic algorithm for our method passes through three main steps [3].

1 **The target event:** The algorithm starts from a given event, called the 'target event'. The target event (e.g., sending a message) is identified by the designer based on the protocol evaluation criteria, e.g., overhead.

2 **The search:** Three steps are taken in the search: *(a) Identifying conditions:* The algorithm uses the protocol transition rules to identify transitions necessary to trigger the target event and those that prevent it, these transitions are called *wanted transitions* and *unwanted transitions*, respectively. *(b) Obtaining sequences:* Once the above transitions are identified, the algorithm uses backward and forward search to build event sequences leading to these transitions and calculates the times of these events as follows. **(i) Backward search** is used to identify events preceding the wanted and unwanted transitions, and uses implication rules that operate on the protocol's transition table. Section 4.2 describes the implication rules. **(ii) Forward search** is used to verify the backward search. Every backward step must correspond to valid forward step(s). Branches leading to contradictions between forward and backward search are rejected. Forward search is also used to complete event sequences necessary to maintain system consistency [4]. *(c) Formulating inequalities:* Based on the transitions and timed sequences obtained in the previous steps, the algorithm formulates relations between timer values and network delays that trigger the wanted transitions and avoid the unwanted transitions.

[2]Evaluation criteria may be related to performance or correctness. In this paper we address performance criteria, correctness has been addressed in previous studies [9, 10].

[3]Our algorithm is a variant of the fault-oriented test generation (FOTG) algorithm presented in [10]. It includes the topology synthesis, the backward search and the forward search stages. Here, we describe those aspects of our algorithm that deal with timing and performance semantics.

[4]The role of forward search will be further illustrated in the response time analysis in Section 6.

3 **Task specific solution:** The output of the search is a set of event sequences and inequalities that satisfy the evaluation criteria. These inequalities are solved mathematically to find a topology or timer configuration, depending on the task definition.

We apply our method to two kinds of tasks: **Topology syntehsis** is performed when the timer values are given, and the objective is to identify the delay matrix that produces the best or worst case behavior. **Timer configuration** is performed when the topology or delay matrix is given, and the timer values that cause the best and worst case behavior are to be determined.

4. TIMER SUPPRESSION (TSM)

In this section, we present a simple description of TSM, then present its model, used thereafter in the analysis. TSM involves a request q and one or more responses p. When a system Q detects the loss of a data packet it sets a request timer and multicasts a request q. When a system i receives q it sets a response timer (e.g., randomly), the expiration of which, after duration Exp_i, triggers a response p. If the system i receives a response p from another system j while its timer is running, it suppresses its own response.

4.1. EVALUATION CRITERIA

We use two performance criteria to evaluate TSM: (1) Overhead of response messages, where the worst case produces the maximum number of responses per data packet loss. As an extreme case, this occurs when all potential responders do indeed respond and no suppression takes place. (2) The response delay, where the worst case scenario produces maximum loss recovery time.

4.2. TIMER SUPPRESSION MODEL

Protocol states ($\mathcal{S}$). Following is the state symbol table.

State	Meaning
R	original state of the requester Q
R_T	requester with the request timer set
D	potential responder
D_T	responder with the response timer set

Stimuli or Events. *(1) Sending/receiving messages:* transmitting response (p_t) and request (q_t), receiving response (p_r) and request (q_r). *(2) Timers and other events:* the events of firing the request timer *Req* and response timer *Res* and the event of detecting packet loss L.

Notation. Following are the notations used in the transition table.

- An event subscript denotes the system initiating the event, e.g., p_{t_i} is response sent by system i, while the subscript m denotes multicast reception, e.g., p_{r_m} denotes reception of a multicast response by all members of the group if no loss occurs. When system i receives a message sent by system j, this is denoted by the subscript i, j, e.g., $p_{r_{i,j}}$ is system i receiving response from system j.
- The state subscript T denotes the existence of a timer, and is used by the algorithm to apply the 'timer implication' to fire the timer event after the expiration period.
- A state transition has a start state and an end state and is expressed in the form $startState \to endState$ (e.g. $D \to D_T$). It implies the existence of a system in the $startState$ (i.e., D) as a condition for the transition to the $endState$ (i.e., D_T).
- An *effect* in the transition table may contain state transition and stimulus in the form ($startState \to endState$).$stimulus$, which indicates that the condition for triggering $stimulus$ is the state transition. An effect may contain several transitions (e.g., 'Trans1, Trans2'), which means that out of these transitions only those with satisfied conditions will take effect.

Transition Table. Following is the transition table for TSM.

Symbol	Event	Effect	Meaning
loss	L	$(R \to R_T).q_t$	loss detection causes q_t and setting of request timer
tx_req	q_t	q_{r_m}	transmission of q causes multicast reception of q
rcv_req	q_r	$D \to D_T$	reception of q causes setting of response timer
res_tmr	Res	$(D_T \to D).p_t$	response timer expiration causes p and change to D
tx_res	p_t	p_{r_m}	transmission of p causes multicast reception of p
rcv_res	p_r	$R_T \to R$, $D_T \to D$	reception of p causes timer suppression
req_tmr	Req	q_t	request timer expiration causes transmission of q

The model contains one requester Q and several potential responders (e.g., i and j). [5] Initially, the requester Q exists in state R and all potential responders exist in state D. Let t_0 be the time at which Q sends the request q. The request sent by Q is received by i and j at times $d_{Q,i}$ and $d_{Q,j}$, respectively. When the request q is sent, the requester transitions into state R_T by setting the request timer. Upon receiving a request, a potential responder in state D transitions into state D_T, by

[5] Since there is only one requester, we simply use q_t instead of q_{t_Q}, and q_{r_i} instead of $q_{r_{i,Q}}$.

setting the response timer. The time at which an event occurs is given by $t(event)$, e.g., q_{r_j} occurs at $t(q_{r_j})$.[6]

Implication Rules. The following cause-effect implication rules are used by the backward search:

1 Transmission/Reception (**Tx_Rcv**): By the reception of a message, the algorithm implies the transmission of that message – without loss– sometime in the past (after applying the network delays). An example of this implication is $p_{r_{i,j}} \Leftarrow p_{t_j}$, where $t(p_{r_{i,j}}) = t(p_{t_j}) + d_{j,i}$.

2 Timer Expiration (**Tmr_Exp**): When a timer expires, the algorithm infers that it was set *Exp* time units in the past, and that no event occurred during that period to reset the timer. An example of this implication is $Res_i.(D_i \leftarrow D_{T_i}) \Leftarrow D_{T_i}$, where $t(Res_i) = t(D_{T_i}) + Exp_i$, and Exp_i is the duration of the response timer Res_i.[7]

3 State Creation (**St_Cr**): A state is created from another by reversing the transition rules and going towards the *startState* of the transition. For example, $D_{T_i} \Leftarrow (D_{T_i} \leftarrow D_i)$.

In the following sections we use the above model to synthesize worst case scenarios according to protocol overhead and response time.

5. PROTOCOL OVERHEAD ANALYSIS

In this section, we conduct worst case performance analysis for TSM with respect to the number of responses. We assume no loss of request or response messages until recovery, and that the request timer is high enough that the recovery will occur within one request round [8]. Worst-case analysis aims to obtain scenarios with maximum number of responses per data loss [9]. In this section we present the algorithm to obtain inequalities that lead to worst-case scenarios. These inequalities are a function of network delays and timer expiration values.

Target event and conditions. Since the overhead in this case is measured as the number of response messages, the designer identifies

[6] The time of a state is when the state was first created, so $t(D_{T_i})$ is the time at which i transited into state D_T.
[7] We use the notation *Event.Effect* to represent a transition.
[8] The case of multiple request rounds is discussed in [11].
[9] For best-case analysis see [11] [12].

the event of triggering a response p_t as the target event, and the goal is to maximize the number of response messages.

The search. Following, we apply to the case study our search algorithm as described in section 3.

- **Identifying conditions:** The algorithm searches for the transitions necessary to trigger the target event, and their conditions, recursively. These are called *wanted transitions* and *wanted conditions*, respectively. The algorithm also searches for transitions that nullify the target event or invalidate any of its conditions. These are called *unwanted transitions.* In our case the target event is the transmission of a response (i.e, p_t). From the transition table described in Section 4.2, the algorithm identifies transition *res_tmr* $[Res.(D_T \rightarrow D).p_t]$ as a *wanted transition* and its condition D_T as a *wanted condition.* Transition *rcv_req* $[q_r.D \rightarrow D_T]$ is also identified as a *wanted transition* since it is necessary to create D_T. The *unwanted transition* is identified as transition *rcv_res* $[p_r.D_T \rightarrow D]$ since it alters the D_T state without invoking p_t.

- **Obtaining sequences:** Using backward search, the algorithm obtains sequences and calculates time values for the following transitions: (1) wanted transition, *res_tmr*, (2) wanted transition *rcv_req*, and (3) unwanted transition *rcv_res*, as follows: (1) To obtain the sequence of events for transition *res_tmr*, the algorithm applies implication rules (see Section 4.2) Tmr_Exp, St_Cr, Tx_Rcv in that order, and we get $Res_i.(D_i \leftarrow D_{T_i}).p_{t_i} \Leftarrow q_{r_i}.(D_{T_i} \leftarrow D_i) \Leftarrow q_{t_Q}$. Hence the calculated time for $t(p_{t_i})$ becomes $t(p_{t_i}) = t_0 + d_{Q,i} + Exp_i$, where t_0 is the time at which q_{t_Q} occurs. (2) To obtain the sequence of events for transition *rcv_req* the algorithm applies implication rule Tx_Rcv, and we get $q_{r_i}.(D_{T_i} \leftarrow D_i) \Leftarrow q_{t_Q}$. Hence the calculated time for $t(q_{r_i})$ becomes $t(q_{r_i}) = t_0 + d_{Q,i}$. (3) To obtain sequence of events for transition *rcv_res* for systems i and j the algorithm applies implication rules Tx_Rcv,Tmr_Exp, St_Cr, Tx_Rcv in that order, and we get $p_{r_{i,j}}.(D_i \leftarrow D_{T_i}) \Leftarrow Res_j.(D_j \leftarrow D_{T_j}).p_{t_j} \Leftarrow q_{r_j}.(D_{T_j} \leftarrow D_j) \Leftarrow q_{t_Q}$. Hence the calculated time for $t(p_{r_{i,j}})$ becomes $t(p_{r_{i,j}}) = t_0 + d_{Q,j} + Exp_j + d_{j,i}$.

- **Formulating Inequalities:** Based on the above wanted and unwanted transitions the algorithm avoids transition *rcv_res* while invoking transition *res_tmr* to transit out of D_T. To achieve this, the algorithm automatically derives the following inequality (see [11] for more details): $t(p_{t_i}) < t(p_{r_{i,j}})$. (1). Substituting expressions for $t(p_{t_i})$ and $t(p_{r_{i,j}})$ previously derived, we get: $d_{Q,i} + Exp_i < d_{Q,j} + Exp_j + d_{j,i}$. Alternatively, we can avoid the unwanted transition *rcv_res* if the system did not exist in D_T when the response is received. Hence, the algorithm automatically derives the following inequality (see [11] for more details): $t(p_{r_{i,j}}) < t(q_{r_i})$. (2).

Substituting, we get: $d_{Q,i} > d_{Q,j} + Exp_j + d_{j,i}$. [10] Figure 1 (a) and (b) show equations (1) and (2), respectively.

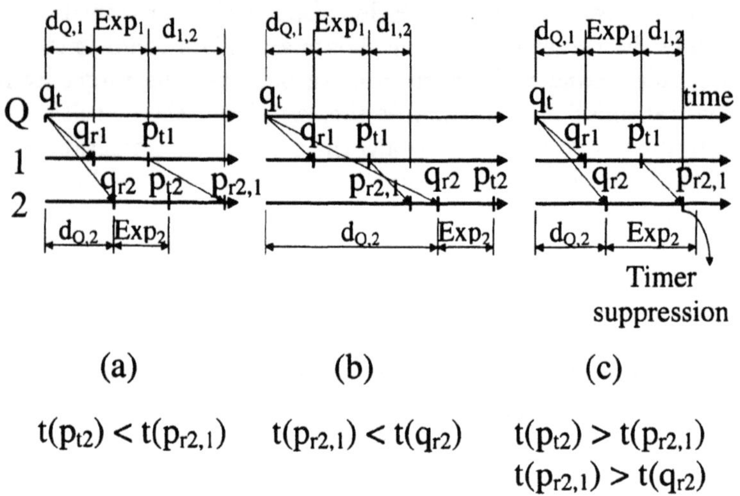

Figure 1 (a) & (b) sequences do not lead to suppression, (c) leads to suppression.

Task specific solutions.

- **Topology synthesis:** Given the timer expiration values or ranges, we want to find a feasible solution for the worst-case delays. A feasible solution in this context means assigning positive values to the delays $d_{i,j} \forall i, j$. In equation (1) above, if we take $d_{Q,i} = d_{Q,j}$ [11], we get: $Exp_i - Exp_j < d_{j,i}$. [12]

[10] Note that equations (1) and (2) are general for any number of responders, where i and j are any two responders in the system.

[11] The number of inequalities (n^2, where n is the number of responders) is less then the number of the unknowns $d_{i,j}$ ($n^2 - n$), hence there are multiple solutions. We can obtain a solution by assigning values to n unknowns (e.g., $d_{Q,i}$) and solving for the others.

[12] These inequalities put a lower limit on the delays $d_{j,i}$, hence, we can always find a positive $d_{j,i}$ to satisfy the inequalities. Note that, the delays used in the delay matrix reflect delays

- **Timer configuration:** Given the delay values or ranges (i.e., bounds), we want to obtain timer expiration values that produce worst-case behavior.

  We can obtain a range for the relative timer settings (i.e., $Exp_i - Exp_j$) using equation (1) above.

The solution for the system of equations given by (1) and (2) above can be solved in the general case using linear programming (LP) techniques (see [11] for more details). Section 7 uses the above solutions to synthesize simulation scenarios. [13]

6. RESPONSE TIME ANALYSIS

In this section, we conduct the performance analysis with respect to the response time. For our analysis, we allow selective loss of a single response message during the recovery phase [14]. In this case, transition rules are applied to only those systems that receive the message.

The algorithm obtains possible sequences leading to the target event and calculates the response time for each sequence. To synthesize the worst case scenario that maximizes the response time, the sequence with maximum time is chosen.

6.1. TARGET EVENT

The response time is the time taken by the mechanism to recover from the packet loss, i.e., until the requester receives the response p and resets its request timer by transitioning out of the R_T state. In other words, the response interval is $t(p_{r_Q}) - t(q_{t_Q}) = t(p_{r_Q}) - t_0$. The designer identifies $t(p_{r_Q})$ as the target time, hence, p_{r_Q} is the target event.

6.2. THE SEARCH

We present in detail the case of single responder, then discuss the multiple responders case.

over the multicast distribution tree. In general, these delays are affected by several factors including the multicast and unicast routing protocols, tree type and dynamics, propagation, transmission and queuing delays. One simple topology that reflects the delays of the delay matrix is a completely connected network where the underlying multicast distribution tree coincides with the unicast routing.

[13]Note, however, that it may not be feasible to satisfy all these constraints, due to upper bounds on the delays for example. In this case the problem becomes one of maximization, where the worst-case scenario is one that triggers maximum number of responses per packet loss. This problem is discussed in [11].

[14]Without loss of response messages this problem becomes one of maximizing the round trip delay from the requester to the first responder.

- **Backward search:** As shown in Figure 2 (a), the backward search starts from p_{r_Q} and is performed over the transition table (in Section 4.2) using the implication rules in Section 4.2, yielding [15]:

  $D_j.p_{r_Q}.(R_Q \leftarrow R_{T_Q}) \Leftarrow p_{t_j}.(D_j \leftarrow D_{T_j}).Res_j.R_{T_Q} \Leftarrow q_{r_j}.(D_{T_j} \leftarrow D_j).R_{T_Q}$

  At which point the algorithm reaches a branching point, where two possible preceding states could cause q_{r_j}:

  - The first is transition *loss* $[D_j.q_{t_Q}.(R_{T_Q} \leftarrow R_Q)]$ and since the initial state R_Q is reached, the search ends for this branch.
  - The second is transition *req_tmr* $[D_j.Req_Q.q_{t_Q}.R_{T_Q}]$. Req_Q indicates the need for a transition to R_{T_Q}, and the search for this last state yields eventually $D_j.q_{t_Q}.(R_{T_Q} \leftarrow R_Q)$.

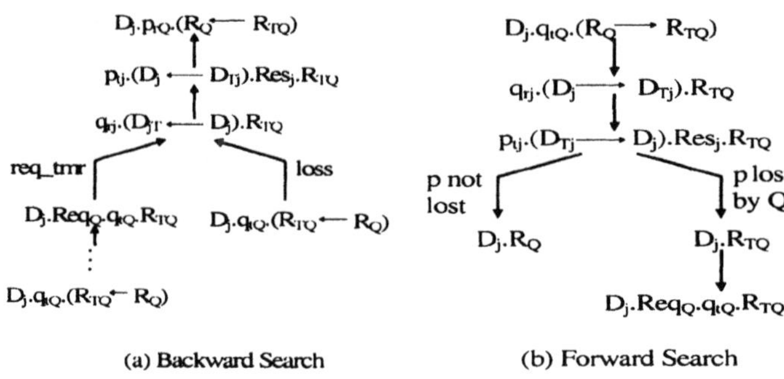

(a) Backward Search (b) Forward Search

Figure 2 Backward search for response time analysis.

- **Forward search:** The algorithm performs a forward search and checks for consistency of the GFSM. The forward search step may lead to contradiction with the backward search, causing rejection of that branch. For example, as shown in Figure 2 (b), one possible forward sequence from the initial state gives:

  $D_j.q_{t_Q}.(R_Q \rightarrow R_{T_Q}) \Rightarrow q_{r_j}.(D_j \rightarrow D_{T_j}).R_{T_Q} \Rightarrow p_{t_j}.(D_{T_j} \rightarrow D_j).Res_j.R_{T_Q}$

  The algorithm then searches two possible next states:

[15]The GFSM may be represented by composition of individual states (e.g., $State_1.State_2$ or $transition_1.State_2$).

- If p_{t_j} is not lost, and hence causes p_{r_Q}, then the next state is $D_j.R_Q$. But the original backward search started from $D_j.q_{t_Q}.Req_Q.R_{T_Q}$ which cannot be reached from $D_j.R_Q$. Hence, we get contradiction and the algorithm rejects this sequence.
- If the response p is lost by Q, we get $D_j.R_{T_Q}$ that leads to $D_j.Req_Q.q_{t_Q}.R_{T_Q}$. The algorithm logs this feasible sequence.

Calculating the time for each feasible sequence, the algorithm identifies the latter sequence as one of maximum response time.

For **multiple responders**, the algorithm automatically explores the different possible selective loss patterns of the response message. The search identifies the sequence with maximum response as one in which only one responder triggers a response that is selectively lost by the requester. To construct such a sequence, the algorithm creates conditions and inequalities similar to those formulated for the best-case overhead analysis with respect to number of responses (see [11]).

7. SYSTEMATIC SIMULATION

We have conducted a set of simulations for the scalable reliable multicast (SRM) [4] based on our worst-case analysis. We tied our method to the network simulator (NS). The output of our method, in the form of inequalities (see Section 5), is solved using a mathematical package (LINDO). The solution, in terms of a delay matrix, is then used to generate the simulation topologies for NS automatically.

We measured the number of responses triggered for each data packet loss. We have conducted two sets of simulations, each using two sets of topologies. The simulated topologies included topologies with up to 200 nodes. The first set of *stress* topologies was generated according to the overhead analysis presented in this paper. The second set of *random* topologies was generated by the GT-ITM topology generator [13].

The first set of simulations was conducted for the SRM deterministic timers [16]. The results of the simulation are shown in Figure 3. The number of responses triggered for all the *stress* topologies was $n-1$, where n is the number of nodes in the topology (i.e., no suppression occurred). For the *random* topologies, the number of responses triggered was almost 20 responses in the worst case.

[16]SRM response timer values are selected randomly from the interval $[D_1.d_r,(D_1+D_2).d_r]$, where d_r is the estimated distance to the requester, and D_1, D_2 depend on the timer type. For deterministic timers $D_2=0$ and $D_1=1$.

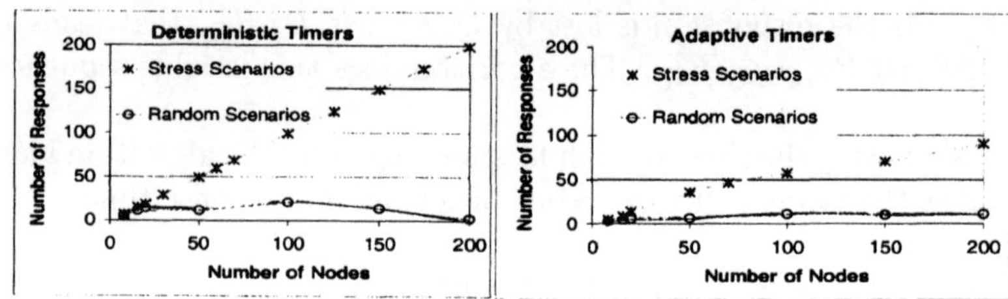

Figure 3 Simulation results for deterministic and adaptive timers over *stress* and *random* topologies.

Using the same two sets of topologies, the second set of simulations was conducted for the SRM adaptive timers [17]. The results are given in Figure 3. For the *stress* topologies almost 50% of the nodes in the topology triggered responses. Whereas *random* topologies simulation generated almost 10 responses in the worst case.

These simulations illustrate how our method may be used to generate consistent worst-case scenarios in a scalable fashion. It is interesting to notice that worst-case topologies generated for simple timers also experienced substantial overhead for more complicated adaptive timers. It is also obvious that *stress* scenarios are consistent when used to compare different mechanisms; the performance gain for adaptive over deterministic timers is very clear under *stress* scenarios.

8. RELATED WORK

Verification systems typically aim to detect violation of well-defined properties (e.g., safety). In general, the two main approaches for protocol verification are theorem proving and reachability analysis [14]. Theorem proving systems define a set of axioms and relations to prove properties.

[17] Adaptive timers adjust their interval based on the number of duplicate responses received and the estimated distance to the requester.

These systems are useful in many applications. However, they do not synthesize network topologies and do not address performance issues per se. Reachability analysis algorithms [15] try to inspect reachable protocol states, and suffer from the 'state space explosion' problem. Reduced reachability analysis has been used in the verification of cache coherence protocols [16], using a global FSM model. We adopt a similar FSM model and extend it for our approach in this study. However, our approach differs in that we address end-to-end protocols, that encompass rich timing, loss and performance semantics.

In [10] we developed a fault-oriented test generation (FOTG) for multicast routing. In [10], FOTG was used to study correctness of a multicast routing protocol on a LAN. Here, we extend FOTG to study performance of end-to-end multipoint mechanisms. In [9] we proposed a simulation-based stress testing framework based on heuristics. However, that method does not provide automatic topology generation, nor does it address performance issues. The VINT [17] tools provide a framework for Internet protocols simulation with a library of protocols and a set of validation test suites. However, it does not provide a generic tool for generating these tests automatically. Work in this paper is complementary to such studies, and may be integrated with network simulation tools as we do in Section 7.

9. CONCLUSION AND FUTURE WORK

We have presented a methodology for scenario synthesis for performance evaluation of multipoint protocols. We used a virtual LAN model to represent the underlying network topology and an extended global FSM model to represent the protocol mechanism. We adopted the fault-oriented test generation algorithm for search, and extended it to capture timing/delay semantics and performance issues for end-to-end multipoint protocols. Our method was applied to performance evaluation of the timer suppression mechanism; a common building block for various multipoint protocols. Two performance criteria were used for evaluation of the worst and best case scenarios; the number of responses per packet loss, and the response delay. Simulation results illustrate how our method can be used in a scalable fashion to test and compare reliable multicast protocols. We hope that similar approaches may be used to identify and analyze other protocol building blocks.

Our future work includes applying our methodology to sensor networks. These networks, similar to ad-hoc networks, assume dynamic topologies, lossy channels, and deal with stringent power constraints, which differentiates their protocols from Internet protocols [18]. Possible research directions in this respect include: (i) Extending the topology representation or model to capture dynamics, where delays vary with time. (ii) Defining new evaluation criteria that apply to the specific problem domain, such as power usage. (iii) Investigating the algorithms and search techniques that best fit the new model or evaluation criteria.

References

[1] D. Estrin, D. Farinacci, A. Helmy et al. Protocol Independent Multicast - Sparse Mode (PIM-SM): Protocol Specification. *RFC 2117*, March 1997.

[2] D. Estrin, D. Farinacci, A. Helmy et al. Protocol Independent Multicast - Dense Mode (PIM-DM): Protocol Specification. *Proposed RFC*, September 1996.

[3] W. Fenner. Internet Group Management Protocol, Version 2. *Internet-Draft*, November 1997.

[4] S. Floyd, V. Jacobson, C. Liu, S. McCanne, and L. Zhang. A Reliable Multicast Framework for Light-weight Sessions and Application Level Framing. *IEEE/ACM Transactions on Networking*, November 1996.

[5] R. Govindan, H. Yu, and D. Estrin. Large-scale weakly consistent replication using multicast. *USC-CS-TR 98-682*, September 1998.

[6] M. Handley. The Address Allocation Protocol. *Internet-Draft*, August 1998.

[7] E. Amir et al. An active service framework and its application to real-time multimedia transcoding. *ACM SIGCOMM'98*, September 1998.

[8] A. Reddy, D. Estrin, and R. Govindan. Fault Isolation in Multicast Trees. *ACM SIGCOMM*, August 2000.

[9] A. Helmy and D. Estrin. Simulation-based STRESS Testing Case Study: A Multicast Routing Protocol. *IEEE MASCOTS*, July 1998.

[10] A. Helmy, D. Estrin, and S. Gupta. Fault-oriented test generation for multicast routing protocol design. *FORTE/PSTV*, November 1998.

[11] A. Helmy et al. Performance Evaluation of Multipoint Protocols Using Systematic Scenario Synthesis. *USC-CS-TR 00-726*, March 2000.

[12] Ahmed Helmy. Systematic Test Synthesis for Multipoint Protocol Design. *USC-CS-TR 99-716, Ph.D. Dissertation*, August 1999.

[13] K. Calvert, M. Doar, and E. Zegura. Modeling Internet Topology. *IEEE Communications*, June 1997.

[14] E. Clarke and J. Wing. Formal Methods: State of the Art and Future Directions. *ACM Wkshp on Strategic Directions in Computing Research*, December 1996.

[15] F. Lin, P. Chu, and M. Liu. Protocol Verification using Reachability Analysis. *Computer Communication Review, Vol. 17, No. 5*, 1987.

[16] F. Pong and M. Dubois. Verification Techniques for Cache Coherence Protocols. *ACM Computing Surveys*, March 1996.

[17] L. Breslau, D. Estrin, K. Fall, S. Floyd, J. Heidemann, A. Helmy et al. Advances in Network Simulation. *IEEE Computer*, May 2000.

[18] D. Estrin, R. Govindan, J. Heidemann, and S. Kumar. Next century challenges: Scalable coordination in sensor networks. *ACM MobiCom*, August 1999.

SIMULATING MULTICAST TRANSPORT PROTOCOLS IN ESTELLE[1]

A re-usable IP network module

Justin Templemore-Finlayson
Intitut National des Télécommunications
Evry, France
templemo@hugo.int-evry.fr

Eugen Borcoci
University POLITEHNICA of Bucharest,
Bucharest, Romania
eugenbo@first.elcom.pub.ro

Key words: Multicast Transport Protocols, Formal Description Techniques, Estelle, Formal Simulation, Component re-use.

Abstract: State of the art multicast research points to a future architecture in which multicast transport protocols provide end-to-end control over the IP multicast datagram service, similar to TCP/IP. However, many aspects of these protocols are still to be resolved, and are active areas of research. We present a simulation architecture for applying the Formal Description Technique Estelle to the development of multicast transport protocols. A key component of this architecture is the IP network (IPN) module, which is the focus of this paper. The IPN module reproduces complex medium behaviours specific to multicast transport protocols and not provided by existing IP medium models, including: IGMP services, network heterogeneity, address-based routing, and multicast delivery. The sophistication of the IPN module presents a realistic simulation environment for the formal investigation of multicast transport protocols. The module has in addition been designed to be independent of the upper-layer protocol being investigated, and as such presents a re-usable component for the simulation of any IP-based protocols.

[1] This work has been partially supported by the Copernicus/IDEMCOP project.

Tommaso Bolognesi and Diego Latella (Eds.), *Formal Methods for Distributed System Development.*

1. INTRODUCTION

Multicast is a key technology to the future of the Internet. Multicast protocols enable *group communication* - communication with more than two participants - which increases the functionality of the Internet without significant negative effect on network resources [10].

Applications which can benefit from a native multicast protocol include electronic file distribution, collaborative environments and multimedia streaming, to name but a few. In addition, important network-related functions will migrate to multicast: network announcements, replication, mirroring and discovery protocols can all benefit from multicast.

The basis of multicast applications will most likely be a TCP/IP - like stack, in which multicast transport protocols provide end-to-end reliability and services using the IP multicast datagram service [7].

Design of multicast transport protocols is ostensibly simple when a native IP multicast service is present, but it has proved difficult to create protocols that scale well on a heterogeneous wide area network, such as the Internet. The growing body of multicast research has also shown that multicast transport protocols, especially reliable protocols, can pose a threat to the overall functioning of the Internet [10], as multicast transport protocols must co-operate with each other, and TCP. These factors make it important to understand the operation of these protocols in the Internet, prior to their deployment.

In this paper we present a simulation architecture for applying the Formal Description Technique (FDT) Estelle [1] to the analysis of multicast transport protocols. This FDT provides a rigorous method for communication protocol design and analysis by simulation, which can help our understanding of multicast transport protocols.

A key element of this architecture is an Estelle IP network (IPN) medium module.

Creation of a medium module is a common technique for simulating Estelle-based protocols, but the IPN module includes substantial novelty through its support of multicast transport protocols: The module provides multicast delivery, and Internet Group Management Protocol (IGMP) services, and emulates heterogeneous wide area network behaviour. These characteristics are essential for realistic simulation of multicast transport protocols.

In addition, the IPN module was designed and implemented completely independently of upper-layer protocols. It presents a definitive, re-usable component for Estelle-based simulation of any transport or application protocols, unicast or multicast, based on IP.

To the author's knowledge, the IPN module is the first multicast-capable medium module to be developed for a Formal Description Technique. Such medium models have been developed during research using informal simulators, namely ns and Opnet, but none have been completely separated from the upper-layer protocol, and are therefore not useful to the broader simulation community.

Section 2 presents the IP Network module requirements, including the service interface which is a key element of re-use.

Section 3 describes the internal functioning of the module in detail.

Section 4 presents simulation experiments of the Reliable Multicast Transport Protocol (RMTP-II) [9],[12], a proposed multicast transport protocol for the Internet. A simulation architecture for RMTP-II using the IPN module is presented, and the use of the IPN module to produce network behaviours is discussed.

2. THE IP NETWORK MODULE REQUIREMENTS

The IPN module is a completely re-usable Estelle module which emulates a wide area, multicast-capable IP network.

In Estelle, it is common practice to specify a system environment using a medium module when simulating a system. This provides the user with control over the environment in which the protocol is being tested. A medium module is usually configured with the static resources of the network, and interactive simulation allows the provocation of dynamic network situations and events.

The IPN module is based on the *generic medium module*, an Estelle medium module successfully used in for simulation of several large protocols (e.g. SSCOP [2], OSI-TP [5], XTP [6]).

The generic medium module is a generic unicast/broadcast transfer medium which produces network events for packet loss, duplication and re-ordering. It models a homogeneous network in that network event probabilities and channel bit rate for each sender-receiver pair are the same. It provides transfer of variable-size packets in unicast and broadcast modes, with transmission delay proportional to the packet size.

Within the EDT simulator, it is possible to change the configured event probabilities during simulation, thereby reproducing dynamic network conditions.

This model has limited applicability for simulation of multicast transport protocols, as it lacks a multicast service and does not model the heterogeneous behaviour exhibited by wide area networks. The IPN module reproduces the behaviour of the generic medium module for an IP network, and adds a native IP multicast service and heterogeneous behaviour between different sender-receiver node pairs in the network.

The key elements of the new IPN module are :

- IP unicast and multicast addressing
- unicast and multicast packet delivery
- IGMP services
- heterogeneous network behaviour

The IPN module multicast model faithfully reproduces that of IP multicast.

Heterogeneous network behaviour is specified by defining event probabilities for each individual pair of nodes connected to the IPN module. The following events can be configured probabilistically:

- packet loss
- packet duplication
- packet re-ordering
- transmission delay

In addition, where the generic medium module requires editing of the internal Estelle code of the module for re-use, the IPN module is encapsulated, and entirely configured through a standard external interface, described in the following section. This configuration by the higher-layer protocol does not add any complexity to the protocol specification, as it is simply a formalisation of the IP network interfaces.

Note that the IPN module simplifies the generic medium module by using a transmission delay independent of packet size. This is a conscious design choice which enables us to encapsulate and re-use the IPN module without change. If desired, transmission delays based on packet-size can be re-introduced by modifying the IPN module.

2.1 IPN module interface

The IPN module implements a simplified interface of actual IP. This interface permits upper-layer entities to dynamically change their relationship with the communication medium, within a static Estelle system structure. This service interface is depicted in Figure 1.

The *bind* service presents the corresponding IP/UDP function to start listening on a given network address. It eliminates the need for static configuration of the interfaces at the IPN module : the parameterised address (*my_address*) is used to perform packets routing and delivery.

The *packet* service performs multicast or unicast delivery (subject to non-deletory network events), based on the destination address (*dest_addr)* specified. The IPN module recognises IP multicast addresses and performs multicast or unicast delivery appropriately. The IPN module does not examine the *PDU* field. By using an external variant record *PDU_type* containing the upper-layer PDU definitions, the IPN is able to transfer any *PDU* without reconfiguration.

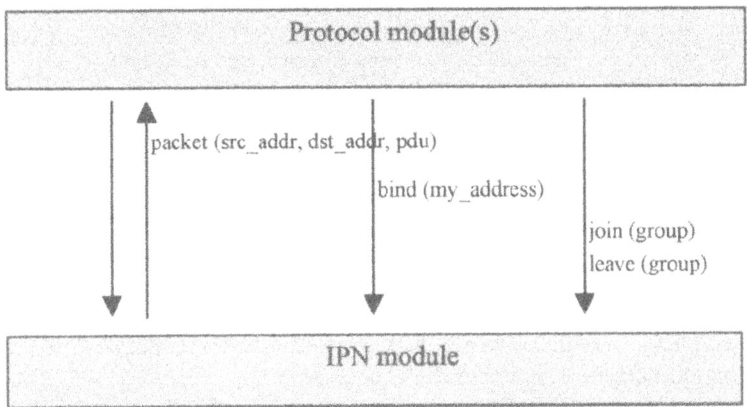

Figure 1. IP Network (IPN) service interface

The *join* and *leave* services correspond to the basic IGMP services for subscribing to and unsubscribing from a multicast group [7].

3. IPN MODULE FUNCTIONAL DESCRIPTION

3.1 Implementing the IPN service interface at the IPN-user protocol

The IPN service interface is defined in an Estelle channel definition, reproduced in Figure 2. To use the IPN module as a medium in a simulation, a protocol needs to extend this channel definition with its own PDU definitions.

There is a one-to-one correlation between the interactions defined in the channel, and the services described in Section 2.1.

Customisation of the channel to carry specific PDUs simply requires the definition of the `PDU_type` as a union record with a case for each PDU.

```
channel IPN_access (PROVIDER,USER);
  by USER:
    join_group (group:network_address_type);
    leave_group (group:network_address_type);
    bind (my_address:network_address_type);
  by PROVIDER,USER:
    pdu (src_addr:network_address_type;
         dst_addr:network_address_type;
         pdu:PDU_type);
```

Figure 2. The Estelle channel defining the IPN module service interface.

3.2 Packet addressing and delivery

The IPN module accepts packets with three fields: *source address*, *destination address*, and *protocol data unit.*

The source and destination address are defined as UDP addresses consisting of an IP address and port number. The IPN module examines the *destination address* IP address to determine the destination host(s). The UDP port numbers are preserved for the upper-layer protocol or an intermediate UDP module to handle.

The choice of whether to use unicast or multicast delivery is based on the *destination address* specified in a packet. IP multicast reserves Class D addresses (224.0.0.0 to 239.255.255.255) for multicast. The IPN module recognises these multicast addresses and delivers using multicast as appropriate.

3.3 Heterogeneous network behaviour

The *generic medium module* presents a medium with homogeneous transmission and event characteristics. It allows the user to configure the following characteristics of the network:

- packet loss probability
- packet duplication probability
- transmission delay based on packet length

These values are shared for all host pairs in the simulated network.

The IPN module extends this model by allowing these characteristics to be configured per host pair connected to the medium. This flexibility enables the heterogeneous conditions of a wide area network to be modelled.

The IPN module makes a simplification to the trasfer delay modelling, calculating delay between two hosts using a normal distribution function around a specified mean delay. Transfer delay is therefore independent of packet length. Measuring the packet length in Estelle would require knowledge of the upper-layer protocol data unit structure at the IPN module, which conflicts with the goal of encapsulation of the IPN module. A protocol designer may of course alter the IPN module to introduce packet-size dependent delays.

These medium characteristics are accessible during interactive simulation. In tool-sets such as EDT [3],[4], the characteristics of the path between two or a group of hosts can be changed to produce situations such as local congestion or single link failures. Construction of medium scenarios as described in [6] can be used to create a realistic Internet environment in which to test a multicast transport protocol.

3.4 Internal architecture

The internal architecture of the IPN module is depicted in Figure 3. The module operates in the following way:

1. A packet placed on the medium by the ith host is received at interaction point SAP[*i*].
2. The time of receipt is recorded (*timeIn*) and stored with the packet in the Input list.
3. If the packet *dst_addr* field is a not Class D address, i.e. unicast, the destination interaction point, SAP[*j*] is looked up in the address tables.
4. The transfer events *loss* and *duplication*, and the *transmission delay* for the host pair (*i*, *j*) are generated. If the packet is duplicated, an independent transmission delay is generated for it.
5. Any packet not lost is placed in the Output list. The scheduled output time is stored in *timeOut* and the destination interaction point index *j* in *dstHost*.
6. If the incoming packet *dst_addr* was a Class D address, i.e. multicast, the SAP of each host joined to that multicast group is looked up in the address tables, and steps 4 and 5 are performed for each host.
7. An output scheduler periodically increments a time counter and searches the Output list for packets whose *timeOut* equals the current time. The referenced packet (*src_addr*; *dst_addr*; *pdu*) stored in the Input list is then output on SAP[*dstHost*].
8. A packet is removed from the Input list when there are no more references to it in the Output list.

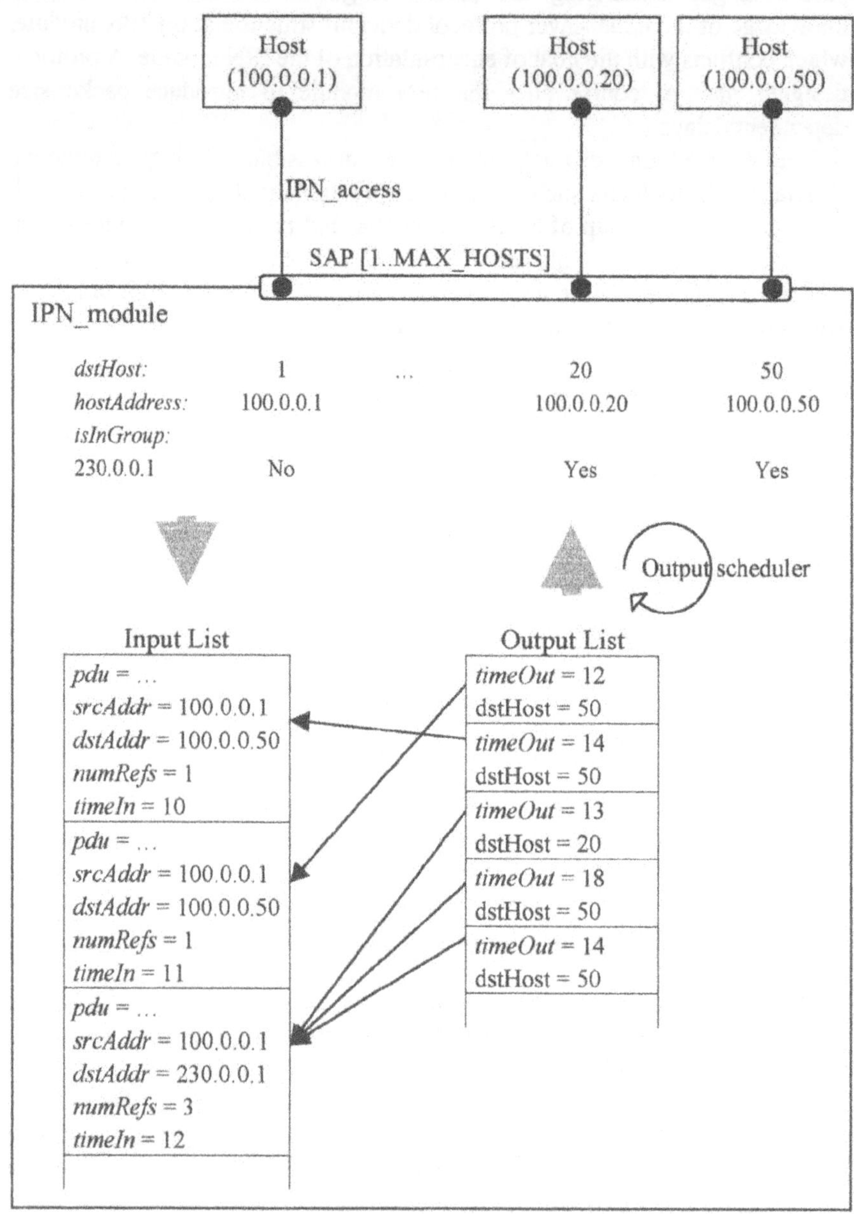

Figure 3. IPN module internal architecture

Within this scheme, re-ordering occurs when a packet to be routed between two hosts experiences a shorter delay than a previous packet scheduled between those two hosts.

WAN heterogeneity is also experienced. Packets travelling between the same two hosts experience varying delays, and probabilistic loss or duplication. Packets sent between different host pairs may experience radically different conditions. Furthermore, packets sent to a group will experience different conditions for each different receivers within that group.

The use of separate Input and Output lists makes the IPN module scalable to a large number of connected hosts: If a packet is multicast to a group with one thousand members, only a single copy of the packet is kept (in the Input list), and the memory requirements for storing a thousand references to that packet are very small.

3.4.1 Example

In Figure 3 the current state of an IPN module gives examples of unicast delivery with re-ordering, and multicast delivery to multiple receivers are given. For brevity, we refer to each host by its last byte only (i.e. 100.0.0.1 becomes simply 1).

In this example, the current network configuration shows host 1 bound to SAP[1], host 20 bound to SAP[20] and host 50 bound to SAP[50]. Hosts 20 and 50 are also all members of the multicast group 230.0.0.1.

All the packets in the Input list have been transmitted by host 1. The first two packets in the list are destined for host 50. None have been lost or duplicated (there is a single reference to each in the Output list), but the transition delay for the first packet is 4 time units, while for the second, transmission delay is only 1 unit. As the second packet was received only one time unit after the first, they will be scheduled for output in the reverse order, thus re-ordered.

The third packet will be multicast, as it has a Class D destination address (230.0.0.1). Entries for each member of the group (hosts 20 and 50) are inserted into the output table. Loss, duplication and transmission delays are calculated separately for each receiving host. None are lost, but the packet destined for host 50 has been duplicated, and each of the resulting three packets in the Output list are scheduled to be output at different times.

4. USE OF THE IPN MODULE TO SIMULATE THE RMTP-II PROTOCOL

The IPN module has been developed within a project to formally specify and validate the Reliable Multicast Transport Protocol-II (RMTP-II) [12]. An Estelle specification of the RMTP-II has been written [11], and is in the process of being simulated. The IPN module is a vital component in these simulations, enabling the wide area network behaviour experienced by this protocol to be reproduced.

4.1 The RMTP-II protocol and specification

The Reliable Multicast Transport Protocol (RMTP-II) defines a transport layer protocol for the reliable multicasting of data from one or relatively few senders, to a large group of receivers. RMTP-II uses the IP multicast service to distribute data, and IP unicast to exchange control traffic between protocol nodes.

RMTP-II provides a multicast equivalent of TCP: a sender-reliable, connection-oriented multicast service. The main challenge to be met by the protocol is the scalable processing of control traffic. RMTP-II solves this by constructing a tree of protocol nodes rooted at the sender and spanning all the receivers. Control traffic from receivers is aggregated as it flows towards the sender, thus avoiding packet implosion.

The Estelle specification of RMTP-II is implementation-oriented, which will be used to generate a working implementation compatible with other RMTP-II implementations.

4.2 Application of the IPN module

Use of the IPN module in the simulation of the RMTP-II protocol does not require any static configuration of the IPN module. However, the RMTP-II protocol must be designed to correctly use the IPN module interface. The following conventions must be observed in the RMTP-II protocol :

1. The channel `IPN_access` (see Section 3.1), must be defined in the specification module, and the type `PDU_type` declared as a variant record containing all the RMTP-II PDUs.
2. A `bind` interaction output is added to the RMTP-II protocol modules initialisation part in order to identify themselves to the IPN module.

3. Packet handling at the protocol modules must be implemented using IP addressing and the interaction definition formats of the channel `IPN_access`.
4. `Join` and `leave` interactions must be explicitly output to subscribe to / unsubscribe from multicast groups.

As the RMTP-II specification is an implementation-oriented specification, which has to deal with underlying IP protocol details, these steps did not add any complexity to the specification, simply syntactic changes.

In the following we present two examples of simulations of the E-RMTP protocol. In these simulations, the IPN module is used in conjunction with a network behaviour scenario to reproduce dynamic WAN conditions in which to test the protocol.

4.3 Simulation experiments with RMTP-II and the IPN module

Figure 4 presents the RMTP-II simulation architecture in Estelle, including the IPN module. In Figure 4 the RMTP-II protocol is defined by the collection of shaded modules, the *Sender Node (SN)*, the *Top Node (TN)* and a group of *Receiver Nodes (RN)*. Each of these performs a specialised function in the protocol.

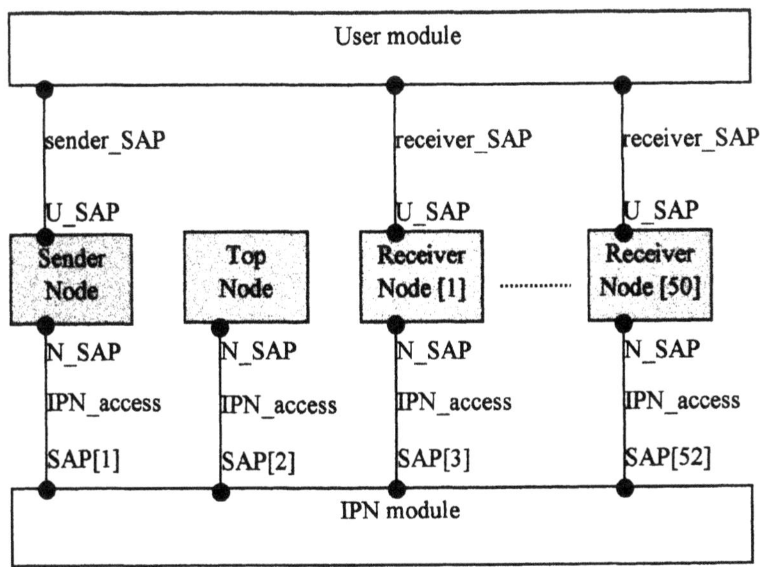

Figure 4. RMTP-II simulation architecture

The SN performs data transmission, the RN data reception, and the TN acts as a co-ordinator between the group members. The *IPN module* provides the IP wide area network environment, and the *User module* simulates sender and receiver user behaviour.

Simulations are performed by a method of guided simulation, used in the validation and performance evaluation of other complex Estelle specifications in [5] and [6].

In our simulations of RMTP-II, we use this technique to define independent user and medium scenarios, which enables us to test different protocol behaviours (stimulated by the user scenario) under different network conditions (stimulated by the medium scenario).

In the following we present an example simulation of RMTP-II using the IPN module. Note that this simulation does not show us anything new or unexpected about the RMTP-II protocol, apart from offering a formal confirmation of a well-known phenomenon. However, these measurements would not be possible without the IPN module. They demonstrate how the IPN module can be used to simulate the complex network conditions experienced by multicast protocols, which can be used to search for solutions to problems faced by multicast transport protocols.

4.3.1 Investigating the "crying-baby" problem

The "crying-baby" problem [10] describes the decrease in throughput observed by a member of a group due to the presence of a slower receiver, or "crying-baby", in that group. This is a common example of the difficulties faced by multicast protocols in a heterogeneous environment.

In this experiment, we create a crying-baby medium scenario for RMTP-II by gradually increasing the error rate observed between the *Sender Node* and an arbitrary *Receiver Node* (the “baby”), while maintaining constant error rates between the sender and other receivers.

We measure the effect of this scenario on protocol throughput by recording and calculating the average of transfer times seen at each receiver. In this experiment, the transfer time is an indicative value of the number of retransmission cycles required to deliver all the packets in a data stream.

Figure 5 shows two results. Firstly the expected rise in transfer time to the crying-baby. Secondly and more interestingly, it shows how the average transfer time experienced by receivers closely shadows that of the crying-baby. This is the expected result in the case of RMTP-II, which uses a flow control window that only advances once all receivers in the group have acknowledged a packet.

The simulation therefore confirms the observation that throughput for the RMTP-II protocol is only as great as the throughput of the slowest receiver in the group.

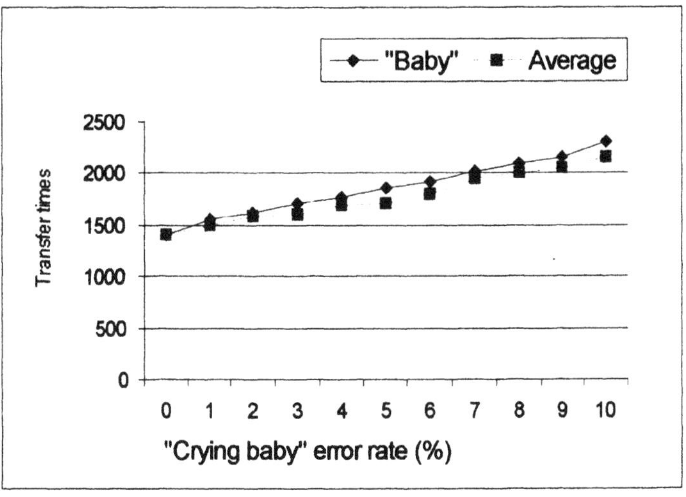

Figure 5. RMTP-II "crying-baby" simulation experiment.

5. CONCLUSIONS

Realistic validation and performance evaluation of a protocol in Estelle requires a realistic model of the environment in which that protocol is designed to operate.

The IPN module presents a Estelle module which reproduces the complex behaviour of a wide area, multicast-capable IP network. The module is self-contained and re-usable, and as such it presents a "ready-to-use" IP network medium, which relieves the protocol designer from writing such a medium module themselves.

The IPN module has been used as a medium module in the Estelle simulations of a multicast transport protocol, RMTP-II. A simulation example presented in Section 4 shows that the IPN module is capable of reproducing the complex heterogeneous environment which these protocols experience. These simulations are part of an ongoing project to formally investigate the many problems faced by Internet multicast transport protocol designers, and in which the IPN module plays an important role.

REFERENCES

[1] ISO/IEC, 1989. Information Processing Systems. Estelle - A Formal Description Technique based on Extended State Transition Model, IS 9074.

[2] E Borcoci, S Budkowski, 1999. Traffic capability of the ATM-SSCOP protocol - Simulation Study, *Proceedings of the European Simulation Multiconference 1999*, Warsaw, Poland.

[3] S Budkowski, 1992. The Estelle Development Toolset, *Computer Networks and ISDN Systems Journal*, Special Issue on FDT Concepts and Tools, Vol.25, No.1.

[4] S Budkowski et al., 1998. *The Estelle Development Toolset*, Institut National des Télécommunications, Evry, France. http://www-lor.int-evry.fr/edt/

[5] O Catrina and S Budkowski, 1998. *Integration of subtransactions in OSI-TP. Validation of the Estelle specification using EDT*, Research Report, Institut National des Télécommunications, France, December 1998.

[6] O Catrina and E Borcoci, 1996. Estelle specification and validation of XTP 4.0, *Copernicus Project (COP62) Deliverable for Workpackage 2, Task 2.1*, University of Bucharest, 1996.

[7] S Deering, 1989. *Host Extensions for IP multicasting, RFC 1112*, August 1989.

[8] A Mankin, A Romanow, S Bradner, V Paxson. *IETF Criteria for Evaluating Reliable Multicast Transport and Application Protocols*, *RFC 2357*, June 1998.

[9] P Sanjoy, K Sabnani, JC Lin and S Bhattacharya, 1997. Reliable Multicast Transport Protocol (RMTP), *IEEE Journal on Selected Areas in Communications*, Vol. 15, No. 3.

[10] B Quinn and K Almeroth, 1999, *IP multicast applications : Challenges and Solutions*, Internet Draft, 26 February 1999.

[11] J Templemore-Finlayson, 2000, *An Estelle Specification of the Reliable Multicast Transport Protocol II*, Research Report N° 00001-LOR, Institut National des Télécommunications, France, January 2000.

[12] B Whetten, M Basavaiah, S Paul, T Montgomery, N Rastogi, J Conlan and T Yeh. *Reliable Multicast Transport Protocol - II*, Internet Draft, April 1998.

GENERATION OF REALISTIC SIGNALLING TRAFFIC IN AN ISDN LOAD TEST SYSTEM USING SDL USER MODELS

Thomas Steinert[1], Georg Rößler[2]
[1] *Univ. of Stuttgart, Institute of Communication Networks and Computer Engineering, Pfaffenwaldring 47, D-70569 Stuttgart, Germany, steinert@ind.uni-stuttgart.de*
[2] *Tenovis GmbH & Co. KG, Kleyerstraße 94, D-60326 Frankfurt am Main, Germany, Georg.Roessler@tenovis.com*

Abstract This paper describes the use of SDL for defining User Models for a test and measurement system whose purpose is to perform load tests of the control unit of an ISDN-PBX with realistic signalling traffic. The generation of realistic signalling traffic is based on the emulation of real users by User Models. The functional behaviour of a User Model is defined by an SDL process using a subset of the SDL syntax. The use of this type of User Models allows, besides the creation of messages according to distribution functions, to process messages received from the system under test and therefore to react appropriately on them, like a real user would do it. Normally several User Models are used together for a load test, so that the generated traffic is a realistic mixture of a variety of different user categories. The User Models are transformed to data structures for execution by the load test system. These data structures are optimized for efficiency in order to reach the target call rate of at least 30000 calls per hour on a standard workstation. The temporal behaviour of the User Models is defined separately by means of Event Generators which send signals to the User Models according to statistical distribution functions.

Keywords: Environment simulation, ISDN-PBX, performance testing, SDL, traffic generator

1. INTRODUCTION

Telephony is still the base function in today's PBX (Private Branch Exchange) systems, but customers expect much more than just telephone service. Some of these additional functions are implemented in the PBX as extensions of the switching software, like the support of mobile users with DECT (Digital Enhanced Cordless Telecommunica-

Tommaso Bolognesi and Diego Latella (Eds.), *Formal Methods for Distributed System Development.*

tions) technology. Other functions are realized by applications running on dedicated servers which are connected to the PBX. Call centers are the most prominent systems where server applications and the PBX together realize the desired functionality.

The main motivation to build the load test system is that the overload protection in the switching software has to be tested with realistic traffic. A second motivation for the load test system is that call centers require extensive testing under high and correlated load. Aspects that need to be tested are the interaction between the PBX and the servers as well as the correct handling of service requests by the call distribution according to the customer specific call flow scripts. Furthermore, the load test system allows to verify that a call center can handle the load which has been specified by a customer.

While there are quite a number of call generators commercially available, many of them are not able to generate the required call rate of at least 30000 calls per hour. Moreover, these call generators often lack some features which we think are important. The simulated behaviour must be the same as the behaviour of real users. This implies that all durations follow given distributions instead of generating calls with fixed interarrival times. Furthermore, the load test system must be able to react appropriately on the actions taken by the PBX, for example by repeating a call attempt after a call has been rejected. Finally, the load test system must be capable to produce load peaks like those arriving at call centers when a phone number is shown in a television commercial.

The focus for the load test system is on the switching software, whereas physical interfaces and the message forwarding by line cards can be tested using other tools. Hence the basic function is to generate and terminate calls at high rates with statistically correct behaviour in time. Statistical correctness means that the functional and temporal characteristics of the generated traffic are the same as from real users. An important feature is flexibility in the sense that it has to be easy to modify the simulated user behaviour and to introduce new signalling messages or even a new signalling protocol in the load test system. This flexibility has to be achieved without requiring significant changes to the load test system software.

This paper is structured as follows: Section 2 describes the developed method for the specification of realistic user behaviour by means of User Models where the definition of the functional behaviour is separated from the definition of the temporal behaviour. The architecture of the load test system and a basic test scenario are presented in section 3 and section 4 concludes the paper with some final remarks about the application of the load test system at Tenovis.

2. SPECIFICATION OF USER BEHAVIOUR

The main purpose of the load test system is to generate realistic signalling traffic, which allows to determine the behaviour of the system under test in heavy load situations. This leads to the question, how this realistic signalling traffic can be specified. The approach chosen for this load test system is to emulate each single user by a so called "User Model" which acts independently from the other emulated users. The resulting traffic of all User Models together represents a realistic signalling traffic load, as it would be created by a multitude of real users.

This section describes the developed method for specifying User Models emulating real user behaviour. Section 2.1 motivates the general principle for the method, 2.2 presents the description technique for the functional behaviour of a User Model which is based on the Specification and Description Language (SDL) defined by the ITU-T in [1] and 2.3 describes the way for specifying the temporal behaviour of a User Model, so that realistic traffic can be generated.

2.1. GENERAL PRINCIPLE

The generation of realistic signalling traffic of the load test system is, as described above, based on User Models. A User Model emulates the behaviour of a real user, i.e. it sends signals for triggering signalling scenarios and reacts appropriately on incoming signals depending on distribution functions.

To emulate the behaviour of a real user concerning signalling the scenario of a simple phone call is analysed. Its phases are sketched in Figure 1. There are two types of events occurring at the user during the phone call: "User Actions" representing actions triggered by the user himself and "User Receptions" caused by the PBX.

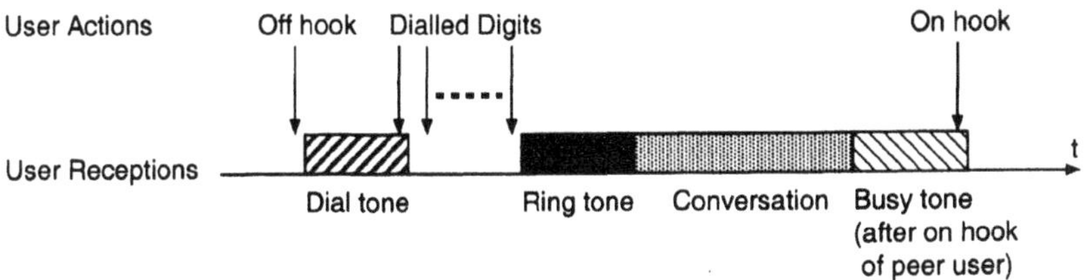

Figure 1 Phases of a phone call from user's point of view

Figure 2 contains a state-transition diagram of the FSM resulting from the scenario of the simple phone call. Again, there is the distinction between the "User Actions", which are sketched above the state symbols, and the "User Receptions", which are sketched under the state symbols. The state-transition diagram uses the same "User Actions" as those of Figure 1, but different "User Receptions". Instead of the continuous signals like e.g. "Dial tone", atomic signals for the start of a tone and for the end of a tone are applied.

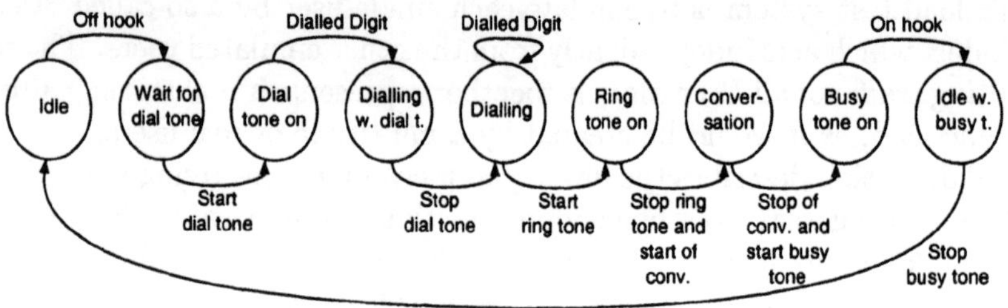

Figure 2 Modelling a user via a Finite State Machine (FSM)

It is important to understand that the "User Actions" are essential for generating realistic user behaviour, whereas the "User Receptions" are input signals from the environment. These "User Actions" require that the User Model is augmented by a specification of the temporal behaviour. The reason is that in order to obtain realistic signalling traffic, the time between the entry in a state and the arriving of a User Action has to be a random variable with a given distribution. The temporal behaviours of multiple User Model instances are independent of each other, although these instances may have the same statistical properties.

There are quite a number of publications ([2] provides a nice overview) about augmented FSMs which describe both functional and temporal behaviour. However, they mostly focus on the performance analysis and not on the emulation of behaviour in real-time. One exception is the paper by Lemppenau and Tran-Gia [3] which describes also a system for traffic generation.

Our approach is to separate the specifications of functional and temporal aspects as far as possible. The User Model FSM describes only the functional behaviour of users. The temporal behaviour is defined by means of so called "Event Generators", which create signals called "Events" corresponding to the "User Actions". These Events are one type of input signals of the User Model FSM. The principle of Event Generators is described in detail in section 2.3.

The second type of input of the User Model FSMs are "Signalling-SDUs" (Service Data Units) caused by the signalling messages sent by the PBX. These Signalling-SDUs (in the remainder of this paper referred to as SDUs) result from the layer 3 protocol of an ISDN signalling system. This protocol is realized as defined in the respective standard, but the interface to the upper layer is adapted to the needs of the load test system. Because several ISDN signalling systems, e.g. Q.931 [4] and 1TR6 [5], with different layer 3 protocols are supported by the PBX, the interface to the User Model FSMs has to be independent of the signalling protocol. Therefore protocol independent SDUs, whose definitions are adopted from Annex A of Q.931, are applied at the interface to the User Model FSMs. This type of signal is also used by the User Model FSMs to communicate in direction to layer 3 of an ISDN signalling system, so that the User Model FSM is able to control the underlying signalling protocol. In Figure 3 the interfaces of a User Model FSM are depicted: A User Model FSM receives Events from Event Generators to trigger actions, like setting up a phone call. It sends SDUs to the Signalling FSM for initiating signalling actions, e.g. for establishing a signalling connection to the PBX and it receives SDUs sent by the Signalling FSM indicating a state of a signalling connection.

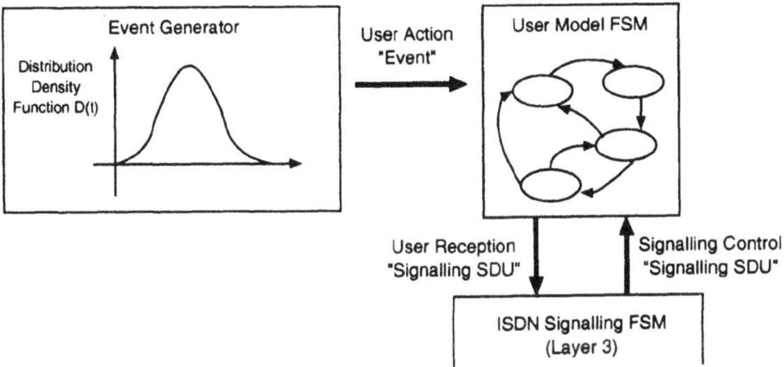

Figure 3 Input- and output signals of a User Model FSM

In the following section 2.2 the definition of User Model FSMs representing the functional behaviour of an emulated user by means of SDL is described.

2.2. SPECIFICATION OF THE FUNCTIONAL BEHAVIOUR USING SDL

Emulation of real user behaviour involves both the generation of signals according to a certain distribution and the reaction on the signals received from the PBX, depending on the current state of a signalling connection. Furthermore, the reaction could also depend on some special variables, e.g. the number of already repeated call attempts allowing to limit the number of repeated call attempts until a user would give up.

SDL is used to specify the User Model FSMs for several reasons. SDL is widely used in the software developing process in the telecommunication sector. Furthermore, commercial SDL tools can be used to create SDL specifications and to verify that their syntax and the static semantics are correct.

The load test system transforms the SDL specifications of User Model FSMs into data structures. During the execution of a load test these data structures are interpreted by the runtime system of the load test system. The data structures address implemented C procedures whose implementations are mostly short and simple. The main advantage of this approach in contrast to the generation of C code by the aformentioned SDL tools is the flexibility concerning the integration of new resp. the adaptation of existing User Model FSMs without a compilation process. Furthermore, the C procedures are part of the load test system and are therefore optimized concerning the performance, so that the real time requirements are fulfilled even if a large number of User Model FSM instances is active simultaneously. For complexity and particularly for performance reasons only a subset of the complete SDL grammar is used for specifying the User Model FSMs.

2.2.1 Applied SDL subset. For specifying the functional behaviour of a User Model the SDL subset contained in Figure 4 is applied.

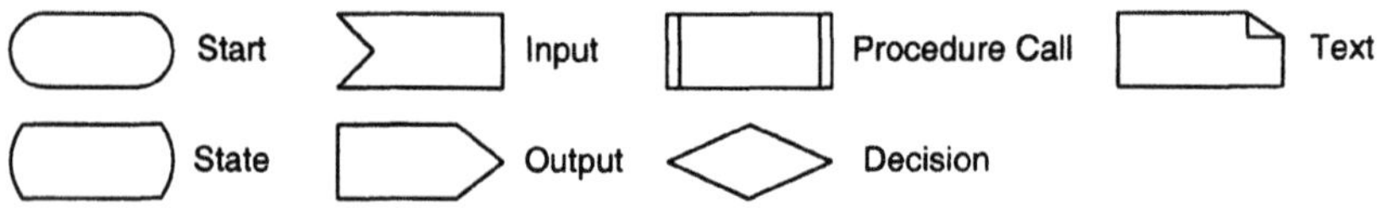

Figure 4 Applied SDL subset

The elements Start, State, Input, and Output are applied as usual in the context of User Model FSMs. The meaning of the other elements of the SDL subset is as follows:

Procedure Call: This element is applied for calling specific procedures. The indicated procedure name addresses a compiled procedure. There is a set of compiled procedures for the specification of the User Model FSMs, which can easily be expanded.

Decision: The decision statement allows to decide which branch of a transition has to be executed. The decision is performed by a procedure. The procedure name in this SDL statement addresses, as in the Procedure Call statement, a compiled procedure. Depending on the return value of this procedure, a branch of the transition is chosen for further execution.

Text: The text element is used for defining variables for the User Model FSM. Within the text symbol the variables are defined by applying the SDL DCL statement ("Declare"). Only the data types defined for the User Model FSMs can be utilized, but new data types can easily be added. The processing of variables of these types is performed in the compiled procedures which can be called by means of the Procedure Call statement. The utilization of variables within User Model FSMs is explained in detail in the following section.

2.2.2 Procedures and variables. The use of procedures and variables in the specification of a User Model FSM allows to introduce more sophisticated features. They permit to hide complex algorithms behind a simple SDL Procedure Call, e.g. for determining the current location of an emulated DECT user. Furthermore, it is possible to define application dependent data types, e.g. for the ISDN Signalling Information Elements, which are used within the ISDN signalling for negotiating the call parameters of an ISDN call.

To realize procedures and variables for User Model FSMs, there is a supporting library, the "FSM Runtime Library". This library contains the definitions of data types, which can be used within the SDL specification of User Model FSMs. Each data type of a User Model FSM has to correspond to a definition included in the FSM Runtime Library. Furthermore, the library contains the procedures applied in the SDL Procedure Call statements and the SDL Decision statements. The library is implemented in the C programming language and can be expanded by applying some simple interface rules. The relation between the SDL specification of a User Model FSM and the FSM Runtime Library is depicted in Figure 5.

A special data type is the one representing ISDN Signalling Information Elements. These Information Elements are used for specifying the parameters of an ISDN connection and for the application of supplementary services. They are realized within the User Models so that different services can be addressed without changing the implementation of the Signalling FSM.

2.2.3 Interfaces. This section describes the interfaces of a User Model FSM and particularly goes into detail to the relation between User

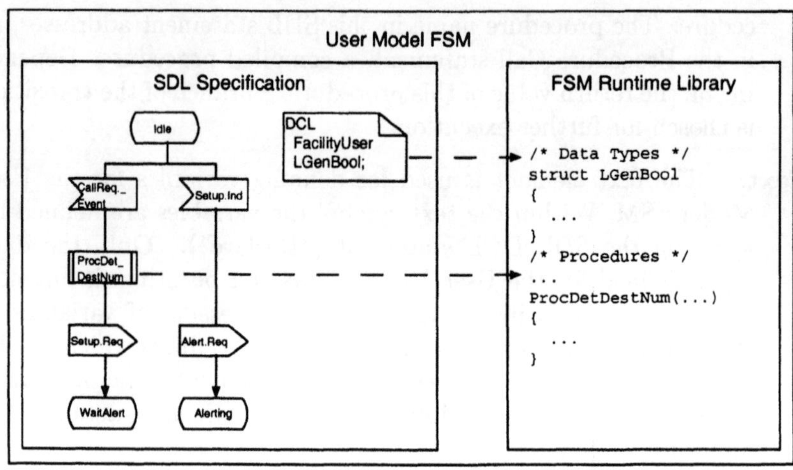

Figure 5 Relation between SDL specification and FSM Runtime Library

Model FSM and Signalling FSM. As described in the previous section, a User Model FSM disposes of interfaces based on messages to the Event Generators and to the Signalling FSM. The messages which are sent by the Event Generators are called "Events" and, therefore, are denoted in an SDL input statement with the suffix Event. The SDUs exchanged between User Model FSMs and Signalling FSMs follow the principle of the Abstract Service Primitives (ASP) defined in the ISO/OSI Basic Reference Model [6]: SDUs sent by the User Model FSMs to the Signalling FSMs have either the suffix Req or Resp, those sent by Signalling FSMs to User Model FSMs either Ind or Conf.

Figure 6 contains a signalling sequence diagram of a simple call with all signals exchanged between Event Generators and User Model FSM, User Model FSM and Signalling FSM and the resulting Signalling PDUs.

As it can be seen in Figure 6, a User Model FSM has a detailed view on the underlying Signalling FSM. However, the Signalling FSM contains all protocol internal procedures, like Timers or error recovery procedures. Therefore, and because of the protocol independent SDUs the User Model FSMs represent a high level of abstraction concerning the applied signalling protocols.

There are many SDUs sent by a Signalling FSM to a User Model FSM indicating a received layer 3 Signalling PDU which do not require a reaction of a User Model FSM, like MoreInfo.Ind in Figure 6. A User Model FSM can ignore these SDUs which results in simple User Model FSMs.

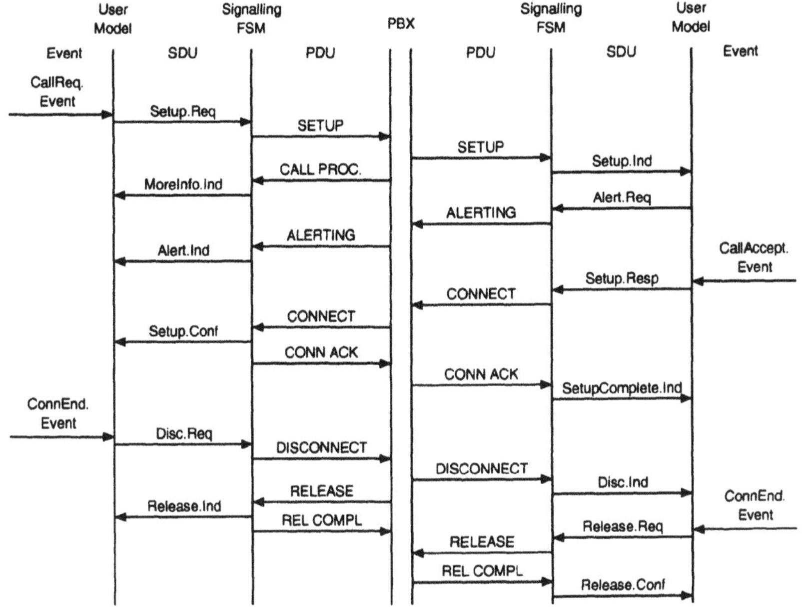

Figure 6 Complete signalling sequence diagram of a simple call

2.3. SPECIFICATION OF THE TEMPORAL BEHAVIOUR

As described in section 2.1, the specification of User Models for emulating real user behaviour is separated into two sections: The functional behaviour which indicates what an emulated user does, specified by means of an SDL process, and the temporal behaviour which defines when an action of a User Model has to be performed.

The realization of the method for specifying the temporal behaviour of a User Model is based on Event Generators. For each input signal of a User Model FSM of type Event there is an Event Generator defined. An Event Generator models a stochastic process by generating Events whose interarrival times are statistically distributed according to a defined distribution density function. The load test system disposes of an expandable set of distribution density functions, which are applied by the Event Generators for generating Events.

One purpose of the load test system is to generate statistically correct signalling traffic. If we analyse again the phases of a phone call illustrated in Figure 1, it can be seen, that only two types of Event Generators are sufficient to describe real user behaviour in an appropriate way: The "State Dependent" Event Generators which are related to a state of a User Model and the "Rate" Event Generators which are used for specifying the frequency of occurrence of an Event, independent from the state of the User Model. These types are presented in the following sections.

2.3.1 State Dependent Event Generator. Event Generators of the State Dependent type are related to the state of a User Model FSM, where the respective Event is specified as input signal. This means they are only active during the time the User Model FSM is in the respective state. Each time a state is entered, the respective Event Generators are started. This means, they determine the time when the Event should be delivered to the User Model FSM according to the configured distribution density function. If this time is reached, the Event will be delivered and the Event Generator will be stopped. If another input signal causes a transition of the User Model FSM, the Event Generator will be stopped before the Event will be delivered. An example for an Event generated by a State Dependent Event Generator in Figure 7 is the initiation of the "User Action" for accepting a call.

2.3.2 Rate Event Generator. In contrast to the State Dependent Event Generators are the Rate Event Generators not only active during a state of a User Model FSM. They are active as long as the corresponding User Model is active. This means they generate continuously Events, independently of the current state of the User Model FSM. This type of Event Generator is applied for specifying the call attempt rate of a User Model.

An Event of a Rate Event Generator is delivered to the User Model FSM at the time determined according to the configured distribution density function. But in contrast to the State Dependent Event Generator the time for the next Event is determined immediately after the delivery. If a User Model FSM cannot process an Event generated by a Rate Event Generator, because this Event is not specified as an input signal of the current state, the Event will be buffered until it can be processed. The motivation for this approach is that the load test system must be capable to drive the PBX into saturation. Some inaccuracies concerning the adjusted rate of Events can be tolerated. Normally, the configuration of the Event Generators of a User Model should be coordinated in that way, that no or only very short buffering is necessary.

3. LOAD TEST SYSTEM

The load test system "LGen" is implemented in software and runs on standard workstations. In the PBX, the real Interface Control Units

(ICUs) which terminate subscriber and trunk lines are replaced by a LAN ICU (LICU) which emulates Virtual ICUs (VICUs) on behalf of LGen. The messages from LGen are the same as those from real ICUs, thus the switching software running on the Generic Control Unit (GCU) cannot distinguish between real ICUs and VICUs. The LICU forwards these messages between the LAN and the PBX internal CBus.

The switching fabric is also involved in tests because the switching software requests B-channels to be through-connected.

3.1. ARCHITECTURE

This section presents the architecture of the LGen software, which is depicted in Figure 7. The architecture consists of the following building blocks:

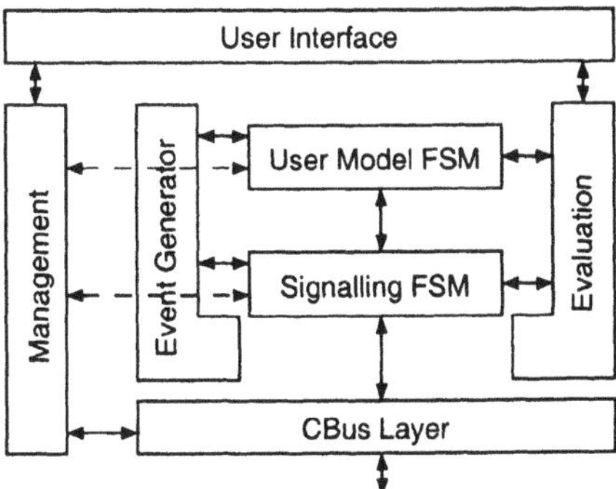

Figure 7 Software architecture of LGen

User Model FSM: This block contains an instance of a User Model FSM for each emulated user. The FSM Runtime Library is used for execution of the SDL processes. Each instance of a User Model is connected with an instance of a Signalling FSM. User Models communicate with these realizations of layer 3 protocols of ISDN signalling systems by means of protocol-independent SDUs. Within a load test multiple User Model FSMs can be applied simultaneously.

Signalling FSM: The block Signalling FSM contains the realizations of layer 3 protocols of different ISDN signalling systems. The messages exchanged between this block and the CBus Layer are Signalling PDUs, which are built according to the respective standard. The Signalling FSMs are implemented in the same way as the User Model FSMs, i.e. they are specified in SDL by applying the same method as described in section 2.2. LGen transforms these specifications into optimized data structures which are executed during a load test. Only the FSM Runtime Library differs slightly from the one of the User Model FSMs. This results in a higher flexibility concerning the supported signalling systems. As for the User Models it is also possible to run multiple Signalling FSMs with different layer 3 protocols simultaneously.

Event Generation: This block contains the Event Generators for the User Model FSMs generating Event messages according to defined distribution density functions. Furthermore, the Timers for the Signalling FSMs are realized within this block. To deliver the Event and Timer messages at the specified time, a Calendar module is applied. This Calendar is realized in a similar way than those for event-driven simulations, but in the case of LGen the messages are processed in real time.

Evaluation: This block contains modules for evaluating a load test. The required information is provided by the User Model FSMs and particularly by the Signalling FSMs. One type of information are the measurement values concerning signalling connections determined by the Signalling FSMs, like e.g. "Setup Answer Delay" or "Number of Accepted Setup". These values are passed to the Evaluation block, to determine results as e.g. mean value and standard deviation. Another type of information are the exchanged messages itself and the time when they were occurring. Therefore, SDUs and PDUs can be recorded during a load test in a file. Because of the large amount of recorded data it is possible to filter this message trace by user-selectable filters. This allows to examine in detail the exchanged messages or to determine the signal flow delay from one emulated user to another emulated user.

CBus Layer: The purpose of the CBus Layer is to put the received layer 3 PDUs in CBus messages and to send them through a TCP connection to the LICU. In the other direction it has to extract layer 3 PDUs from CBus messages received from the LICU. To perform these actions, the CBus Layer disposes of a table assigning CBus addresses of VICUs and ports to emulated users. A second task of the CBus Layer is to exchange the control messages between LGen and LICU for activating and deactivating VICUs and ports.

Management: The management block contains the modules for managing the entities of LGen. It administers the emulated users, i.e. it activates and deactivates the respective User Model and Signalling FSM instances, according to the user-defined test sequence configuration. Furthermore, it initiates the activation and deactivation of VICUs and controls the evaluation, e.g. by setting filters for the message trace.

User Interface: This block allows the interaction between the tester and LGen during a load test. A graphical user interface, which can be connected to this module, allows to initiate actions, like activating or deactivating emulated users, starting and stopping a load test or setting of filters for the message trace. Furthermore, it displays intermediate results of the evaluation of measurement values.

The communication between these blocks is mainly based on the exchange of LGen internal messages. These messages dispose of a fixed header containing information like sender and receiver address and message type. Depending on the type of the message there are supplementary message parts, like a complete Signalling PDU. This message based communication allows to decouple the execution of the different signal processing entities.

To take advantage of running LGen also on multiprocessor workstations, the developed program uses multiple threads to process the messages concurrently. The model applied is very similar to the "peer model" (also known as "workcrew mode"), where all existing threads work concurrently on their task, i.e. process LGen internal messages. This load sharing principle allows LGen to adapt to the underlying hardware platform in an optimized way. For a further description of thread based programming we refer to [7] and [8].

3.2. BASIC TEST SCENARIO

This section describes how the signalling traffic generation is realized with the architecture presented in the previous section. This is done by explaining how a load test of LGen takes place and how the blocks presented above interact during a load test.

After LGen is started, the Management is responsible for the control of a load test. It initiates the activation of the VICUs by sending a message to the CBus Layer. The CBus Layer exchanges then the control messages with the LICU through a TCP connection to activate the VICUs and ports. The activated ports are announced to the Management, so that only those users are activated, whose ports are already active.

The Management activates the users according to the test sequence configuration, which is provided at the configuration phase of LGen. To activate a user, the Management sends a message to the User Model FSM instance and to the respective Signalling FSM instance. The User Model

FSM instance starts the Rate Event Generators, performs the specified start transition and starts the State Dependent Event Generators of the first state. The Signalling FSM instance performs also the start transition.

If a User Model FSM instance receives an Event, it stops all active State Dependent Event Generators, performs the specified transition to the next state and starts then the respective State Dependent Event Generators. Normally, during a transition caused by an Event, an SDU is sent to the related Signalling FSM instance. Upon reception of an SDU the Signalling FSM instance performs the specified transition. If there is a PDU to send, the Signalling FSM instance builds the PDU, as defined in the respective standard, and sends it to the CBus Layer. The CBus Layer builds a CBus message containing the PDU and, among other information, the CBus address of the port, where the emulated user is connected. This CBus message is transmitted by means of TCP to the LICU, which sends the message via the CBus to the GCU, where it is processed. For the GCU there is no difference between CBus messages created by LGen and those of real ICUs.

The GCU transmits the messages for the ICUs (virtual and real ones) on the CBus. The LICU recognizes those messages which have to be forwarded and sends them through TCP to LGen. At LGen they arrive at the CBus Layer, which extracts the PDU, determines the emulated user by means of the CBus address and sends the PDU to the corresponding Signalling FSM instance. The Signalling FSM instance performs the transition which is specified for this PDU. If there is no transition specified, it will be ignored and the Signalling FSM instance remains in the current state. In most cases a PDU causes an SDU to be sent to the User Model FSM instance. The User Model FSM instance checks whether a transition is specified for this SDU. Only if there is a transition, the State Dependent Event Generators are stopped, the specified transition is performed and the State Dependent Event Generators of the new state are started.

To all SDUs and PDUs exchanged within LGen the filter rules of the message trace are applied. If a message has to be recorded, it is sent to the Evaluation, where the message and the current time are written to a file. The Signalling FSM instances perform the measurements during a signalling connection. At the end of the connection they send the measured values to the Evaluation, where the statistical evaluation takes place.

Besides the use of the test sequence configuration, it is also possible to control a load test by means of the User Interface. Upon user action the User Interface sends a message to the Management which interprets the message. It then determines the destination of the request and composes the appropriate internal message for e.g. the activation of emulated users, the setting of filters of the message trace or the stop of the load test. During the load test the Evaluation includes intermediate results in messages and sends them to the User Interface, where these results are displayed in an appropriate manner.

As described, LGen emulates real user behaviour as specified by means of User Models. As LGen is capable to process a large number of emulated users simultaneously with very low latency, the superposition of the signalling messages represents realistic traffic load. The concept of activating and deactivating emulated users is a natural way for emulating load peaks. It is sufficient to activate at a time several users, whose User Models are specified in that way that they start to initiate a call within a short time frame.

4. CONCLUSION

The LGen load test system is used mainly in the development departments and in the test floor. The first application was to verify that the existing overload protection mechanism worked as expected and to test an enhanced overload protection mechanism. In real systems, the observed fluctuations in call arrivals and other durations are due to the statistical nature of real traffic sources. LGen has demonstrated that it is important for meaningful load tests to use realistic traffic instead of using call generators which can only produce deterministic arrival patterns.

The target rate of 30000 calls per hour is clearly exceeded by LGen. The processor load is then still low which makes sure that the simulated behaviour conforms to the expected distributions. On a Sun Ultra 1 workstation, call rates of more than 100000 calls per hour can be reached. This very high call rate is useful for stress tests, even if the simulated timing behaviour deviates from the target distributions. This deviation is due to the non-negligible delays at LGen internal queues when the processor load becomes high.

A second important application area for LGen are call centers and other CTI (Computer Telephony Integration) applications. There are two main reasons why call centers have to be tested under high loads. First, some branches in the call flows which are executed by the external call distribution application are only used when the load is high. An example is a call flow which connects callers to an announcement for the estimated waiting time until an agent will answer the call. Another motivation for tests under high load is the experience that multiple systems from different vendors like in a call center do not always interact smoothly, especially if some systems operate close to their maximum load. LGen is used now in several areas in the context with call centers, and it has also been possible to demonstrate the performance of an operational call center at a customer site.

Work has started now to extend LGen to use it in the DECT context. LGen will be used to generate DECT-specific load peaks which are caused by high call rates and also high roaming and handover rates. A typical situation where such a scenario occurs is at the end of a big meeting when all participants leave the meeting room and many of them try to make calls.

LGen has proven to be a useful tool for tests during development and in the integration test floor. The key features are the realistic generated traffic and the flexible modelling of user behaviour. Both features result from the approach to model the behaviour in SDL and to use Event Generators for the execution in real-time.

References

[1] ITU-T Recommendation Z.100: SDL+ methodology: Use of MSC and SDL (with ASN.1)

[2] Mitschele-Thiel, Müller-Clostermann: Performance engineering of SDL/MSC systems, Computer Networks 31 (1999)

[3] Lemppenau, Tran-Gia: A Universal Environment Simulator for SPC Switching System Testing, Proceedings 11th International Teletraffic Congress (ITC), Kyoto 1985

[4] ITU-T Recommendation Q.931: ISDN user-network interface layer 3 specification for basic call control

[5] FTZ Richtlinie 1TR6: Kennzeichenaustausch zwischen DIVO(ISDN)-Vermittlungsstellen und ISDN-Teilnehmereinrichtungen – ISDN-D-Kanal-Protokoll

[6] ISO 7498, Basic Reference Model for Open Systems Interconnection

[7] IEEE 1003.1c-1995, Standard for Information Technology, Portable Operating System Interface (POSIX): System Application Program Interface (API) Amendment 2: Threads Extension (C Language)

[8] Nichols, Buttlar, Farrell: Pthreads Programming, O'Reilly&Associates, 1996

Part Six

EXHAUSTIVE AND PROBABILISTIC TESTING

SATISFACTION UP TO LIVENESS

Verification based on Exhaustive Testing

Ulrich Ultes-Nitsche
University of Southampton
Department of Electronics and Computer Science
Southampton, SO17 1BJ, United Kingdom
uun@ecs.soton.ac.uk

Abstract An approximation to the usual linear satisfaction of temporal properties is discussed in this paper. It is called *satisfaction up to liveness* as it only differs from the linear satisfaction relation on liveness but not on safety properties. From the point of view of observation, satisfaction up to liveness and linear satisfaction are indistinguishable. Roughly speaking, by observing all *finite* behaviours (exhaustive testing), we do not know whether a system satisfies a property up to liveness or linearly. Being indistinguishable from linear satisfaction in terms of complete observations, satisfaction up to liveness offers an alternative approach to model-checking.

Keywords: Model-Checking, Exhaustive Testing, Fairness, Safety & Liveness.

1. INTRODUCTION

One checks the satisfaction of linear-time temporal properties by checking whether the set of behaviours of a system is a subset of the property [3]. Both, the behaviour and the property, are ω-languages (languages of infinite word-length [15]). Practically, these ω-languages are derived from descriptions written in a high-level specification language. To describe properties, for instance, linear-time temporal logics [7] are very frequently used. Set-inclusion is decidable for regular ω-languages (ω-languages representable by finite-automata [15]). However it is more complex than for regular languages (with finite word-length) [15], mainly because the class of regular ω-languages accepted by deterministic Büchi-automata [5; 15] is only a proper subclass of the regular ω-languages accepted by non-deterministic Büchi-automata.

Tommaso Bolognesi and Diego Latella (Eds.), *Formal Methods for Distributed System Development.*

Using the notion of *relative liveness* [9; 14] and viewing it as a satisfaction relation for temporal properties [13; 14] allows to reduce the satisfaction relation to deciding set-inclusion for finitary languages (i.e. languages of finite word-length). Obviously, by turning to considering finitary languages, we lose some information. It will be argued that the resulting *satisfaction up to liveness* relation, as it will be called, does not ignore any practically important information about the system. To underpin the arguments, different aspects of satisfaction up to liveness will be discussed in order to document the usefulness of the general concept. The main point in the argumentation will be that, by observing all finite behaviours of a system (i.e. by testing it exhaustively), linear satisfaction and satisfaction up to liveness cannot be distinguished from one another.

The paper is structured as follows: It starts with a first definition of satisfaction up to liveness in Section 2. Since throughout this paper only languages and ω-languages appear, but no state-based representations like automata is used, Section 3 contains an outline of how the local language represents an automaton entirely. Section 4 reveals more about the structure of satisfaction up to liveness and how it relates to truth under fairness. Section 5 explains that satisfaction up to liveness is in fact an approximation to the linear property satisfaction. The formal apsects of satisfaction up to liveness and their practical consequences are summarised in Section 6. Section 7 concludes the paper, including an outlook on future work.

2. THE BASIC DEFINITION OF SATISFACTION UP TO LIVENESS

Before introducing the basic definition of satisfaction up to liveness, we review linear satisfaction of properties briefly. Let $\mathcal{A}$ be the finite set of actions that a system can perform. By $\mathcal{A}^\omega$ we denote the set of all infinitely long sequences of actions in $\mathcal{A}$. A behaviour b of the system is an element of $\mathcal{A}^\omega$, i.e. $b = a_1 a_2 a_3 \ldots$ such that all a_i are in $\mathcal{A}$. The set B of all behaviours of the system is therefore a subset of $\mathcal{A}^\omega$.[1] A property P, from an abstract point of view, is itself a subset of $\mathcal{A}^\omega$, representing all correct behaviours with respect to the intuitive meaning of the property. Hence the system, represented by its set of behaviours B, satisfies property P (written: $B \models P$) if and only if all behaviours in B are correct with respect to P and thus are in P. This

[1] In terms of formal language theory, b is an ω-word and B is an ω-language on the alphabet $\mathcal{A}$.

linear satisfaction of properties is defined by

$$B \models P \iff B \subseteq P.$$

Since B can only be a subset of P if $B \cap P = B$, we can rewrite the above definition as

$$B \models P \iff B = B \cap P.$$

Let $\mathcal{A}^*$ denote the set of all finitely long sequences of actions in $\mathcal{A}$. For a subset S of $\mathcal{A}^\omega$ let $pre(S)$ denote the set $\{w \in \mathcal{A}^* \mid \exists x \in \mathcal{A}^\omega : wx \in S\}$ of finite prefixes of behaviours in S. For an element s of $\mathcal{A}^\omega$, let $pre(s) = pre(\{s\})$. Instead of saying that B satisfies P if and only if B is equal to $B \cap P$, we relax this definition to requiring only that the finite prefixes of B have to equal the finite prefixes of $B \cap P$. This relaxation leads us to the definition of satisfaction up to liveness. We write $B \Vdash P$ for saying B satisfies P up to liveness. We define:

$$B \Vdash P \iff pre(B) = pre(B \cap P).$$

What can we see by looking at this definition? Firstly, $B \models P$ always implies $B \Vdash P$ but not vice versa. An example for $B \Vdash P$ not implying $B \models P$ is $B = \{a, b\}^\omega$ and $P = \{a, b\}^* \cdot \{b\} \cdot \{a, b\}^\omega$.[2] Secondly, by observing systems, we cannot distinguish between a system satisfying a property linearly and a system satisfying the same property up to liveness: If we monitor all actions a system performs for an arbitrary period of time, we always see a (prefix-closed) set of prefixes of B. Since all these prefixes are prefixes of $B \cap P$, both, when P is satisfied by B up to liveness or linearly, linear satisfaction and satisfaction up to liveness give us the same view on the observable behaviour of a system.

In other words: From a set of behaviours B satisfying a property P up to liveness but not linearly, we can derive a set of behaviours B' by removing all behaviours from B that do not satisfy P linearly. B' satisfies P linearly. But by observing their finite behaviours, B cannot be distinguished from B', i.e. $pre(B) = pre(B')$. Their differences could only be detected by observing systems for an infinite amount of time, being by that practically undetectable. An observer would consider a system with behaviour B to be equivalent to a system with behaviour B' even though they are distinguishable by the linear satisfaction relation. Since we would have to observe a system's finite behaviour completely to determine whether it satisfies a property up to liveness, we can say that satisfaction up to liveness is based on exhaustive testing.

[2] P is in this example the temporal property "eventually b"; in temporal logic $P \equiv \Diamond b$.

Because the test whether $pre(B) = pre(B \cap P)$ can be reduced to checking whether $pre(B) \cap (\mathcal{A}^* \setminus pre(B \cap P))$ is empty, and the emptiness test as well as intersection and complementation is decidable for regular languages/ω-languages [6; 15], this give us the algorithm to check satisfaction up to liveness automatically (model-checking) [7].

3. HOW THE LOCAL LANGUAGE ENCODES AUTOMATA

In this paper, we always consider languages and never talk about automata (with this section being an exception from the rule). That restriction seems to be relatively far away from practical systems where one would talk of events, states, and so on. The definitions of this paper, however, can be formulated more concisely on languages. A very common view is that by considering languages one loses (state-) information compared to the automaton level. In fact, this view is not correct as an automaton can be characterised completely by its *local language* [6]. We explain this argument briefly in this section.

Consider a finite automaton with Q being its set of states, Σ being its alphabet, $\delta \subseteq Q \times \Sigma \times Q$ being its transition relation, $q_0 \in Q$ being its initial state and $F \subseteq Q$ being its set of final states. A finite automaton accepts sequences of letters in Σ that lead it from its initial state to some final state only taking transitions in δ. For a sequence of letters, we usually do not know in which state we may end up (due to non-determinism). So, by looking at the language, we lose state information. If, instead of keeping track of the letters that are the transition labels, we keep track of the triples *(state, letter, successorstate)*, we do not lose state information. The language consisting of transition triples is called the local language of the automaton. It can be represented by the same automaton except that the alphabet becomes $\Sigma' = \delta$ and the transition relation becomes $\delta' \subseteq Q \times (Q \times \Sigma \times Q) \times Q$ such that $(q, (q, a, p), p)$ is in δ' if and only if (q, a, p) was in δ. The so constructed automaton accepts exactly the local language of the former one.

From its local language a (reduced[3] non-empty) automaton can be reconstructed entirely. From the transition triples that occur as letters in the local language we obtain δ, Q, and Σ. The initial state q_0 is the first component of the leftmost transition triple of any word in the local language. The set of final states contains all right components of righmost transition triples of words in the local language. For instance, from

[3]reduced = all states are reachable from the initial state and from all states a final state is reachable.

knowing that $(q,a,p)(p,b,p)(p,a,r)$ is a word in the local language of an automaton, we learn that q, p, and r are some of its states, a and b are letters in its alphabet, (q,a,p), (p,b,p), and (p,a,r) are transitions of the automaton, q is its initial state, and r is one of its final states. Applying the projection homomorphism, that projects on the middle component of transition triples, to the local language gives us the language accepted by an automaton.

So, in fact, considering only languages in this paper is not a restriction. By considering the local language, we can take the concepts that we develop in this paper to the automaton level, without having to do any proofs on automata, but only more concisely on languages.

4. SATISFACTION OF THE PROPERTY IS ALWAYS POSSIBLE

In this section, we review the definition of satisfaction up to liveness. By giving an alternative, equivalent definition to the one in Section 2, we explore some more aspects of properties satisfied up to liveness. This second definition occurs under the name *relative liveness* in different contexts in [4; 9; 14], where it is mainly a derivate of *machine-closed structures* [1; 2] to classify properties with respect to other properties. A dual type of properties, *relative safety* properties can also be defined [9; 14]. In [14] relative liveness is used as a satisfaction relation as well as in [13], where it is named approximate satisfaction.

Considering our first definition of satisfaction up to liveness,

$$B \Vdash P \iff pre(B) = pre(B \cap P),$$

we recognise that $pre(B \cap P)$ is always a subset of $pre(B)$, because $B \cap P \subseteq B$. Therefore

$$B \Vdash P \iff pre(B) \subseteq pre(B \cap P),$$

emphasising that $B \Vdash P$ states something about the finite prefixes of behaviours in B and their possible infinite continuation. Before being able to define what that "something" is that $B \Vdash P$ says about finite prefixes of B, we have to formalise the continuation of finite prefixes.[4] Let $b \in \mathcal{A}^*$ (remember that $\mathcal{A}$ is the set of actions occuring in the behaviours in B, i.e. $B \subseteq \mathcal{A}^\omega$). Then

$$cont(b, B) = \{x \in \mathcal{A}^\omega \mid bx \in B\}.$$

[4]To define the contiunation of finite prefixes, we borrow the notion *left-quotient* of a language by a word from formal language theory [6]. We use the different notation $cont(w, L)$ rather than $w^{-1}(L)$ because it appears to be more intuitive in our context.

So x is in $cont(b, B)$ if an only if b can be continued with x resulting in a behaviour bx that is in B. Using this notation, we re-define satisfaction up to liveness and prove the new definition to be equivalent to the first one.

Lemma 1

$$B \Vdash P \iff \forall b \in pre(B) : \exists x \in cont(b, B) : bx \in P.$$

Informally this lemma states that a behaviour satisfies a property up to liveness if and only if "all finite behaviours can be continued to an infinite behaviour satisfying the property".

Proof *If $pre(B) \subseteq pre(B \cap P)$, then all b in $pre(B)$ are also in the set $pre(B \cap P)$. Therefore, for all $b \in pre(B)$, there exists an $x \in cont(b, B)$ such that $bx \in P$. If for all $b \in pre(B)$ there exists $x \in cont(b, B)$ such that $bx \in P$, then $b \in pre(B \cap P)$ for all $b \in pre(B)$ and thus $pre(B) \subseteq pre(B \cap P)$.* □

To be able to explain why the discussed satisfaction relation is called satisfaction *up to liveness*, we have to review the definition of safety and liveness properties [3], adapting it to the notation used in this paper. Note that any property is the intersection of a safety and a liveness property [3].

A property $P \subseteq \mathcal{A}^\omega$ is a safety property if and only if

$$\forall b \in \mathcal{A}^\omega : b \not\models P \Rightarrow \exists v \in pre(b) : \forall c \in \mathcal{A}^\omega : vc \not\models P.$$

A property $P \subseteq \mathcal{A}^\omega$ is a liveness property if and only if

$$\forall v \in \mathcal{A}^* : \exists c \in \mathcal{A}^\omega : vc \models P.$$

We show that for safety properties satisfaction up to liveness and linear satisfaction coincide:

Lemma 2 *If $P \subseteq \mathcal{A}^\omega$ is a safety property, then for all $B \subseteq \mathcal{A}^\omega$:*

$$B \Vdash P \iff B \models P.$$

Proof *Because $B \models P$ always implies $B \Vdash P$, we only have to prove that $B \Vdash P$ implies $B \models P$. Recall that, if $B \Vdash P$, then $\forall b \in pre(B) : \exists x \in cont(b, B) : bx \in P$. Assuming $B \models P$ does not hold, we show a contradiction to $B \Vdash P$.*

Assume $B \not\models P$. Then, because P is a safety property, there exists $v \in pre(B)$ such that $vc \not\models P$, for all $c \in \mathcal{A}^\omega$. So, in particular, we have $vc \not\models P$, for all $c \in cont(v, B)$. Since v is in $pre(B)$, this implies that

$\forall b \in pre(B) : \exists x \in cont(b, B) : bx \in P$ does not hold, contradicting that $B \Vdash P$. □

So, for safety properties, there is no difference between satisfaction up to liveness and linear satisfaction. Only when checking liveness properties has one to distinguish between the two concepts. This fact led to the name *up to liveness*, stating that for checking whether a liveness property is satisfied linearly, knowing that it is satisfied up to liveness, we would have to perform additional checks.

5. SATISFACTION UP TO LIVENESS APPROXIMATES LINEAR SATISFACTION

To make the point again that the only difference between the usual linear satisfaction relation and satisfaction up to liveness can be revealed if one observes a system for an infinite amount of time, it is not surprising that satisfaction up to liveness can be regarded as an approximation to linear satisfaction. To make this precise, we present the characterisation of satisfaction up to liveness within the Cantor topology, i.e. the topological space over Σ^ω compatible with the following metric [6]. (For topological notions see [10].)

Definition 3 *Let $common(a, b)$ designate the longest common prefix of two behaviours a and b in $\mathcal{A}^\omega$. We define the metric $d(a, b)$ by*

$$\forall a, b \in \mathcal{A}^\omega, a \neq b : d(a, b) = \frac{1}{|common(a, b)| + 1}$$

$$\forall b \in \mathcal{A}^\omega : d(b, b) = 0.$$

What does it mean in the given context that a satisfaction relation approximates another one? It means that the correctness criterion of the first satisfaction relation must imply that all behaviours of the system are arbitrarily close to meeting the second satisfaction relation's correctness criterion. Being arbitrarily close to meeting the correctness requirement must be translated into, in our context, that the correct behaviours of the system must be dense within the set of all behaviours. We can indeed derive this fact from knowing that a property is satisfied up to liveness, as stated by the following lemma [14]:

Lemma 4 *Let $B \subseteq \mathcal{A}^\omega$ and $P \subseteq \mathcal{A}^\omega$ be the set of behaviours of a system and the property it should meet, respectively. Then $B \Vdash P$ if and only if $B \cap P$ is a dense set in B.*

Proof *Let $B \Vdash P$ and let $b \in B$. Then $pre(B) = pre(B \cap P)$. Thus, $pre(b) \subseteq pre(B \cap P)$, implying $\forall w \in pre(b) : \exists a \in B \cap P : w \in pre(a)$. So, for all $b \in B$ and all $\varepsilon > 0$ (ε is related to $\frac{1}{|w|+1}$), there is $a \in B \cap P$ such that $d(a,b) < \varepsilon$. So $B \cap P$ is a dense set in B.*

Let $B \cap P$ be a dense set in B. Then, for all $b \in B$ and all $\varepsilon > 0$, there exists $a \in B \cap P$ such that $d(a,b) < \varepsilon$. Let $w \in pre(b)$ and $\varepsilon = \frac{1}{|w|+1}$. Let $a \in B \cap P$ such that $d(a,b) < \varepsilon$. Then $w \in pre(a)$. Because w was chosen arbitrarily, $pre(B) \subseteq pre(B \cap P)$, implying $pre(B) = pre(B \cap P)$. So $B \Vdash P$. □

If we agree that a coarser satisfaction relation is, from a practical point of view, as good as a more exact one, if we could only distinguish the two relations by observing a system for an infinitely long time, then we can define:

Definition 5 *A satisfaction relation is a reasonable satisfaction relation for linear temporal properties whenever the satisfaction of property P by behaviour B implies that $B \cap P$ is dense in B.*

The condition "$B \cap P$ is dense in B" just states that, if $B \cap P$ can be distinguished from B, then only by observing B until infinity. Obviously, linear satisfaction itself is *reasonable* with respect to the definition above: $B \models P$ is equivalent to $B \cap P = B$, which implies that $B \cap P$ is dense in B.

According to Lemma 4, satisfaction up to liveness is the weakest possible *reasonable* satisfaction relation for linear temporal properties: $B \vdash P$, for any *reasonable* satisfaction relation "$\vdash$", always implies $B \Vdash P$. So, alternatively, we could define a *reasonable* satisfaction relation to be one that implies $B \Vdash P$.

6. DISCUSSION

We discuss in this section whether satisfaction up to liveness offers a reasonable alternative to the usual satisfaction relation of properties that is called linear satisfaction in this paper. Not surprisingly is the author in favour of satisfaction up to liveness. However, we have to weigh the advantages and disadvantages at this point.

Obviously, satisfaction up to liveness and linear satisfaction are different satisfaction relations. But to distinguish a system that satisfies a property linearly from one that does only satisfy it up to liveness we may have to observe the two systems for an infinite amount of time. So even though there are differences these differences do not have an impact on the system behaviour as we will be able to observe it. If we cannot observe the difference, it appears that the difference between the two concepts does not matter in practical terms. By arguments given in Section 5 we may regard linear satisfaction as the strongest and satisfaction up to liveness as the weakest *reasonable* satisfaction relation for linear-time temporal properties.

The advantage of satisfaction up to liveness is that by its definition, language inclusion has only to be checked on the level of finite-word languages rather than on ω-languages as in the case of linear satisfaction. Satisfaction up to liveness comes with an exact notion of abstraction, namely weakly continuation-closed homomorphisms [14]. Linear satisfaction does not have such a clear and simple notion of abstraction.

The major disadvantage of satisfaction up to liveness is that it is not closed under intersection. If a system satisfies properties P and P' up to liveness, it does not necessarily satisfy $P \cap P'$ up to liveness. This fact states a lack of compositionality. But if either P or P' is a safety property, then intersection is preserved. In contrast, linear satisfaction is preserved under intersection in any case.

A big bonus of satisfaction up to liveness is that it includes an inherent fairness condition [14]. If a property is satisfied up to liveness then it will be satisfied linearly under an abstract notion of fairness that can be made concrete to *strong fairness* [8] for particular implementations of the system [14]. Being able to ignore extreme execution scenarios without having to deal with fairness explicitly is a definite advantage of satisfaction up to liveness, having a strong impact on applying model-checking to practical problems [12]. We consider here a very simple example taken from [14]:

Consider a system that on receiving a *request* may produce either a *result* of some sort or *rejects* the request. The decision is taken internally (*yes* or *no*) before its outcome is reported to the requesting client. The decision is dependent on the availability of a resource that is handled by a resource manager and may be either *locked* or *free*. The system is described as a Petri net in Figure 1.

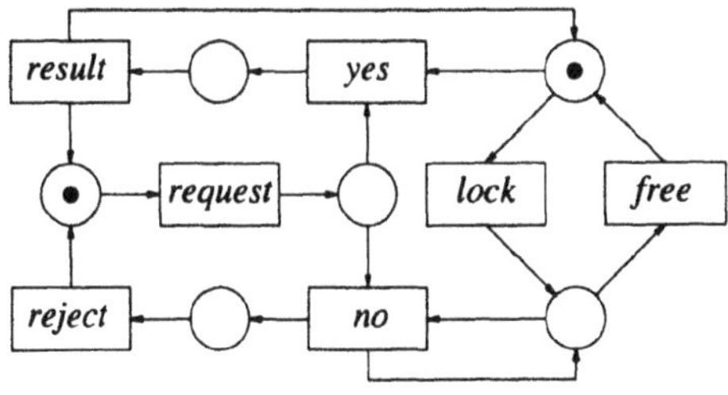

Figure 1 A small system

The possible behaviours of the system are represented by the finite labelled transition system shown in Figure 2 (the reachability graph of the Petri net). The initial state is shaded grey.

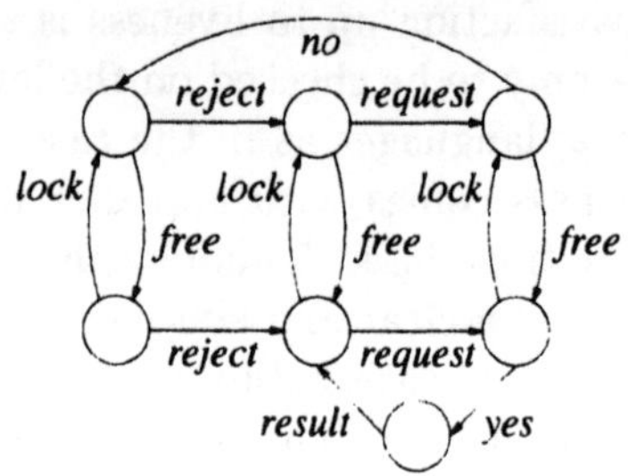

Figure 2 The behaviours of the small system

From Figure 2, it is easy to see that our system does not satisfy linearly the propositional linear-time temporal logic [7] property $\Box\Diamond(result)$. This property requires that infinitely often a result will be produced by the system. Let $\mathcal{A} = \{request, reject, result, yes, no, lock, free\}$. Then the language representation of this property is $\mathcal{A}^* \cdot (result \cdot \mathcal{A}^*)^\omega$.

The property is violated because, for the example, whenever a request is made the resource could be unavailable. For instance, $lock \cdot (request \cdot no \cdot reject)^\omega$ is such a behaviour violating the linear satisfaction of the property. However, such a behaviour is highly unfair [8] and the specification in Figure 1 itself is very reasonable. One way of dealing with this situation would be to add additional fairness constraints to the specification which would make the design process of the system more complicated. A less complicated way of dealing with it would be using the inherent fairness condition of satisfaction up to liveness.

The property $\Box\Diamond(result)$ is satisfied by the system of Figure 1 up to liveness, because satisfaction up to liveness ignores extreme execution scenarios such as $lock \cdot (request \cdot no \cdot reject)^\omega$. Quite frequently, in communication protocols and Intelligent Network service specifications, such "extreme execution scenarios" are part of the specification ("whenever A calls B, B's line is busy") but should be ignored by the satisfaction relation [12]. Satisfaction up to liveness does exactly this job without additional effort, making it useful in practice [12].

To summarise, satisfaction up to liveness is at least an alternative, worth considering way of checking properties of reactive systems that is observationally indistinguishable from linear satisfaction.

7. CONCLUSIONS AND FUTURE WORK

Different aspects of satisfaction up to liveness as an alternative satisfaction relation to the traditional linear satisfaction have been discussed in this paper. One of the major points is that systems satisfying a property up to liveness cannot be distinguished from systems satisfying the property linearly by observing (even all of) their finite behaviours. The major contribution of the paper is to give alternative presentations of satisfaction up to liveness in a unified framework and to interpret their mathematical foundation with respect to the observability of system behaviours. By this it is possible to compare satisfaction up to liveness with the usual concept of (linear) satisfaction, discussing their similarities and differences. In addition, a criterion for a relation to be a *reasonable* satisfaction relation for linear-time temporal properties has been given (again with respect to observability).

The concept of satisfaction up to liveness of non-safety properties is relatively close to linear satisfaction of safety properties (in the case of safety properties it is the identical concept). In both cases, one has to look at finitary languages (finitely long words). So in order to model-check satisfaction up to liveness, improvements might be gained by considering the particular aspects of model-checking safety properties [11]. Exploring possible improvements in that direction will be part of future work.

In this paper we were talking constantly about observability, knowing that the system satisfies a property (up to liveness). The notion of observing a system is understood by means of exhaustive testing. It will therefore also be part of the future work to explore in detail the link to testing and test case generation from specifications verified up to liveness.

Probabilistic satisfaction relations [17; 18] appear to lie somewhat in between linear satisfaction and satisfaction up to liveness. It will be part of future research to see how they could be embedded into the definition of a *reasonable* satisfaction relation as given in this paper.

References

[1] M. Abadi and L. Lamport. The existence of refinement mappings. SRC Report 29, DEC System Research Center, July 1988.

[2] M. Abadi and L. Lamport. Composing specifications. SRC Report 66, DEC System Research Center, October 1990.

[3] B. Alpern and F. B. Schneider. Defining liveness. *Information Processing Letters*, 21(4):181–185, October 1985.

[4] R. Alur and T. A. Henzinger. Local liveness for compositional modeling of fair reactive systems. In P. Wolper, editor, *Computer Aided Verification (CAV) '95*,

volume 939 of *Lecture Notes in Computer Science*, pages 166–179. Springer, 1995.

[5] J. Büchi. On a decision method in restricted second order arithmetic. In E. Nagel et al., editors, *Proceedings of the International Congress on Logic, Methodology and Philosophy of Science 1960*, pages 1–11. Stanford University Press, 1962.

[6] S. Eilenberg. *Automata, Languages and Machines*, volume A. Academic Press, New York, 1974.

[7] E. A. Emerson. Temporal and modal logic. In van Leeuwen [[16]], pages 995–1072.

[8] N. Francez. *Fairness*. Springer Verlag, New York, first edition, 1986.

[9] T. A. Henzinger. Sooner is safer than later. *Information Processing Letters*, 43:135–141, 1992.

[10] J. L. Kelley. *General Topology*. Van Nostrand, Princeton, 1955.

[11] O. Kupferman and M. Y. Vardi. Model-checking of safety properties. In N. Halbwachs and D. Peled, editors, *CAV'99*, volume 1633 of *Lecture Notes in Computer Science*, pages 172–183, Trento, Italy, 1999. Springer Verlag.

[12] U. Nitsche. Application of formal verification and behaviour abstraction to the service interaction problem in intelligent networks. *Journal of Systems and Software*, 40(3):227–248, March 1998. ISSN:0164-1212.

[13] U. Nitsche and P. Ochsenschläger. Approximately satisfied properties of systems and simple language homomorphisms. *Information Processing Letters*, 60:201–206, 1996. ISSN: 0020-0190.

[14] U. Nitsche and P. Wolper. Relative liveness and behavior abstraction (extended abstract). In *Proceedings of the 16th ACM Symposium on Principles of Distributed Computing (PODC'97)*, pages 45–52, Santa Barbara, CA, 1997.

[15] W. Thomas. Automata on infinite objects. In van Leeuwen [[16]], pages 133–191.

[16] J. van Leeuwen, editor. *Formal Models and Semantics*, volume B of *Handbook of Theoretical Computer Science*. Elsevier, 1990.

[17] M. Y. Vardi. Automatic verification of probabilistic concurrent finite-state programs. In *Proceedings of the 26th IEEE Symposium on Foundations of Computer Science*, pages 327–338, Portland, October 1985.

[18] M. Y. Vardi and P. Wolper. An automata-theoretic approach to automatic program verification. In *Proceedings of the 1st Symposium on Logic in Computer Science*, Cambridge, June 1986.

TESTING IP ROUTING PROTOCOLS – FROM PROBABILISTIC ALGORITHMS TO A SOFTWARE TOOL

Ruibing Hao, David Lee*, Rakesh K. Sinha, and Dario Vlah

Bell Laboratories, Lucent Technologies

Murray Hill, New Jersey 07974

**Bell Laboratories Research China*

Beijing, China

{rhao,lee,rks1,dv}@research.bell-labs.com

Abstract We present probabilistic algorithms for testing IP routing protocols. Different than commercial testing tools, which select their test cases in an ad-hoc manner, our technique chooses test cases with a guaranteed fault coverage. Our algorithms are applied to testing RIP, OSPF, and BGP routing protocols. We test the routing table correctness, database information consistency with network topology and packet forwarding behaviors. We provide an analysis of test execution time and fault coverage. The algorithms are part of a software tool SOCRATES, developed at Bell laboratories. SOCRATES also creates a testing environment so that the generated test cases can be executed in real time on high speed routers.

Keywords: IP protocol, RIP, OSPF, BGP, router fault detection, fault coverage

1. INTRODUCTION

Internet is a packet-switching network that enables its attached computers to exchange information by sending packets to each other. As packets are sent from source to destination, routing decisions are made by special purpose computers, called IP routers. To route packets correctly and efficiently, routers follow a set of rules, which are implemented as distributed algorithms and are called routing protocols [5]. New sophisticated functions and services, such as service differentiation, QoS routing and multicasting, are introduced on a regular basis. Consequently, the design and implementation of routing protocols become

Tommaso Bolognesi and Diego Latella (Eds.), *Formal Methods for Distributed System Development.*

more complex and less reliable. Testing is indispensable for the correct functioning and performance of the routers.

In this work we study testing of router's network layer IP routing protocols. We present algorithms and a software tool for testing Routing Information Protocol (RIP) [4], Open Shortest Path First (OSPF) protocol [7], and Border Gateway Protocol (BGP) [9]. Unfortunately, there is no precise specifications for these routing protocols. The only design requirements are RFCs (Requests for Comments), which tend to be incomplete. Furthermore, vendors often make their own implementation decisions for their routers, which are being deployed throughout the Internet. Consequently, it is out of question to test whether routing protocols are implemented in conformance to the "specification". On the other hand, there are commonly agreed upon "expected behaviors", such as correct routing table, packet forwarding, and network topology information, which are invariant with implementations. There is a practical need and also it is feasible to test whether a router behaves as expected.

Detecting faults in IP router has been an open challenge for both router vendors and network administrators. Most of the systems and tools for detecting router faults can be classified as either passive monitoring tools or active testing tools.

Passive monitoring tools such as RouteMonitor [6] deploy a collection of monitoring devices in a production network to observe the routing traffic exchanged between routers. These monitors do not inject any route update into the network. Based on the observed routing information, these monitors imitate the operation of a real router, execute the routing protocol and compute a routing table. Faults can be found through the comparison of routing tables and analysis of the statistics for each route in the routing table.

Active testing is a more effective approach for fault detection compared with monitoring. Most testing tools test the router in an isolated environment and check the conformance of router's behavior in respect to the RFCs. A main function of these tools is to generate a set of tests that exercise each requirement in the design/RFC for "typical" network configurations/topologies. Because it is computationally infeasible to consider all possible network configurations, any tool can only consider a small subset of all possible configurations. The existing commercial tools pick this subset in an ad-hoc manner so that even if a router passes all their tests, there is no guarantee of the fault coverage.

Our technique is also active testing. The key strength of our approach is that we employ a probabilistic algorithm to select a subset of network configurations. This has several benefits:

1 We can guarantee a high fault coverage.

2 The existing commercial tools use a *fixed* set of network configurations. So if the fault is outside their chosen set, they have absolutely no chance of uncovering the fault. We, on the other hand, consider any configuration with nonzero probability.

3 There is little (if any) benefit of repeating the tests with an existing commercial tool. In contrast, each independent run of our tool picks a different random set of configuration and increases the chances of uncovering a fault. The same effect can be achieved with commercial tools if the user handcodes a different set of configuration each time, but this becomes quite cumbersome especially if the tests need to be repeated multiple times. After all, one of the main goals of test generation tools is to minimize human intervention.

4 We test the router behavior in dynamic environment when routers and networks are constantly going up and down. Again, the choice of which router/network goes up/down is made probabilistically. It is very difficult to manually encode such a huge traffic pattern.

For the test execution on real routers, a testing environment is created so that a Router Under Test (RUT) "perceives" that it is interacting with a real network of routers.

Section 2 introduces a mathematical model of network topologies. Section 3 presents a general probabilistic algorithm for testing routing protocols. Sections 4, 5 and 6 specialize the general algorithm to RIP, OSPF and BGP testing, respectively. Section 7 describes a software tool SOCRATES for automatic testing of routing protocols and reports experimental results. We omit all the proofs and the interested readers are referred to [3].

2. A MATHEMATICAL MODEL OF IP NETWORK TOPOLOGIES

We consider two different graph models of the IP network topologies.
First Model: If routers are connected by point-to-point link (e.g., T1 lines), we represent the topology as a weighted graph, where nodes model routers and edges model the link between routers. We use this model for BGP testing.
Second Model: For multiaccess networks where each router can interface with more than one network and vice versa (e.g., routers connected by Ethernet), we model the topology as a weighted bipartite graph on router and network nodes. We use this model for OSPF testing. Since

this model is more general than the previous one (if each network node is restricted to have degree exactly two, this model becomes similar to the first model) and we will be stating all our algorithms on this model, we describe it in more detail.

A network of internet connections is modeled by a directed graph $G =< V, W, E >$ where V is a set of router-nodes, W is a set of network-nodes, and E is a set of directed edges [1]. Router-nodes represent routers and network-nodes represent networks that connect the routers. The interface between the routers and networks is represented by edges in E where each edge has one end node in V and the other in W. Specifically, an edge (v, w) from a router-node v in V to a network-node w in W represents a router v interfacing with a network w with a cost $c(v, w) > 0$, and an edge (w, v) from a network-node w in W to a router-node v in V represents a network w interfacing a router v with a cost $c(w, v) = 0$. A pair of edges (v, w) and (w, v) correspond to a link between router v and network w. We can replace each such pair by an undirected edge $[v, w]$ to obtain an undirected graph. This undirected graph contains two sets of nodes V and W, and there are no edges between nodes in V (W). Thus it is a bipartite graph, denoted by G_b. Both directed graph G and the equivalent bipartite graph G_b model IP network topologies, and we shall use them interchangeably.

Suppose that we have $|V| = n$ routers and $|W| = m$ networks interconnected. One important question we address is what is the smallest value of m (as a function of n) needed in order to represent all possible network topologies? What we mean is that given any bipartite graph G_b, we can construct an auxiliary graph $G^* =< V, E^* >$ on router-nodes so that two router nodes are connected by an edge iff they connect to a common network node in G_b. Essentially $G^* =< V, E^* >$ represents the "connectivity-pattern" among routers in G_b. Then we determine the smallest value of m such that for each possible G^* on n nodes, there is a G_b with m network nodes that will correspond to this particular G^*. This is a basic problem for other applications such as network simulation. If a chosen m is too small, we miss some network topologies. On the other hand, if m is too large, it introduces redundancy and wastes resources. We present the result:

Theorem 1 *Any network topology can be represented by a bipartite graph $G_b =< V, W, E_b >$ with n router-nodes V and $\lfloor \frac{n}{2} \rfloor \lceil \frac{n}{2} \rceil$ network-nodes W. Conversely, for any bipartite graph $G_b =< V, W, E_b >$ with n router-nodes V and $\lfloor \frac{n}{2} \rfloor \lceil \frac{n}{2} \rceil$ network-nodes W, there is a network topology, which is represented by G_b.*

Therefore, we have a complete representation of network topologies by bipartite graphs with a matching lower and upper bounds on the numbers of network-nodes needed for the representation. Our probabilistic test generation algorithm and fault coverage analysis are based on this representation and its completeness.

3. A PROBABILISTIC TESTING ALGORITHM

We present a general probabilistic method for testing routing protocols. Starting with an empty network topology graph, we probabilistically insert and delete edges and nodes until the graph becomes complete. After each network topology update, we check the RUT for its network topology database, routing table, and packet forwarding behaviors. We first describe the overall probabilistic testing strategy and then summarize in a generic algorithm with an analysis of its run time. In later sections, we specialize the algorithm to testing RIP, OSPF and BGP.

3.1. TESTING STRATEGY

Ideally, we want to check the RUT for all possible network topologies, however, it is impossible since there are too many ($2^{n\lfloor\frac{n}{2}\rfloor\lceil\frac{n}{2}\rceil}$). Instead, we test on a small portion of network topologies, which are generated probabilistically, and we shall show that our approach guarantees a high fault coverage.

Initially we start with a graph $G_b = < V, W, E_b >$ with only one router-node $V = \{v_0\}$, which is the RUT, and $W = E_b = \emptyset$. Depending on the networks where a router is to be deployed, we set an upper bound n on the number of router-nodes V in the network. The number of network-nodes W is bounded above by $c(n) = \lfloor\frac{n}{2}\rfloor\lceil\frac{n}{2}\rceil$ as in Theorem 1. We repeat the following steps until G_b becomes a complete bipartite graph with n router-nodes in V and $c(n)$ network-nodes in W.

1 Randomly insert or delete an edge. An edge insertion means: a new link is added between a router and a network or a down-link has come back. An edge deletion means: a link between a router and a network is down.

2 Randomly insert or delete a router-node (network-node). To insert a router-node (network-node) means: a new router (network) is added, or a crashed router (network) comes back with all its previous links before crash restored. To delete a router-node (network-node) means: a router (network) crashes with all its links to net-

works (routers) down; we remove them from the graph and save them for later restoration.

3 Probability of edge insertion, node insertion, edge deletion, and node deletion are $0 \leq p_1, p_2, p_3, p_4 \leq 1$ respectively; $p_1 + p_2 + p_3 + p_4 = 1$.

4 Keep track of the set $V_0 \subseteq V$ of neighboring router-nodes of v_0, which are connected by a network-node with v_0. Specifically, a router-node v is in V_0 if and only if there exist a network-node w such that $[v_0, w], [w, v] \in E_b$.

5 For each network topology generated above, check the RUT in node v_0:

(a) Network topology database and routing table.
For each neighboring router-node v in V_0:

i Compute its routing update information because of topology change;
ii Send the computed information to v_0;
iii Obtain network topology database and routing table information from RUT v_0;
iv Compute the expected network topology database and routing table of router-node v_0;
v Compare information from Items iii and iv. A discrepancy indicates a fault.

(b) Packet forwarding behavior.
For each neighboring router-node v_i in V_0:

i Find all the router-nodes v in V such that the chosen path by the routing protocol under test - usually a shortest path - from v_i to v contains v_0, and determine the first node u in V_0 which is on the path after v_0; we have a chosen path $[v_i, v_0, u, \ldots, v]$.
ii Construct and send an IP packet P from v_i to each such router-node v;
iii Router-node u is to receive packet P from RUT v_0; otherwise, there is a fault.

3.2. A TESTING ALGORITHM

The testing strategy described above can be summarized as a generic algorithm that applies to routing protocols, including RIP, OSPF and BGP.

For clarity, we do not include parameters and variables in subroutines except for the RUT v_0. We will also use the notation "send an IP packet to router-node v" for "send an IP packet to *a stub network attached to router-node v*".

Algorithm 1
input. $n, 0 \leq p_1, p_2, p_3, p_4 \leq 1$.
output. implementation fault or conformance.

```
1  repeat
2       construct initial network topology graph G_b with
        V = {v_0}, V_0 = W = E_b = ∅;
3  while (G_b is not complete)
4       UPDATE(v_0);
5       if ROUTE(v_0) =FALSE or FPACKET(v_0) =FALSE;
6            return "faulty";
7  return "conforms"
```

Figure 1 Testing Algorithm

The Algorithm is probabilistic in nature. In the next subsection, we will derive the expected number of iterations of the **while**-loop in line 3 (Proposition 1). Each iteration of the **while**-loop guarantees a small fault-coverage. Repeating the test inside the **while**-loop increases fault-coverage. The **repeat**-loop in line 1 runs for a sufficient number of times for a desired fault coverage. The exact number of repetitions needed will be computed later for specific protocol testing. Line 2 constructs an initial network topology graph with only one router-node under test: v_0. Loop in Line 3 continues until a complete bipartite graph is obtained. Subroutine $UPDATE(v_0)$ in Line 4 gets a new network topology, as described in Item 1-3. This subroutine will be described in Section 3.3 with a run time analysis of the algorithm. Subroutines $ROUTE(v_0)$ and $FPACKET(v_0)$ in Line 5 check the routing table and packet forwarding behavior, respectively, of the RUT, as in Items 4(a) and 4(b). If any faults are detected, we abort the process and report "faulty" in Line 6. Otherwise, we declare conformance in Line 7 with a good confidence in the topologies and router behaviors that we have tested. Subroutine $FPACKET(v_0)$ will be discussed in Section 3.4, and $ROUTE(v_0)$ will be described in Section 4, 5 and 6 for RIP, OSPF and BGP, respectively, since it checks different functions for different protocols.

3.3. SUBROUTINE UPDATE(V_0) AND RUN TIME ANALYSIS

We repeat the loop in Line 3 of Algorithm 1 until we obtain a complete bipartite graph. Each repetition of the loop runs the Subroutine $UPDATE(v_0)$ in Line 4, which generates a network topology for a test on the router.

Subroutine *UPDATE*(v_0)
parameters: $n, m = c(n), 0 \leq p_1, p_2, p_3, p_4 \leq 1$ with $p_1 + p_2 + p_3 + p_4 = 1$.
variables: $G_b =< V, W, E_b >, V_0$.
1 **switch** (p)
2   **case** 'p_1': **if** ($|E_b| < |V| \cdot |W|$) /* graph is not complete */
3   insert an edge u.a.r. in E_b;
4   **case** 'p_2': **if** ($|V| + |W| < n + m$) /* nodes below upper bounds */
5   insert a node u.a.r. in $V \cup W$;
6   **case** 'p_3': **if** ($|E_b| > 0$) /* edge set not empty */
7   delete an edge u.a.r. from E_b;
8   **case** 'p_4': **if** ($|V| + |W| > 1$) /* node set not empty */
9   delete a node u.a.r. from $V \cup W$;
10 compute V_0; /* neighboring router-nodes of v_0 */
11 compute *RoutingUpdate*;
    /* obtain protocol specific routing update information*/
12 **return** *RoutingUpdate*

Figure 2 subroutine UPDATE(v_0)

For a network topology update, one of the four operations on edge or node insertion or deletion is performed with probabilities $0 \leq p_1, p_2, p_3, p_4 \leq 1$. We can partition the unit interval into four subintervals of length p_i, $i = 1, 2, 3, 4$, and then sample uniformly at random (u.a.r.) in the unit interval and obtain $0 \leq p \leq 1$. We then "switch" on the value of p in Line 1. The run time is summarized below:

Proposition 1 *The expected number of iterations of the* **while**-*loop in Line 3 of Algorithm 1 is at most:*

$$\begin{cases} (n + c(n) + n \cdot c(n) - 1)^2, & \text{if } p_1 + p_2 - p_3 - p_4 = 0 \\ \dfrac{n + c(n) + n \cdot c(n) - 1}{p_1 + p_2 - p_3 - p_4}, & \text{if } p_1 + p_2 - p_3 - p_4 > 0 \end{cases}$$

where n is the maximal number of router-nodes, $c(n) = \lfloor \frac{n}{2} \rfloor \lceil \frac{n}{2} \rceil$, and p_1, p_2, p_3, and p_4 are the chosen probability of edge insertion, node insertion, edge deletion, and node deletion, respectively. Furthermore, *any network topology with no more than n router-nodes has a non-zero probability of getting tested by Algorithm 1.*

We can choose the four probability distributions so long as $p_1 + p_2 \geq p_3 + p_4$, which guarantees the completion of the algorithm.

3.4. SUBROUTINE FPACKET(V_0) AND PACKET FORWARDING CHECK

Subroutine $FPACKET(v_0)$ at Line 5 in Algorithm 1 tests if the RUT forwards packets correctly. Each packet switched by the router-node under test v_0 must pass through a router-node in the neighboring set V_0, and we only need to check the packet forwarding behavior of v_0 for the packets sent from V_0. For each router-node v_i in V_0, we first find all the router-nodes v in V such that the chosen path by the routing protocol under test - usually a shortest path - from v_i to v contains v_0. We then determine the router-node u in V_0 which is the first node on the path from v_0 to v. Hence, a packet P sent from v_i to v along the path $[v_i, v_0, u, \ldots, v]$ must be received by u in V_0. When we construct and send an IP packet P from v_i to the destination router-node v, the router-node u must receive the packet P from v_0; otherwise, there is a fault and the subroutine returns FALSE.

```
Subroutine FPACKET(v0)
input. variables Gb =< V, W, Eb >, V0.
1  for each router-node vi in V0, i = 1,...,r
2       construct SPT Ti rooted at vi;
3       for each router-node v in subtree of Ti rooted at v0
4            send packet P from vi to v0 with destination v;
5            let router-node u in V0 be ancestor of v in Ti;
6            if node u does not receive packet P from v0
7                      return FALSE;
8  return TRUE
```

Figure 3 subroutine for testing packet forwarding behavior

Assume that a routing protocol uses shortest path route. (For BGP, a similar scheme can be devised with its notion of "preferred routes.") For each router-node v_i in V_0, Line 2 constructs a Shortest Path Tree (SPT) rooted at v_i, and the router-node under test v_0 is a child of v_i. A packet from v_i to a destination router-node v passes v_0 if and only if v is a descendant of v_0 in the SPT T_i. Line 3-7 checks v_0 for its forwarding packet P from v_i to v. The packet P must be sent from v_0 to u, a node in V_0 and an ancestor of v. Otherwise, a fault in packet forwarding is reported in Line 6-7.

4. TESTING RIP PROTOCOL

We now study the subroutine *ROUTE* for testing RIP [4], a simple distance vector protocol. It uses the asynchronous version of Bellman-Ford algorithm to construct shortest paths to all router-nodes connected to the network. For RIP, $c(u, v)$ - the cost of interfacing from the router-node v to the network-node w is always equal to one. For each destination node, the routing table contains the distance to and also the next-hop to route packets to that destination. RIP intends to cope with dynamic networks with nodes and links up and down. To make sure that distance vectors get updated efficiently and also to avoid routing loops, most implementations use various heuristics such as *Triggered update* and *Split horizon* [5].

Algorithm 1 tests RIP for the packet forwarding behavior in subroutine *FPACKET* with dynamic networks generated by subroutine *UPDATE*. It uses subroutine $ROUTE(v_0)$ in Figure 4 to check whether RIP of the RUT responds correctly to the changed network topology, i.e., whether it constructs a correct distance vector.

Subroutine $ROUTE(v_0)$
variables. V_0. /* neighboring router-nodes of v_0 */
1 **for** nodes in V_0 in random permutation: v_i, $i = 1, 2, \ldots, r$
2 construct distance vector D_i of v_i;
3 send to v_0 distance vector: $D_i^* := split\_horizon(D_i)$;
4 construct updated (expected) distance vector of v_0:
$D_0 := D_0 \oplus D_i^*$;
5 obtain distance vector of v_0: $\hat{D}_0$;
/* via RIP protocol interface with v_0 */
6 **if** $(\hat{D}_0 \neq D_0)$
7 **return** FALSE;
8 **return** TRUE;

Figure 4 subroutine ROUTE for RIP

Now we analyze the fault coverage of Algorithm 1. The analysis has two parts. We first claim that if the algorithm reports a fault then there indeed is a fault in the implementation. The second claim is that if the implementation contains a fault then our algorithm is going to catch it with a high probability. We present the results:

Proposition 2 *If Algorithm* 1 *reports a fault then there is indeed a fault in the implementation of Bellman-Ford algorithm.*

We first describe a fault model and then discuss the fault coverage. Because Bellman-Ford algorithm proceeds by applying a series of relaxation steps, a reasonable fault model is to assume that a relaxation step is computed incorrectly. This is also known as a *single-fault* model. A single fault involves three router-nodes v_1, v_2, and v_3. The idea is that given a triangle consisting of nodes v_1, v_2, and v_3, the implementation, in trying to decide the shortest path from source v_0 to node v_2, incorrectly picks the length two path $v_1 \rightarrow v_3 \rightarrow v_2$ over edge $v_1 \rightarrow v_2$. Given that the implementation contains a single-fault, there are many possible ways in which the testing algorithm can detect this fault:

Theorem 2 Given $0 < \epsilon < 1$, with $e^2.n^2.\ln\frac{1}{\epsilon}$ repetitions in line 1, Algorithm 1 catches any single-fault in a RIP implementation with probability at least $1 - \epsilon$.

5. TESTING OSPF PROTOCOL

We now study a more complex routing protocol OSPF [7]. OSPF is a link state routing protocol. Neighboring OSPF routers maintain adjacency relationship by exchanging "Hello" packets. Each OSPF router generates Link State Advertisements(LSAs) to describe its own network connections and routes learned from other routing protocols. For broadcast network, a designated router is responsible for maintaining adjacency relationships with all other routers on this network. These LSAs are sent to adjacent OSPF routers via flooding. Each OSPF router keeps a LSA database that describes the current network topology, and exchanges its database information with all the neighbors so that each node has the same view of the network topology. Based on the network topology information in the LSA database, each router-node constructs a routing table using shortest paths algorithms.

Algorithm 1 uses subroutine *ROUTE* in Figure 5 to check whether OSPF of the RUT responds correctly to the changed network topology with a link or node up or down. Specifically, it checks: (1) After receiving a link-state advertisement *LSA*, RUT v_0 constructs a correct link-state database D_0; and (2) v_0 floods a correct link-state advertisement to each immediate neighbor node v_i in V_0. (3) v_0 construct correct routing table from its LSA database.

Whenever a link is down, the two adjacent nodes detect it, and form an *LSA* to send to all the neighbors. Whenever a router or network node is down, all its neighbors detect it as all the links to that node is down, and flood this information through an *LSA*. For simplicity, for each

```
Subroutine ROUTE(v0)
input. LSA.
variables. V0. /* neighboring router-nodes of v0 */
1  if (changed link not adjacent to v0)
2        select u.a.r. a node vr in V0 reachable from changed links;
3        send LSA from vr to v0;
4  else /* changed links adjacent to v0 */
5        v0 informed of new link status;
6  for each vi, i ≠ r, in V0 /* check LSA flooding from v0 */
7        receive LŜA from v0; /* via protocol interface */
8        if LŜA ≠ LSA
9              return FALSE;
10 compute expected LSD for v0: D0;
11 obtain LSD from v0: D̂0; /* via explicit LSD download */
12 if (D̂0 ≠ D0) /* check LSD of v0 */
13       return FALSE;
14 calculate the expected routing table R0 from this LSD;
15 obtain routing table R̂0 from v0;
16 if (R̂0 ≠ R0) /* check routing table of v0 */
17       return FALSE;
18 return TRUE;
```

Figure 5 subroutine ROUTE for OSPF

node or link up or down, we denote this network update information by LSA. We modify Line 4 of Algorithm 1 and collect this information via $LSA = UPDATE(v_0)$.

6. TESTING BGP

The routers within the Internet have been grouped into administrative units called autonomous systems (AS). RIP and OSPF are examples of routing protocols used *within* an AS. BGP [9] is the routing protocol of choice between ASs. Each router maintains its preferred path (called AS-Path) to all possible destinations. Each BGP router advertises these paths to all its adjacent (peer) routers. A key aspect of BGP is that the path used for routing is not necessarily the shortest path. This is done by specifying a set of policies. Each AS can independently set preferences for its neighboring routers. When this AS receives two different routes for the same destination, it picks the route advertised by the router with the higher preference.

A simple BGP testing algorithm can use a subroutine for routing information checking similar to ROUTE(v_0) for RIP protocol testing, and we omit an explicit description of the subroutine. We comment that the only notable differences is that instead of exchanging distance vectors, we now exchange AS-PATHs to each possible destination. We set the routing policy of the RUT and then we check whether it correctly computes its set of preferred paths to all destinations.

Note that the strategy that we have just described only tests the behavior for the policy that we have chosen. Which policy should we pick? A fault may show up only under certain routing policies. We would like to change the policies while the test is in progress with dynamic network topologies; it could give us a larger fault-coverage with respect to policies. However, changing the routing policies of the RUT is feasible only with the software tool for the routers made by some specific vendors.

Policies, in general, are a complex and ill-understood part of the BGP protocol. For example, each AS can set its policy independently. It was shown [10] that policies which appear reasonable locally may be mutually inconsistent. Furthermore, testing whether a given set of routing policies are mutually consistent is NP-hard [2].

There are other difficulties with BGP testing. A design philosophy of BGP is to maintain route stability, so if the network configuration changes very fast, many vendors design their routers to deliberately ignore the route changes. So we are forced to slow down our network simulator to enable the router to "catch up." (In general, this is not a problem with I-BGP testing.)

7. EXPERIMENTAL RESULTS

The probabilistic algorithm for IP router testing has been implemented and is a part of SOCRATES, a software tool for automatic IP routing protocol testing, developed at Lucent Bell Laboratories. The tool is written in ANSI C and runs under Linux operating system. It is inexpensive and completely portable. The only hardware support we require is a workstation capable of connecting to the RUT via at least two Ethernet. Currently SOCRATES supports the testing of RIP (versions 1,2), OSPF (version 2) and BGP (version 4) protocols.

The architecture of SOCRATES is shown in Figure 6. It consists of four components and several auxiliary utilities. The four components are *network topology generator, test executor, test and traffic log* and *GUI*.

Network topology generator implements the probabilistic test generation algorithm. It uses a graph model to represent the network topology, simulates the network links (routers) up or down by adding or removing edges (nodes). For each network topology change, a test case is generated and sent to the *test executor*. *Test executor* implements the FPACKET

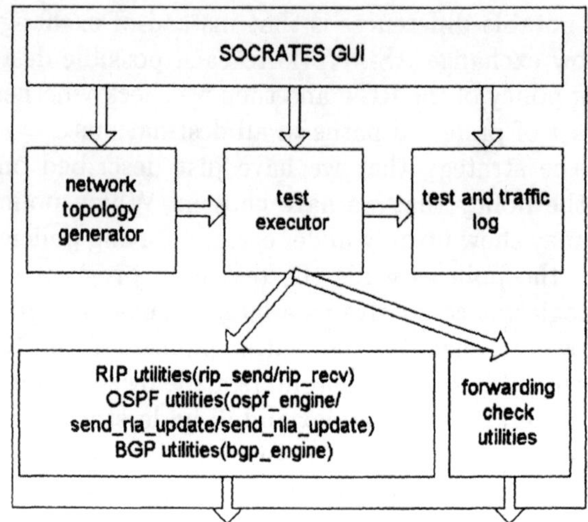

Figure 6 Architecture of SOCRATES system

subroutine and the $ROUTE(v_0)$ subroutine for RIP, OSPF and BGP. It executes the test cases and verifies if RUT updates its routing table correctly and forwards IP packets correctly. *Test executor* uses the utilities discussed below to send topology updates to the RUT, queries the RUT to get the updated routing table, performs consistency check with the calculated expected routing table, performs packet forwarding checking if there is a change in the routing table, and makes final test verdict. *Test and traffic log* records all the generated test cases in a log file. For debugging purpose, we also log the IP level trace of all message exchanges between RUT and SOCRATES.

SOCRATES uses some auxiliary utilities to communicate with the RUT and carry out real testing. Except for BGP utility, which is based on TCP connection for packet exchanging, all other utilities use the networking facilities of the host operating system to read all packets on the network in promiscuous mode, and to generate packets from arbitrary source addresses. This allows the emulation of multiple routers, with different IP addresses, from a single test host.

RIP utilities include *rip_send* and *rip_recv*. Whenever there is a change in the network topology during RIP testing, *rip_send* is used to send a RIP update packet (reflecting the change) to the RUT, and *rip_recv* is used to explicitly query the RUT to get the updated routing table.

OSPF utilities includes *ospf_engine*, *send_rla_update* and *send_nla_update*. An *ospf_engine* is a small OSPF kernel, but can be controlled by the *test executor*. Its function is to maintain the fully adjacent relationship between the RUT and itself. It also synchronizes its local LSA database with RUT and responds to all the flooding to make sure it has the exact image of the LSA database in RUT. Whenever there is a change in the network topology during OSPF testing, a new router-LSA is generated to reflect the change. A new network-LSA may be generated depending on which node is the designated router for the affected network in the graph. These two LSAs are sent to the RUT by using *send_rla_update* and *send_nla_update*, respectively.

bgp_engine is a small program for setting up a TCP connection between the RUT's BGP port and the tester. It is controlled by the *test executor* via SOCRATES control protocol to exchange BGP packets with the RUT. Whenever there is a change in the network topology during BGP testing, *bgp_engine* is directed to send a BGP UPDATE packet to the RUT.

Packet forwarding checking utilities are used to verify the RUT's forwarding behavior whenever there is a change in the RUT's routing table.

We used SOCRATES to test the RIP and OSPF implementation of GATED [8], Lucent PacketStar (release 1.2) and CISCO 7206 Router (IOS 11.3). We ran test sessions on a Linux PC with four network interfaces connected to the RUT.

In order to verify that SOCRATES can detect the implementation errors, we introduced a bug into the RIP source code in GATED [8] by modifying the way it processes any RIP update. Specifically, the buggy implementation ignored the last entry of any distance vector it received. We used SOCRATES to test against this buggy implementation, and found that for a configuration of size 20 routers $\times$ 100 networks, SOCRATES could catch this bug in less than one minute.

The RIP standard document [4] was written to summarize already existing implementations. The vagueness of the document permitted the three RIP implementations to add heuristics, in order to improve the performance of the protocol. Unfortunately, this made testing more difficult. On the other hand, OSPF is a well specified protocol with less room for arbitrary modifications. Testing OSPF implementations resulted in successful test sessions with no interoperability issues. The only minor problem we found is that one vendor's router sends out redundant OSPF DD packets during LSA database synchronization. We also found that compared with GATED, commercial routers do a much strict correctness check before accepting OSPF LSAs into their LSA database.

8. CONCLUSION

Testing is indispensable for the reliability of the routers. We describe a general probabilistic testing algorithm that can guarantee a fault-coverage under fault-models. We apply this general testing algorithm to RIP, OSPF, and BGP testing. We provide a run time and fault coverage analysis. Finally, we describe a software tool that incorporates these algorithms and also creates a testing environment such that generated test cases can be executed in real time on high speed routers.

References

[1] D. E. Comer and D. L. Stevens. *Internetworking with TCP/IP, Vol. II.* Prentice-Hall, 1999.

[2] T. G. Griffin and G. Wilfong. An analysis of BGP convergenece properties. In *ACM SIGCOMM*, pages 277–288, 1999.

[3] R. Hao, D. Lee, R. K. Sinha, and D. Vlah. Testing IP routing protocols - from probabilistic algorithms to a software tool. *Bell Labs Tech Memo*, 2000.

[4] C. Hedrick. RFC 1058. URL = `www.ietf.org/rfc.html`, June 1988.

[5] C. Huitema. *Routing in the Internet.* Prentice-Hall, 1995.

[6] D. Massey and B. Fenner. Fault detection in routing protocols. *Proceeding of International Conference on Network Protocols*, pages 31–40, 1999.

[7] J. Moy. RFC 2178. URL = `www.ietf.org/rfc.html`, July 1997.

[8] Merit Networks. Gated. URL = `www.gated.org`.

[9] Y. Rekhter and T. Li. RFC 1771. URL = `www.ietf.org/rfc.html`, March 1995.

[10] K. Vardhan, R. Govindan, and D. Estrin. Persistent route oscillations in inter-domain routing. Technical Report 96-631, USC/ISI, 1996.

Part Seven

HARDWARE SPECIFICATION, IMPLEMENTATION AND TESTING

VERIFYING AND TESTING ASYNCHRONOUS CIRCUITS USING LOTOS

Ji He and Kenneth J. Turner
Computing Science and Mathematics, University of Stirling, Stirling FK9 4LA, UK
h.ji@reading.ac.uk, kjt@cs.stir.ac.uk

Abstract

It is shown how DILL (Digital Logic in LOTOS) can be used to specify, verify and test asynchronous (unclocked) hardware designs. New relations for (strong) conformance are defined for assessing a circuit implementation against its specification. An algorithm is also presented for generating and applying implementation tests based on a specification. Tools have been developed for automated verification of conformance and generation of tests. The approach is illustrated with three case studies that explore speed independence, delay insensitivity and testing of sample asynchronous circuits.

Keywords:

Conformance Testing, Hardware Description, Hardware Verification, LOTOS

1. INTRODUCTION

DILL (Digital Logic in LOTOS, e.g. [8, 9]) is an approach, a language and a toolset for specifying, analysing and testing digital hardware designs. DILL builds on the theories and tools for LOTOS. The authors have previously developed methods for dealing with synchronous (clocked) circuits. The recent work reported in this paper extends DILL to asynchronous (unclocked) circuits.

Conventional digital circuits operate under control of a master clock for synchronisation, avoiding problems due to component speed variations. However, asynchronous circuits are interesting as they operate at the speed of individual components. Unfortunately asynchronous circuits are much harder to design and analyse. Verification of asynchronous circuits, especially speed-independent and delay-insensitive ones, has been a research topic for some years [1, 2]. Rigorous testing of asynchronous circuits is still in its infancy.

Like DILL, many other asynchronous verification approaches define relations that judge correctness of a circuit design. The relations *confor* and *strongconfor* defined in this paper resemble those introduced by others, e.g. *conformance*

Tommaso Bolognesi and Diego Latella (Eds.), *Formal Methods for Distributed System Development.*

[1], *decomposition* [2] and *strong conformance* [5]. It is not possible to detect deadlocks and livelocks with *conformance* and *decomposition*. Although *strong conformance* can do this, it does not work for non-deterministic specifications.

For validating hardware designs, test cases are in practice manually defined or are randomly generated. More rigorous approaches use traditional software testing techniques or state machine representations. In the DILL approach, tests are derived from higher-level specifications in a novel adaptation of the conformance testing theory developed for protocols.

2. CIRCUIT SPECIFICATION

2.1. ASYNCHRONOUS CIRCUIT CLASSES

Asynchronous circuits exhibit a variety of forms due to the different delay and environment assumptions made. An asynchronous circuit can behave correctly only when these assumptions are met. Some of the better-known design approaches include the following:

DI (Delay-Insensitive) circuits assume unbounded (but finite) delays for components and wires. Correct operation does not depend on actual delays.

QDI (Quasi Delay-Insensitive) circuits augment the delay model of DI circuits with the use of isochronic forks. These are branching connections on which the difference of delay magnitudes is negligible. This is a good compromise for building practical circuits using single-output gates.

SI (Speed-Independent) circuits have gates with unbounded delay, but wires with zero delay. If all gates have just one output, SI and QDI are identical.

Specifying bounded delays needs a formalism that supports quantitative timing specification. This paper studies the classes above since they assume unbounded delays and so are suitable for LOTOS. (Quasi) delay-insensitive designs can be easily changed to speed-independent circuits by inserting artificial delay components. Since unbounded delay plus unbounded delay is still unbounded, most wire delays can be absorbed into the preceding components. Only components with more than one output need special treatment.

The rest of the paper therefore mainly discusses speed-independent designs. Happily these are a good match to the DILL approach since component delays are unbounded, just like the interval between consecutive LOTOS events. In DILL, component ports are connected by synchronising their LOTOS events. This actually assumes that delay on the connecting wires is negligible, an assumption that is also adopted by speed-independent circuits. Speed independence is closely related to the concept of semi-modularity. If new inputs cannot change any pending outputs, the design is termed semi-modular. Semi-modularity is commonly used as a correctness criterion for speed independence, since the violation of semi-modularity causes speed-dependent behaviour.

2.2. MODELLING SPEED INDEPENDENCE

In this section, a basic logic gate is used to illustrate how to model speed independence (really semi-modularity) of asynchronous circuits. In [8, 9], the specification of basic logic gates allows new inputs to pre-empt pending outputs. Some of these input transitions may change the levels of pending outputs, resulting in the violation of semi-modularity. Consider a *Nand2* gate (two-input 'not and') whose inputs and output *Ip1, Ip2, Op* are initially *1, 1, 0.* After *Ip1* changes to *0*, the output should change to *1.* If *Ip1* changes back to *1* before output happens, the output may either undergo the *1* to *0* transition or remain at *0.* This depends on the speed of the gate. To respect semi-modularity, such inputs have to be forbidden. Note that the specification is partial in that inputs are forbidden at certain points. An input offer happens only when there is no potential output, or when input cannot alter potential output.

```
process Nand2 [Ip1,Ip2,Op] (dtIp1,dtIp2,dtOp:Bit) : noexit :=
  let newOp:Bit = dtIp1 nand dtIp2 in                    (* potential output *)
    (
      Ip1 ?newIp1:Bit [(newIp1 ne dtIp1) and             (* first input changes *)
        ((dtOp eq newOp) or                              (* no new potential output *)
          ((dtOp ne newOp) and                           (* there is potential output *)
          ((newIp1 nand dtIp2) eq newOp)))];             (* but no change is needed *)
      Nand2 [Ip1,Ip2,Op] (newIp1,dtIp2,dtOp)             (* continue behaviour *)
    []                                                   (* or *)
      Ip2 ?newIp2:Bit [(newIp2 ne dtIp2) and             (* second input changes *)
        ((dtOp eq newOp) or                              (* no new potential output *)
          ((dtOp ne newOp) and                           (* there is potential output *)
          ((dtIp1 nand newIp2) eq newOp)))];             (* but no change is needed *)
      Nand2 [Ip1,Ip2,Op] (dtIp1,newIp2,dtOp)             (* continue behaviour *)
    []                                                   (* or *)
      Op !newOp [dtOp ne newOp];                         (* new output produced *)
      Nand2 [Ip1,Ip2,Op] (dtIp1,dtIp2,newOp)             (* continue behaviour *)
    )
endproc
```

2.3. SPECIFYING CIRCUIT COMPONENTS

Asynchronous circuit design uses special-purpose components in addition to basic logic gates. To give a flavour of the approach and to show that their specifications in DILL are straightforward, a sampling of the components is specified below. For brevity an abbreviated syntax is used for process definitions. It is also common practice to omit signal levels when specifying asynchronous circuits, since the levels strictly alternate. For example the input sequence *Ip !0, Ip !1, Ip !0* can be represented as *Ip, Ip, Ip*. The following outline specifications respect semi-modularity.

Wires are the simplest components. Delays on wires are assumed zero, so they are not needed for speed-independent circuits. But when a (quasi) delay-insensitive circuit is made speed-independent, wire delays may have to be explicitly specified.

```
Wire [Ip,Op] := Ip; Op; Wire [Ip,Op]
```

Fork components are also necessary when a (quasi) delay-insensitive design is transformed to speed-independent form. A fork has one input *Ip* and two outputs *Op1, Op2*. The value on input *Ip* is fanned out to *Op1* and *Op2*. Because of delays, the two outputs may occur at different times. New input has to wait until both outputs have been produced.

```
Fork [Ip,Op1,Op2] := Ip; (Op1; exit ||| Op2; exit) >> Fork [Ip,Op1,Op2]
```

C-Elements are very important in asynchronous design. A C-Element (named after its conventional output *C*) is used as a transition synchroniser since its output can change only after both inputs have changed. For this reason, it is sometimes also called a join element. A C-Element has two inputs *Ip1, Ip2* and an output *Op*. The output changes to *1* when both inputs change to *1*, and changes to *0* when both change to *0*.

```
C-Element [Ip1,Ip2,Op] := (Ip1; exit ||| Ip2; exit) >> (Op; C-Element [Ip1,Ip2,Op])
```

Merge components copy input signals *Ip1, Ip2* to a single output *Op*.

```
Merge [Ip1,Ip2,Op] := Ip1; Op; Merge [Ip1,Ip2,Op] [] Ip2; Op; Merge [Ip1,Ip2,Op]
```

Selectors take an input *Ip*, and non-deterministically output on *Op1* or *Op2*.

```
Selector [Ip,Op1,Op2] := Ip; (i; Op1; exit [] i; Op2; exit)>> Selector [Ip,Op1,Op2]
```

2.4. INPUT (QUASI-)RECEPTIVENESS

In LOTOS, communication between processes is based on symmetric synchronisation at a gate. However, digital hardware has a clear distinction between inputs and outputs. A hardware component can never refuse inputs, while its outputs can never be blocked by other components.

A specification is said to be input-receptive if every input is allowed in every state. In such a case, the DILL model represents the real circuit faithfully. However input-receptive specifications cannot be written for most asynchronous circuit components since unexpected inputs are not permitted. One way to address this is by explicit deadlock if unexpected inputs arrive. If more accurate analysis is required, input quasi-receptive specifications should be written. Informally, a specification is input quasi-receptive if it can always participate in input events except when deadlocked. An input quasi-receptive specification can be obtained by adding a choice when there is a potential output.

It is not straightforward to transform a specification with more than just sequence and choice operators into input quasi-receptive form. In such a case, a partial specification can be used to generate the corresponding LTS (Labelled Transition System). An input quasi-receptive specification can be obtained by modifying the LTS. For a state that cannot participate in all input events, outgoing edges leading to deadlock are added for missed inputs. This method works very well for LTSs without internal events. But subtle problems can arise for those containing internal events. Suppose state S accepts an input $Ip1$, performs an internal action, and ends up in state S' where $Ip2$ can be accepted. It would be incorrect to add a transition to state S that accepts $Ip2$ and leads to deadlock. That is, S accepts $Ip2$ only through an internal action.

Internal events are less meaningful when considering input receptiveness as they mean the environment has no effect on choices. For this reason, LTSs with internal events are determinised before outgoing edges are added to create input quasi-receptive specifications. An LTS is input quasi-receptive if, after determinisation, all states except terminal ones can accept all inputs.

3. CIRCUIT VERIFICATION AND TESTING

3.1. VERIFYING CIRCUIT DESIGNS

As it is more difficult to specify components in an input (quasi-)receptive manner, verification may be based on components that are not input-receptive. The verification may, however, not be exact in that some problems may not be discovered. Input quasi-receptive specifications result in a larger state space and thus make verification more difficult.

Assumptions about the environment often have to be made. When an environment is not explicit, many methods simply assume the mirror of a specification as the environment of its implementations [1]. If S is the abstract specification and I is the implementation specification, verification means comparing $S \mid\mid I$ with S or checking if a logic formula holds for $S \mid\mid I$. But verifying $S \mid\mid I$ is not always satisfactory. When an implementation can accept more inputs than its specification does, $S \mid\mid I$ restricts the inputs considered to only those in the specification. This assumes that the environment does not provide extra inputs, so inputs that are accepted only by the implementation are ignored when verifying the joint behaviour. This is reasonable, but permits an implementation to produce more outputs than a specification. This is undesirable since an implementation producing unexpected output for legitimate input is normally erroneous. Moreover, when a specification is non-deterministic this method may exclude correct deterministic implementations.

When an implementation is specified in an input (quasi-)receptive way, a distinction is made between inputs and outputs. If its environment is also receptive, it will deadlock on unexpected outputs from an implementation.

However, it is very hard to extract an input quasi-receptive environment from a behavioural specification – especially if this is complicated or contains internal events. An alternative method is therefore adopted for verifying asynchronous circuits. Relations are defined that take into account the difference between input and output. These relations do not require a (quasi-)receptive environment or implementation, and are natural criteria for asynchronous circuit correctness.

3.2. INPUT-OUTPUT CONFORMANCE

Although many relations have been defined to characterise the relationship between two LTSs, they are not very helpful for verifying asynchronous circuits. This is especially true when the environment of a circuit is not explicitly supplied. Two new relations, *confor* and *strongconfor*, are therefore defined to assess (strong) conformance of an implementation to its specification. The inspiration was *ioconf* and *ioco* [11] for protocol conformance testing.

Suppose *Spec* is an abstract specification of a circuit and *Impl* is its implementation specification. *Spec* may be partial in the sense that in some states it does not accept some inputs, i.e. it is not input-receptive. An input is absent if the environment of a circuit does not provide it, if the behaviour of the circuit upon receiving the input is not of interest, or if the behaviour is undefined. Although a circuit may accept all inputs at any time, most specifications are partial to avoid detail. *Impl* may be partial or total in the sense of input receptiveness.

Suppose that *sp* is a state of *Spec* and that *im* is the corresponding state in *Impl*. To define the *confor* relation, consider the input transitions of *sp* and *im*. If input *ip* is accepted by *sp*, it is reasonable that *ip* also be accepted by *im*. But if *im* accepts an input that is not accepted by *sp*, this input and all the behaviour afterwards can be ignored. Since the environment will never provide such an input, or even if it is provided, such behaviour is not of interest. In short, the inputs acceptable in *sp* should be a subset of those acceptable in *im*.

Concerning outputs, if *sp* can produce *op* then a correct implementation should also produce it. If *sp* cannot produce a certain output, neither should its implementation. However when a specification is allowed to be nondeterministic, it is too strong to require *im* to produce exactly the same outputs as *sp* since a deterministic implementation could produce a subset of the outputs. A suitable relation should thus require output inclusion instead of output equality. Unfortunately a circuit that accepts everything but outputs nothing may also be qualified as a correct implementation. The special δ 'action' overcomes this weakness by indicating the absence of output. Like any other output, if δ is in the output set of *im* it must be in the output set of *sp* for conformance to hold. That is, *im* can produce nothing only if *sp* can do nothing.

In the above, *sp* and *im* are not actually LTS states but are all possible situations that a circuit may be in after a certain input-output sequence. Because

δ is also involved in the sequence, the state spaces of both specification and implementation are transformed into automata that are explicitly labelled with δ. The input-output sequences are actually traces of the specification automaton. For implementation $i \in \mathcal{LTS}(L_I \cup L_U)$ and specification $s \in \mathcal{LTS}(L_I \cup L_U)$:

$$
\begin{array}{lll}
i \text{ confor } s & =_{def} & \forall \sigma \in \text{STrace}(s): \ \text{out}(i \text{ after } \sigma) \subseteq \text{out}(s \text{ after } \sigma)\text{and} \\
& & \text{if } i \text{ after}\sigma \neq \emptyset : \text{in}(s \text{ after } \sigma) \subseteq \text{in}(i \text{ after } \sigma) \\
i \text{ strongconfor } s & =_{def} & \forall \sigma \in \text{STrace}(s): \ \text{out}(i \text{ after } \sigma) = \text{out}(s \text{ after } \sigma)\text{and} \\
& & \text{if } i \text{ after}\sigma \neq \emptyset : \text{in}(s \text{ after } \sigma) \subseteq \text{in}(i \text{ after } \sigma)
\end{array}
$$

L_I and L_U refer to the inputs and outputs from the point of view of the implementation. The inputs and outputs of an automaton after some trace are computed by *in* and *out*. The *after* operator yields an automaton after it has executed a given trace. *STrace* generates a suspension trace.

To define suspension traces, the transition relation is extended with output refusals: self-loop transitions labelled with L_U indicate that no output action can occur. Refusal to output can also be expressed using δ, which is regarded as an output action distinct from L_I and L_U. A suspension trace thus contains ordinary actions and δ actions. If L_δ denotes $L \cup \delta$, a suspension trace $\sigma \in L_\delta^\star$.

The *confor* relation requires that, after a suspension trace of *s*, the outputs that an implementation *i* can produce are included in what *s* can produce. If *i* can follow the suspension trace, the inputs that *s* can accept are also accepted by *i*. *strongconfor* has a similar definition except that output inclusion is replaced by output equality. Normally *confor* is used for a deterministic specification and implementation, while *strongconfor* is used when an implementation is more deterministic than a specification. *(strong)confor* is more easily observed if a specification LTS is transformed to a suspension automaton.

A suspension automaton Γ_p of an LTS p is obtained by determinising p and adding the necessary δ transitions. The suspension traces of p coincide with the traces of its suspension automaton Γ_p. In addition, for all $\sigma \in L^\star$, $out(\Gamma_p \ after \ \sigma) = out(p \ after \ \sigma)$; see [11] for the proof. Therefore checking *(strong)confor* can be easily reduced to checking trace inclusion on suspension automata. Only traces without δ transitions are checked for *confor*, while all the traces of a suspension automaton are checked for *strongconfor*.

A verification tool *VeriConf* was developed by the authors to check the *(strong)confor* relations using the programming interface of CADP (Cæsar Aldébaran Development Package [4]). Briefly, CADP is exploited to generate LTSs of both specification and implementation. Then the verifier is used to produce the suspension automata from the LTSs and to compare the automata according to the relations. The verifier has been successfully used in analysing several asynchronous circuits, including the examples in the following sections.

3.3. TESTING ASYNCHRONOUS CIRCUITS

By 'test hypothesis', an implementation is assumed to be modelled by an IOLTS (Input Output Labelled Transition System) with actions L partitioned into inputs L_I and outputs L_U. Input actions are always enabled in any state. This class of system is denoted by $\mathcal{IOLTS}(L_I, L_U) \subseteq \mathcal{LTS}(L_I \cup L_U)$. Specifications, however, are still modelled as LTSs to permit an abstract view. They are interpreted as incompletely specified IOLTSs where some inputs are not specified in some states. Incompleteness might be for implementation freedom, or because the environment will not provide undesirable inputs.

Several implementation relations have been defined by others between LTSs and IOLTSs. Some of these relations, such as the one analogous to testing preorder, are too strong in that they require *specifications* to be IOLTSs. This is obviously impractical in most cases. The *ioco* relation [11] has been defined to support conformance testing; it is very similar to *confor*. The difference is that implementations accepted by *confor* may not be input-receptive, whereas *ioco* assumes that implementations are modelled as IOLTSs (and can thus always accept all inputs). Consequently, the input inclusion condition required by *confor* is always satisfied in the case of *ioco*.

A test suite is a set of test cases with finite behaviour. Test cases must also be deterministic to allow a tester to have control over test execution. As a result a state of a test case is either terminal, offers one input to the implementation, or accepts all possible outputs from the implementation (including the δ action). The terminal states of a test are labelled with *Pass* or *Fail* to yield a verdict. Since an implementation can be non-deterministic, different terminal states can be reached with different test runs of the same test case. Only when an implementation passes all possible test runs is it said to pass the test case.

A test case is modelled as an $\mathcal{LTS}(L_I \cup L_U \cup \{\delta\})$. As before, δ means a state cannot produce any output. Test cases are obtained by a finite recursive application of the following non-deterministic choices: terminate the test case, give a next input to the implementation, or check the next outputs of the implementation. The first choice terminates the generation procedure to ensure a test stops at some point even though the specification may include infinite behaviour. The second choice will never result in deadlock as inputs are always enabled. The third choice ensures failure if an implementation produces an output not belonging to *out*(Γ). This test generation algorithm is sound and complete with respect to *(strong)confor*; see [11] for the proof.

In the DILL approach to testing digital circuit designs, a circuit is specified in LOTOS (whose semantics is given by an LTS). The implementation of the same circuit is described by VHDL (VHSIC Hardware Description Language [7]). The behaviour of a VHDL program is presumed to be modelled by an IOLTS that need not be known explicitly. The test suite for a circuit is generated by an algorithm based on that of [11]. The authors have extended CADP to generate hardware test suites automatically from the suspension automaton of the specification. A VHDL testbench executes and evaluates the test cases.

The test cases generated in the DILL approach have the form of traces rather than trees. This allows easy measurement of test coverage, and automatic execution of test cases. A test suite cannot usually cover the entire behaviour of a specification as this is normally infinite. The strategy is therefore to cover all transitions in a transition tour that addresses the Chinese Postman problem. As a suspension automaton may not be strongly connected, it is not possible to make direct use of conventional transition tour algorithms. Instead the approach of [6] is used because it is suitable for all kinds of directed graphs. Depth-first search is used until an unvisited edge cannot be reached. Breadth-first search is then employed to find an unvisited edge, and then depth-first search recommences. The authors have developed the *TestGen* tool that realises this algorithm using the CADP application programming interface.

The problems caused by non-determinism can be solved by marking contradictory output branches in a suspension automata. This situation arises if an output may not be matched by the implementation under test since other outputs are permitted. This method is not so effective when the behaviour of an implementation is non-deterministic. The problem is that when an inconclusive verdict is reached, a test run is aborted and other test cases are applied. However, the test case could be still useful if other neighbouring outputs can be found so that the test run may continue.

Contradictory branches in the suspension automaton are therefore marked with '$\star$' and the corresponding state label. In this way the branching structure of the automaton tree is reflected in a test case. The transition tour algorithm is able to cover all the transitions in a suspension automaton. If an implementation differs from all the outputs with a certain mark, a fail verdict should be recorded.

Figure 1 is an example of this technique. If an implementation has the behaviour *Ip, Op1, Op2, Op3, ...* it will follow *Ip, Op1, Op2* in a test run. But when output *Op3* fails at *Op1($\star$S4)*, the testbench must look for another output with the same mark to see if *Op3* can be matched. In this case it can find *Op3($\star$S4)* and continue testing. If an implementation behaves as *Ip, Op3, ...* then there will be no output marked with *($\star$S1)* that can match *Op3*; the implementation would be regarded as incorrect.

The testbench would normally search the remainder of a test trace when an inconclusive point is met so that testing can go forward. However such marks sometimes exist only in the previous part of the trace, forcing the search to go backward. This means that loops may arise during testing, so the testbench needs a strategy to avoid this. The testbench also needs to maintain a timer. If there is no output within a certain period, the δ transition can be regarded as having occurred; otherwise a failure verdict must be given. A testbench

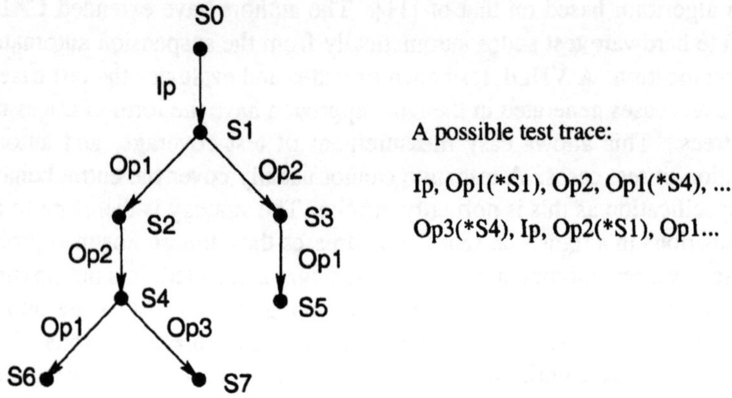

Figure 1 Test Trace for Nodes with Several Outputs

will also have to decide when to provide inputs. For test case *T1* in the later example of figure 3, if *InF !0* is provided too late after the first input *InF !1* then an output may have already been produced. The behaviour should be fully explored by other test cases such as *T2*.

4. CASE STUDY: ASYNCHRONOUS FIFO

As a typical circuit, an asynchronous FIFO (First In First Out buffer) is specified and analysed. The FIFO has two inputs *InT, InF* and two outputs *OutT, OutF*. Its inputs and outputs use dual-rail encoding in which one bit needs two signal lines. The pair of *T/F* (true/false) signal values *1/0* corresponds to data value *1*, while the pair *0/1* corresponds to *0*. A signal of *0* on both lines indicates idle, which means there is no valid data. Lines have to be reset to idle between two transmissions. Suppose a FIFO with one stage is initially empty. It can accept either *1* or *0* on receipt of *InT* or *InF*. The data is delivered to the output lines. After one successful transmission, the raised input and output lines return to *0* to wait for other data. One FIFO stage is specified as:

```
process Stage [InT,InF,OutT,OutF] :noexit :=          (* one FIFO stage *)
  InT !1; OutT !1; InT !0; OutT !0;          (* input/output 1 then idle *)
  Stage [InT,InF,OutT,OutF]                        (* continue behaviour *)
[]                                                               (* or *)
  InF !1; OutF !1; InF !0; OutF !0;          (* input/output 0 then idle *)
  Stage [InT,InF,OutT,OutF]                         (* continue behaviour*)
endproc
```

A possible implementation for a FIFO stage is given in figure 2 (a). Apart from the data path, there are two lines that control data transmission. *Req* comes from the environment of a stage, indicating that environment has valid data to transfer. The *Ack* line goes to the environment, indicating that the stage is empty and is thus ready to receive new data. Both of these control signals are active when *1*. The implementation uses two C-Elements (see section 2.3) and a *Nor2* gate (two-input 'not or'). Initially both *Req* and *Ack* are *1*. When there is valid data on *InT* or *InF*, it is passed to *OutT* or *OutF*. At the same time, *Req* should be reset to *0* until *InT* or *InF* returns to the idle state. *Ack* is reset to *0* when ready to send data on *OutT* or *OutF*, indicating the stage is full. When the data on output lines is fetched, the stage returns to idle and is ready for the next data. The inputs/output of the C-Elements are start as *0,1/0* while those for the *Nor2* gate start as *0,0/1*. The specification of this FIFO cell is:

```
process Cell [InT,InF,OutT,OutF,Req,Ack] : noexit :=
  (CElement [InT,Req,OutT] (0,1,0) |[Req]| CElement [InF,Req,OuTF] (0,1,0))
|[OutT,OutF]|
  Nor2 [OutT,OutF,Ack] (0,0,1)
endproc
```

To ensure a FIFO works correctly, the environment has to be coordinated. For example, it should provide correct input data according to the dual rail encoding. To make things easier, it is convenient to think about the environment in two parts: *EnvF* (front-end) is a provider that is always ready to produce data, while *EnvB* (back-end) is a consumer that can always accept data.

```
process EnvF [InT,InF,Req] : noexit :=              (* data provider *)
  InT !1; Req !0; InT !0; Req !1;                    (* provide 1 *)
  EnvF [InT,InF,Req]                                 (* continue behaviour *)
[]                                                   (* or *)
  InF !1; Req !0; InF !0; Req !1;                    (* provide 0 *)
  EnvF [InT,InF,Req]                                 (* continue behaviour *)
endproc

process EnvB [OutT,OutF,Ack] : noexit :=            (* data consumer *)
  OutT !1; Ack !0; OutT !0; Ack !1;                  (* accept 1 *)
  EnvB [OutT,OutF,Ack]                               (* continue behaviour *)
[]                                                   (* or *)
  OutF !1; Ack !0; OutF !0; Ack !1;                  (* accept 0 *)
  EnvB [OutT,OutF,Ack]                               (* continue behaviour *)
endproc
```

A two-stage FIFO can then be implemented as in figure 2 (b):

```
process Impl [InT,InF,OutT,OutF] : noexit :=        (* FIFO implementation *)
  hide Req,IntT,IntF,IntR,Ack in                     (* internal signals *)
    EnvF [InT,InF,Req]                               (* data provider *)
  |[InT,InF,Req]|                                    (* synchronised with *)
    Cell [InT,InF,IntT,IntF,Req,IntR]                (* first cell *)
  |[IntT,IntF,IntR]|                                 (* synchronised with *)
    Cell [IntT,IntF,OutT,OutF,IntR,Ack]              (* second cell *)
  |[OutT,OutF,Ack]|                                  (* synchronised with *)
    EnvB [OutT,OutF,Ack]                             (* data consumer *)
endproc
```

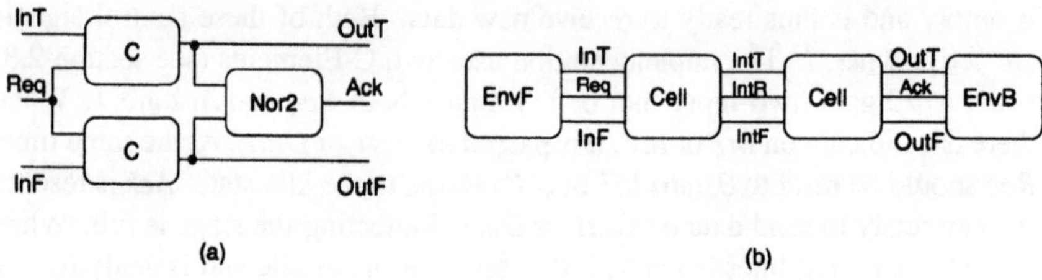

Figure 2 Implementation of A Two-Stage FIFO from Individual Cells

When speed independence needs to be verified, each building block (including the environment) should be specified in the input quasi-receptive style. *Impl_QR* is the corresponding implementation specification. It can use the DILL library for quasi-receptive specifications of the basic building blocks. It also needs the corresponding quasi-receptive specifications *EnvF_QR* and *EnvB_QR*. *EnvF* has no inputs and so is identical to *EnvF_QR*. *EnvB_QR* has the form:

```
process EnvB_QR [OutT,OutF,Ack] : noexit :=          (* receptive data consumer *)
  OutT !1;                                            (* value 1 output by FIFO *)
  (Ack !0;        (OutT !0;         (Ack !1; EnvB_QR [OutT,OutF,Ack]
                                    [] OutT !1; stop [] OutF !1; stop)
                  [] Ack !1; stop [] OutF !0; stop)  (* incorrect ack or FIFO output *)
  [] OutT !0; stop [] OutF !0; stop)                  (* incorrect FIFO output *)
[]                                                    (* or *)
  OutF !1;                                            (* value 0 output by FIFO *)
  (Ack !0;        (OutF !0;         (Ack !1; EnvB_QR [OutT,OutF,Ack]
                                    [] OutT !1; stop [] OutF !1; stop)
                  [] Ack !1; stop [] OutT !0; stop)  (* incorrect ack or FIFO output *)
  [] OutT !0; stop [] OutF !0; stop)                  (* incorrect FIFO output *)
[]
  Ack !0; stop
endproc
```

The specification should exhibit liveness. Using CADP, it was verified that the specification satisfies the following property expressed in ACTL (Action-based Computational Tree Logic [10]). If there is an input of *1*, then output will become *1* eventually: $AG([\text{InT } !1]A[\text{true}_{true}U_{Out!1}\text{true}])$. The formula for data *0* is similar and was also shown to be true. It was verified that *Spec* $\approx$ *Impl* || *(EnvB* |[$\cdots$]| *EnvF)*, where $\approx$ denotes observational equivalence.

To check speed independence, the input quasi-receptive specifications were used. It was also verified that *Spec* $\approx$ *Impl_QR* || *(EnvB_QR* |[$\cdots$]| *EnvF_QR)*, which gives more confidence in the design of the FIFO. The implementation *Impl_QR* || *(EnvB_QR* |[$\cdots$]| *EnvF_QR)* also satisfies the liveness property.

Figure 3 gives the LTS for the FIFO (minimised with respect to observational equivalence), the suspension automaton for the LTS, and several tests. Because the LTS is deterministic, the suspension automaton has almost the

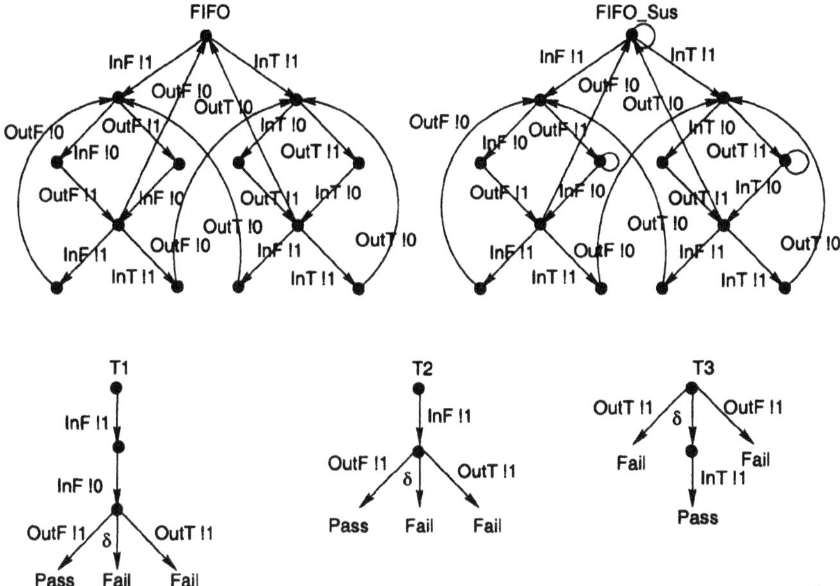

Figure 3 LTS, Suspension Automaton and Several Tests of FIFO

same structure except for the δ transitions, which appear as circles in the figure. Test *T1* provides two inputs and then checks the output of an implementation. If output *OutF* changes, the implementation passes the test. However if *OutT* changes or if there is no output, the implementation fails the test. Similarly, test *T2* checks output after one input is provided. Test *T3* checks output right away. For this test, an output from the initial state is incorrect and results in a fail. Only after δ, meaning that no output is produced, can testing continue.

It was shown that *Impl_QR* || *(EnvB_QR* $|[\cdots]|$ *EnvF_QR) strongconfor Spec* using the *VeriConf* tool. The *TestGen* tool builds a single test case of length 28:

InF !1	InF !0	OutF !1	InF !1	OutF !0	OutF !1	InF !0
InT !1	OutF !0	InT !0	OutT !1	InT !1	OutT !0	OutF !1
δ	InF !0	OutF !0	InT !1	OutT !1	InT !0	InT !1
OutT !0	OutT !1	δ	InT !0	OutT !0	δ	Pass

5. CASE STUDY: *OR-AND* CIRCUIT

This example was introduced in [3] to show the difference between speed independence and delay insensitivity. Although small, it reveals the necessity of using input quasi-receptive specifications. As in section 2.3, events are abbreviated by omitting signal values *!1* and *!0*. The circuit has inputs *Ip1*,

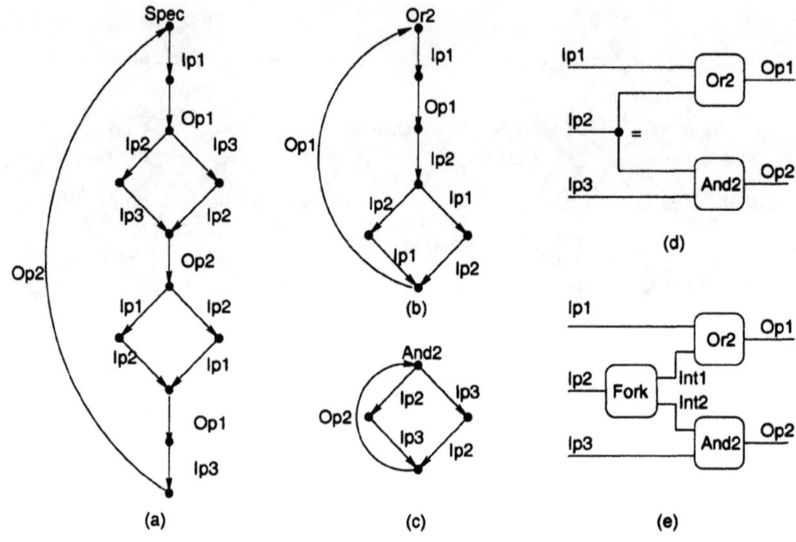

Figure 4 An *Or-And* Circuit

Ip2, Ip3 and outputs *Op1, Op2*. Output *Op1* is the logical *or* of *Ip1, Ip2*, while output *Op2* is the logical *and* of *Ip2, Ip3*.

The abstract circuit LTS is shown in figure 4 (a). The component *Or* and *And* LTSs appear in figures 4 (b) and 4 (c) respectively. The proposed implementation is in figure 4 (d). The verification task is to check if this implementation is delay-insensitive and speed-independent. For analysing delay insensitivity, the circuit is transformed to figure 4 (e), where the isochronic fork (=) in figure 4 (d) is replaced by an explicit Fork element (see sections 2.1 and 2.3 for an explanation of these). The implementations in figures 4 (d) and figure 4 (e) are specified as *Impl1* and *Impl2*:

```
process Impl1 [Ip1,Ip2,Ip3,Op1,Op2] : noexit :=  (* isochronic fork implementation *)
  Or2 [Ip1,Ip2,Op1] |[Ip2]| And2 [Ip2,Ip3,Op2]           (* or plus and gates *)
endproc
process Impl2 [Ip1,Ip2,Ip3,Op1,Op2] : noexit :=   (* ordinary fork implementation *)
  hide Int1,Int2 in                                     (* hide internal signals *)
    (Or2 [Ip1,Int1,Op1] ||| And2 [Int2,Ip3,Op2])          (* or plus and gates *)
  |[Int1,Int2]|                                          (* synchronised with *)
    Fork [Ip2,Int1,Int2]                                       (* fork input *)
endproc
```

The state spaces of *Impl1* and *Impl2* are much larger than that of *Spec*. For example, both can accept *Ip2* and *Ip3* from their initial states but *Spec* cannot. Since no explicit environment is given, a direct verification approach is to compare *Impl1* || *Spec* with *Spec*, i.e. assuming that *Spec* is also the environment of its implementations. It was found that *Impl1* || *Spec* $\approx$ *Spec*

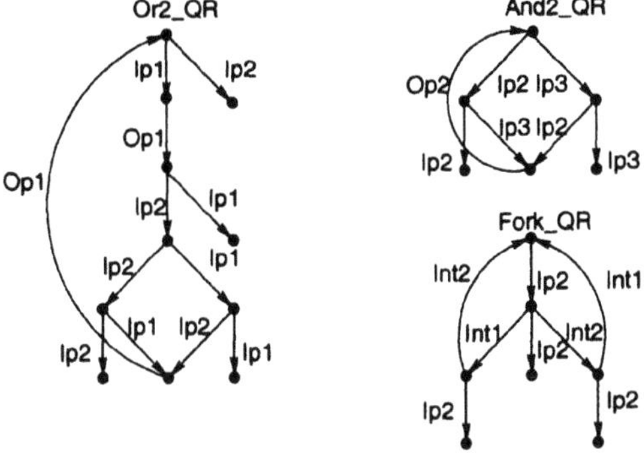

Figure 5 Input Quasi-Receptive Specifications of *Or-And* Components

and *Impl2* || *Spec* ≈ *Spec* by using CADP. This suggests that figures 4 (d) and (e) are both correct implementations of *Spec*.

However, to ensure that both implementations are really speed-independent, the more accurate input quasi-receptive model is needed. Figure 5 shows the revised LTSs. The corresponding implementations are *Impl1_QR* and *Impl2_QR*. It was discovered that *Impl1_QR strongconfor Spec* by using the *VeriConf* tool. Unfortunately *Impl2_QR* does not have the *confor* or *strongconfor* relationship to *Spec*. The tool gives a diagnostic trace: *Ip1, Op1, Ip3, Ip2, Op2, Ip2, Ip1*. By analysing this trace it can be found that after *Op2* is produced, the specification is able to receive *Ip1* and *Ip2*. But for the implementation in figure 4 (e), after *Op2* is produced the fork component may still be in the unstable state since *Int1* has not been produced yet. In this unstable state, an *Ip2* input makes the behaviour of the fork component undefined. This means that figure 4 (e) is not speed-independent, i.e. the correctness of the circuit depends on the speed of *Fork*. Figure 4 (d) is therefore not a delay-insensitive implementation.

6. CASE STUDY: SELECTOR

As a final example, a selector (see section 2.3) allows non-deterministic behaviour in implementations. After a change on input *Ip*, either *Op1* or *Op2* may change depending on the implementation. Figure 6 gives its LTS (minimised with respect to observational equivalence), the suspension automaton of the LTS, and one of the test cases. Sample test *T4* shows that after input *Ip !1*, an implementation producing either *Op1 !1* or *Op2 !1* will pass the test.

Using the method in section 3.3, the *TestGen* tool produces a single test case of length 11 for the selector. This example shows how contradictory branches are marked: after *Ip !1*, *Op !1* is marked with state *S1* since an implementation

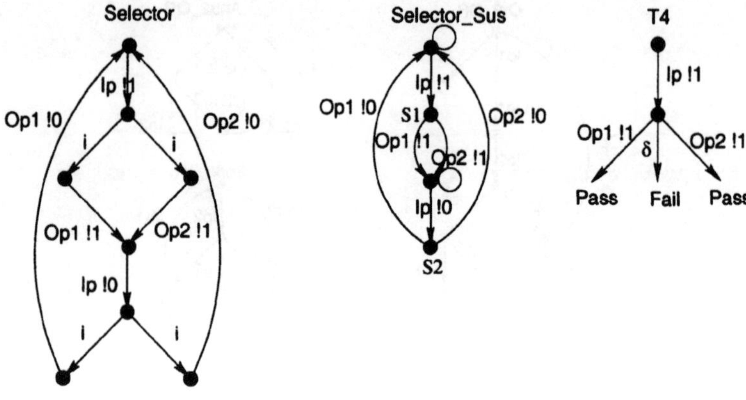

Figure 6 LTS, Suspension Automaton and One Test of Selector

may also do *Op !2*. A selector that insists on sending its input to *Op1* can follow the first row of steps in the test case below. After the sixth step (*Ip !1*), it cycles back to the second step (*Op !1*) – a loop that the testbench must break.

Ip !1	Op1 !1 (⋆S1)	Ip !0	Op1 !0 (⋆S2)	δ	Ip ! 1
Op2 !1 (⋆S1)	δ	Ip !0	Op2 !0 (⋆S2)	Pass	

7. CONCLUSION

An approach to verifying asynchronous circuits has been presented for a range of typical asynchronous components. (Quasi) delay-insensitive circuits are transformed into speed-independent designs. Violations of speed-independence (or rather of semi-modularity) are checked using specifications that are input (quasi-)receptive. New relations *confor* and *strongconfor* have been defined to assess the implementation of an asynchronous circuit against its specification. The *VeriConf* tool has been developed to support the *(strong)confor* relations. The *TestGen* tool generates test suites using transition tours of automata. This allows automatic generation of test suites for reasonable coverage, and also allows testing of non-deterministic implementations.

References

[1] D. L. Dill. *Trace Theory for Automatic Hierarchical Verification of Speed-Independent Circuits.* ACM Distinguished Dissertations. MIT Press, 1989.

[2] J. C. Ebergen, J. Segers, and I. Benko. Parallel program and asynchronous circuit design. In G. Birtwistle and A. Davis, editors, *Asynchronous Dig-*

ital Circuit Design, Workshops in Computing, pages 51–103. Springer-Verlag, 1995.

[3] J. Ebergen. A formal approach to designing delay-insensitive circuits. *Distributed Computing*, pages 107–119, 1991.

[4] J.-C. Fernández, H. Garavel, A. Kerbrat, R. Mateescu, L. Mounier, and M. Sighireanu. CADP (CÆSAR/ALDÉBARAN Development Package): A protocol validation and verification toolbox. In R. Alur and T. A. Henzinger, editors, *Proc. 8th. Conference on Computer-Aided Verification*, number 1102 in Lecture Notes in Computer Science, pages 437–440. Springer-Verlag, Berlin, Germany, Aug. 1996.

[5] G. Gopalakrishnan, E. Brunvand, N. Michell, and S. Nowick. A correctness criterion for asynchronous circuit validation and optimization. *IEEE Transactions on Computer-Aided Design*, 13(11):1309–1318, Nov. 1994.

[6] R. C. Ho, C. H. Yang, M. A. Horowitz, and D. L. Dill. Architecture validation for processors. In *Proc. 22nd. Annual International Synposium on Computer Architecture*, 1995.

[7] IEEE. *VHSIC Hardware Design Language*. IEEE 1076. Institution of Electrical and Electronic Engineers Press, New York, USA, 1993.

[8] Ji He and K. J. Turner. Protocol-inspired hardware testing. In G. Csopaki, S. Dibuz, and K. Tarnay, editors, *Proc. Testing Communicating Systems XII*, pages 131–147, London, UK, Sept. 1999. Kluwer Academic Publishers.

[9] Ji He and K. J. Turner. Specification and verification of synchronous hardware using LOTOS. In J. Wu, S. T. Chanson, and Q. Gao, editors, *Proc. Formal Methods for Protocol Engineering and Distributed Systems (FORTE XII/PSTV XIX)*, pages 295–312, London, UK, Oct. 1999. Kluwer Academic Publishers.

[10] R. D. Nicola and F. Vaandrager. Three logics for branching bisimulation. In *Proc. 5th. Annual Symposium on Logic in Computer Science (LICS 90)*, pages 118–129. IEEE Computer Society Press, 1990.

[11] J. Tretmans. Test generation with inputs, outputs and repetitive quiescence. *Software Concepts and Tools*, 17:103–120, 1996.

HARDWARE IMPLEMENTATION OF CONCURRENT PERIODIC EFSMS

Hisaaki Katagiri[1], Masayuki Kirimura[1],
Keiichi Yasumoto[2], Teruo Higashino[1] and Kenichi Taniguchi[1]
[1] *Graduate School of Eng. Sci., Osaka Univ., Toyonaka, Osaka 560-8531, JAPAN*
[2] *Dept. Info. Proc. & Man., Shiga Univ., Hikone, Shiga 522-8522, JAPAN*
yasumoto@biwako.shiga-u.ac.jp, (kirimura,higashino,taniguchi)@ics.es.osaka-u.ac.jp

Abstract This paper proposes a concurrent periodic EFSMs model and a technique to synthesize hardware circuits from real-time system specifications in this model. In the proposed model, the data exchange by synchronous execution of the same events in multiple EFSMs (multi-way synchronization) can be specified like LOTOS. The executable time range of each event can be specified as a logical conjunction of linear inequalities of the execution time of its preceding events, constants and integer variables with some values input from environments. Here, we assume that every event sequence starting from the initial state in each EFSM returns to the initial state in the specified time interval. The proposed synthesis technique implements only executable combination of branches in a given specification with a schedule table which guarantees that if each I/O event is executed within the specified time range, at least one of the following event sequences remains executable. Some experiments show that the performance and size of the generated circuits are reasonable for practical use.

Keywords: Real-time systems, hardware implementation, scheduling, multi-way synchronization, E-LOTOS

1. INTRODUCTION

For the needs of high-speed computer networks, hardware implementation techniques for communication systems have become important. In general, in development of high performance communication systems, we repeatedly modify parameters such as timing restriction for each task or I/O action as well as internal algorithms and structures of the systems, and simulations are carried out until the expected performance is achieved. Therefore, in order to reduce development costs, rapid pro-

Tommaso Bolognesi and Diego Latella (Eds.), *Formal Methods for Distributed System Development.*

totyping techniques for automatic synthesis of hardware circuits from high-level specifications are essential in early stages of development [11].

In recent years, high-level synthesis techniques using FDT have been stressed. [1, 13] have proposed hardware implementation techniques using SDL and Estelle, respectively, and [8, 10, 14] have proposed techniques based on LOTOS [5]. However, those existing techniques do not handle timing constraints for event execution nor guarantee real-time processing of tasks in the generated circuits. [2] has proposed a technique for guaranteeing such real-time processing. This technique supposes that multiple concurrent processes are given as the corresponding task graphs and calculates an optimal scheduling where the total cost of processors is minimized under the condition that the deadlines of all tasks are met. Unfortunately, it may be difficult to apply this technique to development of real-time communication systems because it does neither handle communication among concurrent processes nor consider execution timings of I/O events.

This paper proposes a model called *concurrent periodic EFSMs* and a technique to synthesize hardware circuits from real-time system specifications described in this model. In the proposed model, like LOTOS the data exchange by synchronous execution of the same events in multiple EFSMs (called *multi-way synchronization*) can be specified, and like E-LOTOS [6] a timing constraint represented as a logical conjunction of linear inequality expressions of the execution time of preceding events, constants and integer variables (where some values are input from external environments) can be specified to decide an executable time range of each event. In the proposed model, we assume that each branch of event sequences starting from the initial state must return to the initial state after executing some events in the specified time interval. Since most of real-time and multimedia systems consist of periodic behavior, this assumption seems reasonable for development of such systems.

The following criteria should be considered when we synthesize circuits from specifications in the proposed model: (1) guarantees for deadlock-free execution of event sequences, (2) fairness in selection of all executable branches, (3) assignment of a possibly wide time range to each I/O event execution. Therefore, the proposed synthesis technique generates circuits where all executable branches in the given specification can be selected, and only schedulable event sequences are dynamically executed depending on the execution time of the preceding events. For such guarantees, the proposed technique derives an appropriate executable time range of each event with a technique for solving integer linear programming problems (ILP). Here, we maximize the total sum of executable time ranges of I/O events.

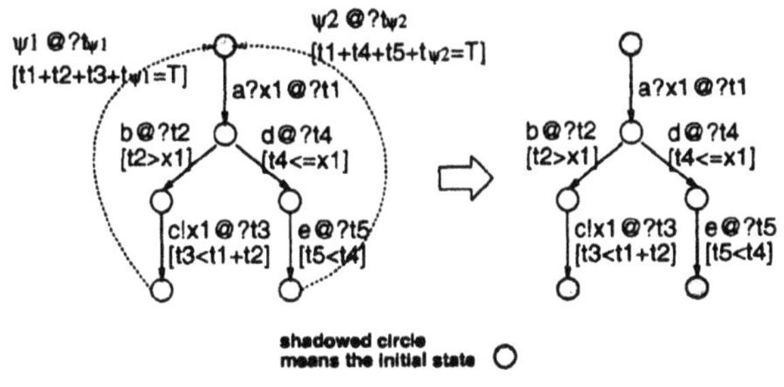

Figure 1 Periodic EFSM

Based on the techniques we previously proposed in [7, 14], a given specification in the proposed model is implemented as a hardware circuit where it consists of sequential circuits corresponding to EFSMs and a combinational logic circuit for the controller of multi-way synchronization among EFSMs. For timing guarantees, we construct a scheduling circuit indicating schedulable event sequences at each time.

To evaluate the proposed technique, we have implemented tools to generate the corresponding RTL-level VHDL descriptions. Through some experiments, we have confirmed that the performance and size of the generated circuits are reasonable for practical use.

2. CONCURRENT PERIODIC EFSMS

EFSM is defined as 5-tuple $< S, I, V, clock, \delta, init >$. Here, S is a finite set of states, I is a finite set of events (i.e., transitions), V is a finite set of variables, *clock* is a counter holding the waiting time since the EFSM has moved to a new state, δ is a finite set of transition rules, and *init* represents the initial state and initial values of all variables.

It is assumed that the execution of each event is completed in a moment. The event which exchanges data between the system and its external environment is called an *I/O event*, and the event which is executed in synchronization with the same one of other EFSMs is called a *synchronous event*. Like the notation of LOTOS, the input and output events are represented as $a?x$ and $b!E$, respectively. Here, a and b denote gate names for input and output, respectively. In addition, in one event, multiple inputs/outputs such as "$a?x!10$" can be specified. Moreover, like E-LOTOS, $a@?t$ denotes that the value of *clock* at the execution time of event a (called *delay time*) is substituted to time variable t. The value of *clock* is treated as an integer.

Each transition rule is defined as 4-tuple $< s_{cur}, s_{next}, a, guard >$ $(s_{cur}, s_{next} \in S, a \in I)$. Here, *guard* is called a transition condition, and it is specified by a logical expression consisting of the input/time variables used in the EFSM. In each EFSM, arbitrary types of variables such as integer, bool and string can be specified. However, the types of all variables used in the transition conditions must be integer. If the value of the transition condition *guard* is true at state s_{cur} and event a is executed, then the EFSM moves to state s_{next}. By using time variables in transition conditions, we can give a restriction for the executable time of each event. For example, $a?x@?t[(2 \leq t)and(0 \leq x)and(x \leq 10)]$ denotes that input event $a?x$ must be executed after waiting at least 2 units of time in the current state, and that an integer satisfying the condition $(0 \leq x)and(x \leq 10)$ is input from gate a and it is substituted into input variable x. As a transition condition, only a logical conjunction of integer linear inequality expressions consisting of time variables representing the delay time of the preceding events and/or input variables can be specified. If multiple events are executable at the current state, only one event is selected and executed non-deterministically.

If a given EFSM satisfies the following restrictions, then the EFSM is called a periodic EFSM.

- All paths (event sequences) from the initial state return to the initial state.
- The transition conditions are specified so that even if any path from the initial state is selected, it can return to the initial state in a certain fixed time interval T.

The graph of the left-hand side in Fig. 1 shows an example periodic EFSM. In this EFSM, at first, a value is input from gate a and substituted to variable $x1$ at time $t1$. Then, until $x1$ units of time has passed, event d can be executed. After $x1$ units of time has passed, event b can be executed. If event b is executed, then the delay time of event b is substituted into time variable $t2$, and event $c!x1@?t3$ becomes executable. In order to make the EFSM return to the initial states in the specified time interval T, we assume that every path (event sequence) from the initial state has the special dummy transition ψ as the last event of the path where the transition condition of the dummy transition ψ gives the timing constraint to make the path return to the initial state in the specified time interval T. In Fig. 1, such dummy transitions ψ_1 and ψ_2 are shown as dotted edges. For the sake of simplicity, hereafter, we will omit such dummy transitions in the graphs and depict the graphs in the form of the right-hand side in Fig. 1.

The synchronization relation among EFSMs is specified as follows using the parallel operators in LOTOS.

$$S ::= S \ |[gate\_list]| \ S \quad | \quad S \ ||| \ S \quad | \quad (S) \quad | \quad EFSM$$

Here, *EFSM* is a name of EFSM, and *gate_list* is a gate list of the events to be synchronized between its operator's both sides of EFSMs.

For example, in $M1|[a,b]|(M2|[b]|M3)$ of Fig. 4, event a must be executed simultaneously either between $M1$ and $M2$ or between $M1$ and $M3$. When synchronous events a of $M1$ and $M2$ are executed simultaneously, the output value $x1-2$ of event "$a!x1-2@?t3$" of $M1$ is substituted to variable $x3$ of event "$a?x3@?t10$" of $M2$.

The details about the transition conditions for synchronous events are omitted here, since they are the same as those of LOTOS [5].

3. BASIC POLICY OF CIRCUIT DERIVATION

Static Scheduling Policy

In order to derive a hardware circuit from a given specification, it is important to decide when we should execute each event so that it can satisfy the timing constraints described in the specification.

For example, consider the following event sequence.

$$s1 \overset{a@?t_a[t_a \geq 2]}{\longrightarrow} s2 \overset{b@?t_b[t_a+t_b \leq 5]}{\longrightarrow} s3$$

Since the timing constraint $t_a \geq 2$ of event a means that event a can always be executed after the delay time at state $s1$ becomes 2, event a can be executed, for example, at delay time 10. However, it does not satisfy the constraint $t_a + t_b \leq 5$ of the following event b in this case.

On the other hand, when event a is executed at delay time 2, event b can be executed at any delay time between 0 and 3. If all events in a given event sequence can be executed at the time satisfying given timing constraints, then we will say that the event sequence is *schedulable.* In order to make each event sequence schedulable, we must decide the execution time of each event so that the succeeding events can satisfy their timing constraints.

Then, the proposed technique derives the range of the execution time of each event which makes the whole event sequence schedulable. Note that generally such a range may become narrow compared with that of the given specification. Moreover, there exists a case where the range of the execution time of the succeeding event is dynamically changed depending on the execution time of its preceding events. In the above example, if event a is executed at delay time h $(2 \leq h \leq 5)$, then event b is executable at any delay time between 0 to $(5-h)$. Since such a dynamic scheduling is not easy to implement as a hardware circuit, we will adopt a static scheduling policy in this paper. Therefore, the proposed technique derives the constant time range for each event where the derived time range guarantees that if each event is executed at any time in the range, the succeeding events can also be executed at any time in the derived time ranges of those events. For example, for the above example, if we give a schedule so that $2 \leq t_a \leq 3$ and $0 \leq t_b \leq 2$ hold, this event sequence is always executable. Using such a static scheduling policy, it becomes possible to judge the executability of each event by using only one global timer representing the current time and some simple logic circuits. Also, we do not need to keep the actual execution time of all events unlike a dynamic scheduling policy.

Requirements for Derived Circuits

Our synthesis technique derives circuits which satisfy the following requirements.

- In order to make given concurrent EFSMs periodic, only schedulable sets of paths (event sequences) in a given specification must be selected and executed.
- It is desirable to derive a possibly wide range of the execution time of each event.

Since each EFSM has some branches as its event sequences where some of them may include synchronous events, the following problems should be considered to satisfy the above requirements.

Like the example in Fig. 2, if there are conditional branches after an I/O event a is executed, the range of executable time of event a may differ depending on which conditional branch is selected. Note that, as pointed out in Fig. 2, as time passes, some of conditional branches may become un-executable (un-schedulable). Moreover, like the example in Fig. 3, synchronous event b of EFSM1 becomes executable after a is executed, and it can synchronize with either of two synchronous events b in EFSM2. Depending on which branch α or β is selected, the range of executable time of the preceding event a may differ. Then, in the proposed technique, we find all possible sets of paths (event sequences) of concurrent EFSMs, and derive the range of the execution time of each event depending on which set of paths is selected. In the derived circuits, at each unit of time, the controller checks which path becomes un-schedulable and it makes only schedulable paths executable.

4. DERIVATION OF EXECUTABLE TIME RANGE OF EACH EVENT

In this section, in order to derive the executable time range of each event, we construct linear inequality expressions from the information about tuples of synchronous events (which are executed simultaneously

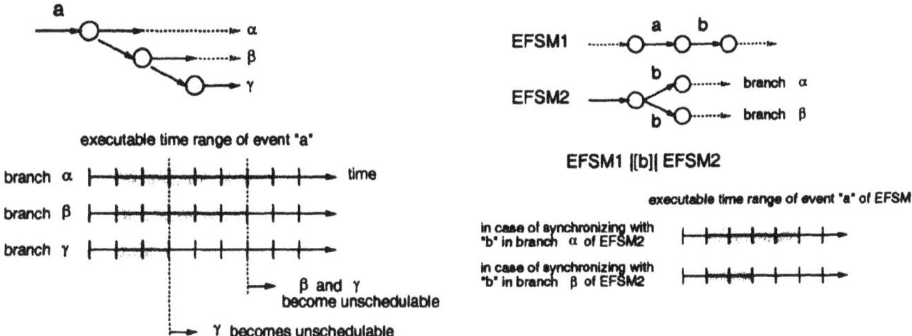

Figure 2 Executable time range of an event followed by branches

Figure 3 Executable time range of an event depending on selection of multi-way synchronization

among EFSMs), temporal ordering of events in each EFSM, and the timing constraint of each event. Since some input variables may influence the execution of event sequences (e.g., the cases that the variables are used in guard expressions or in timing constraints of events), we also derive allowable value ranges for those variables.

4.1. DERIVATION OF TUPLES OF SYNCHRONOUS EVENTS

By tracing executable events in the tuple of all EFSMs sequentially from the initial global state, we can derive the information about in what combination of EFSMs and in what tuple, events must be executed in synchronization with each other. Several techniques for such reachability analysis have been proposed for models including event synchronization among concurrent processes, for example, in [12]. We can use the techniques to derive such information about synchronization of events. However, since reachability analysis generally needs to explore n^m global states in the worst case (here, n and m represent the number of states in each EFSM and the number of synchronizing EFSMs, respectively), computation time grows exponentially as the number of parallel processes increases. Also, in EFSMs, it may take much time to check execution conditions (i.e., boolean expressions) for synchronizations.

In our concurrent periodic EFSMs model, since each EFSM executes one of its event sequences in a specified time interval and all EFSMs have the same time interval, due to restrictions explained in Sect. 2, we can easily derive synchronous event tuples by simply checking whether the synchronous events appear in the same order for each combination of event sequences (due to space limitation, the detailed procedure is omitted). So, in our model, computation time for deriving synchronous event tuples is bound to $N \cdot B^m$ where N is the maximum number of events in each event sequence, and B is the maximum number of branches in each EFSM. In general, $N \cdot B^m \ll n^m$ holds.

4.2. CONSTRUCTION OF ILP PROBLEMS

We use the execution time of each event as a parameter of the ILP problem. We also use the allowable value for each input variable as a parameter. From the set of event sequences ES_i of each $EFSM_i$ and the set of tuples of synchronous events denoted by SYN, we construct the following logical conjunction of linear inequalities as the ILP problem. Hereafter, we denote the execution time of event a by T_a, and the allowable value of variable x by V_x. We assume that each time variable is counted as the time elapsed from the beginning of the current time interval. We use concurrent periodic EFSMs in Fig. 4 as an example.

(1) constraints for temporal ordering of events in each EFSM

Execution time of an event must be less than execution time of the succeeding events. Therefore, for example, for event sequence $P12$ of $M1$ in Fig. 4, the following constraint is derived.

$$T_c < T_e < T_b$$

(2) constraints for tuples of synchronous events

All events of each tuple of synchronous events in SYN must be executed at the same time. So, for example, the following constraint is derived for event b executed synchronously among event sequences {P12, P21, P33} in Fig. 4.

$$T_{b_{M1}} = T_{b_{M2}} = T_{b_{M3}}$$

(3) timing constraints attached to each event

Timing constraints attached to each event are included as it is in the ILP program. However, since each original time variable represents the delay time in the current state, it must be replaced by the corresponding variable representing the time elapsed from the beginning of the current time interval.

For example, the timing constraint of event b in the following event sequence (corresponding to $P12$ in Fig. 4)

$$c?x1@?t1[...]; e@?t4[...]; b@?t5[t5 < t1 + t4]$$

is represented as

$$\begin{aligned}(T_b - T_e) &< T_c + (T_e - T_c) \\ &\Rightarrow T_b - 2T_e < 0\end{aligned}$$

by replacing $t1, t4$ and $t5$ by T_c, $T_e - T_c$ and $T_b - T_e$, respectively.

(4) constraints for the allowable value of each integer variable

Each input variable used in some timing constraints of events should have a certain value which satisfies those constraints.

In $P11$ of Fig. 4, event $c?x1@...$ inputs some value to the variable $x1$, and $x1$ is used in the timing constraint of the next event $d@?t2[t2 > x1]$. Thus, in this case, the following constraint for x is derived.

$$T_d - T_c > V_{x1}$$

In addition, event $a!(x1 - 2)$ of $P11$ is specified to synchronize with either $a?x3...[(x3 > x2)and...]$ of $P22$ or $a?x4...[t15 < x4 + 5]$ of $P32$. So, the following constraint for the variable $x1$ is derived for the case that synchronization of a is executed between $P11$ and $P22$.

$$V_{x1} - 2 = V_{x3}$$

As we explained in Sect. 3, the executable time range of each event may differ depending on what branch is selected and executed. So, in the proposed technique, we construct the above constraint expression for each branch (i.e., each event sequence from the initial state), and solve the corresponding ILP problem to obtain the executable time of each event in the sequence. Similarly, when an event sequence includes a synchronous event, its executable time range may also differ depending on what combination of synchronous events is executed. For example, event $a!(x1 - 2)$ in sequence $P11$ in Fig. 4 can be executed in synchronization with either $a?x3...$ in $P22$ or $a?x4...$ in $P32$, where the executable time range of a in $P11$ differs depending on which a is synchronized. Thus, we derive each case of the executable time range by constructing the constraint expression for each combination.

4.3. DERIVATION OF AN EXECUTABLE TIME RANGE OF EACH EVENT

In the constraint expression constructed in Sect. 4.2, we used variables representing the "execution time" of events. To obtain the "executable time range", we modify the expression as follows.

As we explained in Sect. 3, we derive the upper and lower bounds of executable time of each I/O event as constants. For the purpose, in the constraint expression defined in Sect. 4.2, we replace variable T_a (representing the execution time of event a) by $T_{a_{min}}$ and $T_{a_{max}}$ which represent the lower and upper bounds of execution time, respectively. Also, for each variable which influences the executable time ranges of some events or the selection of event sequences, we similarly derive the lower and upper bounds for its allowable value range. Hereafter, we denote the execution time of a_k by T_k in event sequence $a_1; a_2; ...; a_n$, and the lower and upper bounds of T_k by $T_{k_{min}}$ and $T_{k_{max}}$, respectively (for each variable, we use $V_{k_{min}}$ and $V_{k_{max}}$).

First, we add the following constraint so that the executable time ranges of a series of events in each event sequence does not have common ranges.

$$T_{1_{min}} \leq T_{1_{max}} < T_{2_{min}} \leq T_{2_{max}} < \cdots < T_{n_{min}} \leq T_{n_{max}} \tag{1}$$

Next, the constraint for synchronous event a_j which synchronizes with event a'_k in another sequence, is replaced by the following constraint so that the executable time range of the event is equal in the both event sequences.

$$T_{j_{min}} = T'_{k_{min}} \wedge T_{j_{max}} = T'_{k_{max}} \tag{2}$$

Finally, timing constraints attached to each event is modified as follows.

For example, the timing constraints of events $e@?t4$ and $b@?t5$ of the following event sequence $P12$ in Fig. 4

$$c?x1@?t1[...]; e@?t4[t4 \leq x1]; b@?t5[t5 < t1 + t4] \tag{3}$$

are represented by the following expressions (4) and (5), respectively, with time variables T_c, T_e and T_b and integer variable V_{x1}.

$$\begin{aligned} T_e - T_c &\leq V_{x1} \\ \Rightarrow T_e &\leq V_{x1} + T_c \qquad (4) \\ T_b - T_e &< T_c + (T_e - T_c) \\ \Rightarrow T_b &< 2T_e \qquad (5) \end{aligned}$$

If we assume that the values of T_c and V_{x1} vary between $[T_{c_{min}}, T_{c_{max}}]$ and $[V_{x1_{min}}, V_{x1_{max}}]$, respectively, the minimum value of the right-hand side of (4) is $V_{x1_{min}} + T_{c_{min}}$. Expression (4) shows that the upper bound of T_e must be equal to or smaller than the value. Similarly, expression (1) shows that the lower bound of T_e must be equal to or greater than $T_{c_{max}}$.

According to the above discussion and expression (4), the following constraint expression for event e is finally derived.

$$T_{c_{max}} \leq T_{e_{min}} \leq T_{e_{max}} \leq V_{x1_{min}} + T_{c_{min}} \tag{6}$$

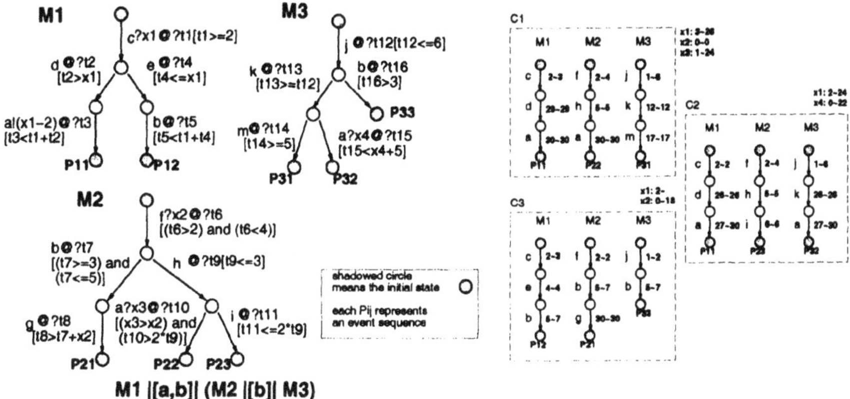

Figure 4 Example of concurrent periodic EFSMs

Figure 5 Executable time range of each event derived from the example in Fig.4

Similarly, for event b, the following constraint expression is derived from expressions (1) and (5).

$$T_{e_{max}} \leq T_{b_{min}} \leq T_{b_{max}} \leq 2T_{e_{min}} \tag{7}$$

By solving a logical conjunction of the above constraint expressions for the following objective function (here, wt_i, wv_j represent weighted coefficients),

$$max \sum wt_i(T_{i_{max}} - T_{i_{min}}) + wv_j(V_{j_{max}} - V_{j_{min}}) \tag{8}$$

we can obtain the executable time range of each event as well as the allowable value range of each integer variable. By setting appropriate values as weighted coefficients of the above objective function, we can derive larger executable time ranges for specified I/O events and/or larger allowable value ranges for some variables.

Here, we applied the proposed technique to the example specification in Fig. 4. We used 30 units of time as the time interval of all EFSMs and used the following objective function for the ILP problem.

$$(T_{a_{max}} - T_{a_{min}}) + (T_{b_{max}} - T_{b_{min}})$$

For three combinations $C1$, $C2$ and $C3$ of synchronous event sequences among $M1$, $M2$ and $M3$, the derived executable time ranges of all events are shown in Fig. 5.

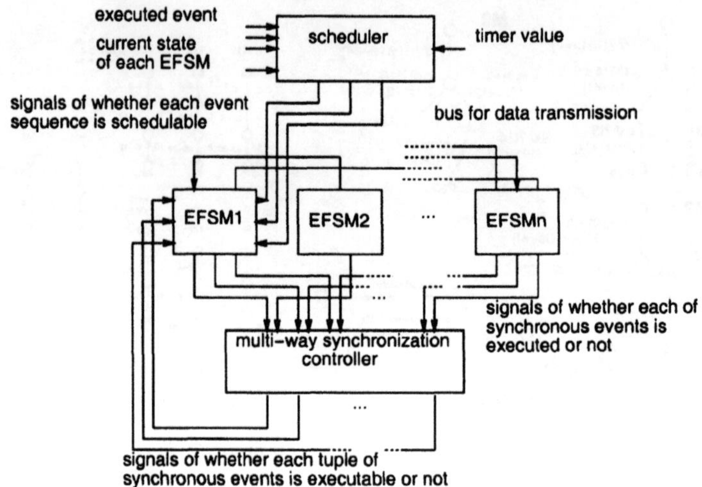

Figure 6 Architecture of the hardware circuit

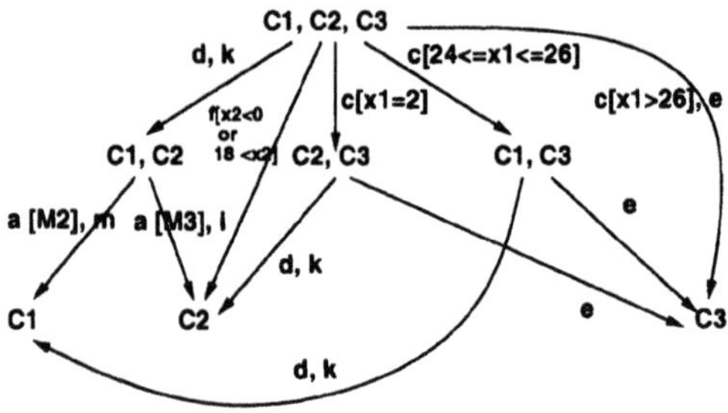

Figure 7 Scheduling tree of concurrent periodic EFSMs in Fig. 4

5. HARDWARE CIRCUIT CONSTRUCTION

5.1. ARCHITECTURE

As shown in Fig. 6, we compose the final circuit of the following three sub-modules: (1) sequential circuits corresponding to EFSMs working with the same clock, (2) a multi-way synchronization controller, and (3) a scheduler to indicate only schedulable event sequences to EFSMs at each time.

The sequential circuit implementing each EFSM has a state register for keeping the current state and a time register for keeping the time elapsed from the beginning of the current time interval (reset to zero in the initial state). The EFSM executes one of events outgoing from the current state at each clock cycle if it is executable. Otherwise, it stays in the same state.

The multi-way synchronization controller is constructed as (i) the executability checking part for checking whether each tuple of synchronous events is executable or not in each clock cycle, and (ii) the conflict avoidance part for selecting one of tuples which become executable but cannot be executed simultaneously. The above part (i) is constructed for each synchronous event tuple. Since each tuple becomes executable when all events in the tuple becomes executable, it is implemented as an AND circuit with n inputs where each input connects to the signal from each EFSM in the tuple (n is the number of events in the tuple). For (ii), although there are several ways for selecting one of conflicting tuples, one of the simplest way is to put priority order among those tuples so that the tuple with the highest priority is selected. In that case, this part can be implemented as a priority encoder.

[7, 14] describe the details of the multi-way synchronization controller.

5.2. SCHEDULER

As we explained in Sect. 4, the executable time range of each event is derived for each event sequence or each combination of synchronous event sequences. So, the time range of an event included in several event sequences or in several combinations of synchronous event sequences is represented as a union of the derived time ranges. The executable time range of each event may change only when the following events are executed: (1) events on a branching state, and (2) synchronous events where each of them can be executed in different combinations of synchronous tuples. Thus, we can know the executable time range of each event by keeping the information about which branch is selected in each EFSM and on what tuple of synchronous event sequences is selected.

In the proposed technique, with the information about branches in each EFSM and the synchronous event tuples derived in Sect. 4.1, we construct the *scheduling tree* as shown in Fig. 7 where each node represents the current schedulable tuples of synchronous event sequences, and each edge represents an event execution which may reduce the number of schedulable tuples. The executable time range of each event and the allowable value range of each variable for each tuple of schedulable event sequences is derived as constant ranges (the lower and upper bounds) using techniques in Sect.4. So, using the ranges and the above scheduling tree, we construct the scheduler module as two parts: (1) a sequential circuit corresponding to a table which outputs the ranges of the executable time of each event and of the allowable value of each variable, depending on the current state in the scheduling tree (such a table can be easily implemented as ROM), (2) and a simple hardware component to check whether the current timer value and the input value are within those ranges.

6. APPLICATION

We have implemented a tool to derive the ranges for execution time of events and variables from a given specification described in E-LOTOS syntax. The tool uses the free-ware called lp_solve [9] to make a solution from the given ILP problems. We have also developed another tool to generate the RTL-level VHDL specification from the original specification and information derived with the first tool. With the tools and usual VHDL synthesis tools on the market, automated synthesis of hardware circuits from high-level specifications in the proposed concurrent periodic EFSMs model becomes possible.

As an example for evaluation, we have described a specification of a video play-back chip for video-conferences which can treat at most two video play-backs in a user terminal, and generated the corresponding hardware circuit with our tools.

In the example, we assume that (1) video streams from other two participants are transmitted to each chip (at each user) by packets (corresponding to video frames) where one packet is sent every 50 time units (each received packet is stored to the buffer in the chip), that (2) time interval between any two succeeding packets changes due to jitter on the network. Each received packet is decoded with an external module such as DSP and the decoded data is output to the display device, that (3) each user can dynamically join and leave the current session of the video-conference, and when a user leave the session, his/her video stream need not be processed in the chip, that (4) the external module can decode a video frame in 35 units of time, and that (5) the time period of all EFSMs is set to 100 units of time.

We modeled the system behavior as six concurrent periodic EFSMs consisting of (1) two displaying parts for drawing video frames in appropriate quality; (2) two admission parts for admitting the joining/leaving requests from two participants; (3) a decoding part for decoding a compressed video frame and outputting the decoded picture data (this operation is actually done by the external module like DSP); and (4) a control part for keeping the current number of participants.

In each EFSM for the displaying part, we have specified three branches where the first branch processes two video frames every time period (high-quality), the second one processes one frame in the same period (low-quality), and the final one does nothing in the case of user's leaving the session. One of the branches is dynamically selected depending on the number of users in the current session. When each EFSM obtains the compressed video frame data from the buffer, it transfers the data to the EFSM of the corresponding decoding part. Since there is only one EFSM for the decoding part and the decoding process takes 35 time units of time, when two displaying parts request decoding at the same time, one of them should be selected alternately (another one is delayed until the decoding of the first one is completed). In that case, the schedulable branch (with lower quality) should be dynamically selected in each EFSM using the proposed technique.

We have generated the corresponding RTL-level VHDL descriptions from the specification of the above example. The total time for the generation was about two seconds where most of the time is consumed by solving the ILP problem to obtain executable time ranges of events. We also synthesized the corresponding hardware circuit using Design Compiler of SYNOPSYS Corp. as the VHDL based synthesis tool on the market where we used the technology of CMOS 0.5 μm. As a result, the size of the synthesized circuit was about 3500 logic gates and its minimum clock period (representing the performance) was 150 nsec.

7. CONCLUSION

In this paper, we have proposed a concurrent periodic EFSMs model suitable to handle parallel and real-time processing, and a hardware implementation technique for the model. In the generated circuits, it is guaranteed that all of branches are schedulable in the initial state, and in any state at least one of the following branches is executable satisfying its timing constraints as long as the preceding I/O events were executed in the time rages calculated in advance. We have also developed tools to generate the corresponding RTL-level VHDL descriptions from given specifications described in E-LOTOS syntax. The generated VHDL specifications can be treated in usual VHDL synthesis tools on the market to synthesize the hardware circuits. Through our experiment for a video playback chip generation, we have confirmed that the proposed model has sufficient expressiveness to describe typical real-time/multimedia systems and that the performance and size of the generated circuit is reasonable for practical use. In [7], we have shown network switches for ATM can be efficiently implemented as hardware circuits with the reasonable size and performance using our concurrent EFSM model where multi-way synchronization can be specified but timing constraints are not treated. The proposed method can also be used to synthesize such network switches requiring some timing guarantees.

Our future work includes integration of cost functions in the ILP problem to efficiently share limited resources among EFSMs.

References

[1] Csopaki, G. and Turner, K. J.: "Modeling Digital Logic in SDL", *Proc. Joint Int. Conf. on 10th Formal Description Techniques and 17th Protocol Specification, Testing and Verification (FORTE/PSTV'97)*, pp. 367-382 (1997).

[2] Dave, B. P. , Lakshminarayana, G. and Jha, N. K. : "COSYN: Hardware-Software Co-synthesis of Embedded Systems", *Proc. 34th Design Automation Conf. (DAC'97)*, pp. 703–708 (1997).

[3] Fernandez, L. S. and Koch, G. and Madrid, N. M. and Vallejo, M. L. L. and Kloos, C. D. and Rosenstiel, W. : "Hardware-Software Prototyping from LOTOS", *Design Automation for Embedded Systems 3*, pp. 117-148 (1998).

[4] IEEE: "IEEE Standard VHDL Language Reference Manual", *IEEE Std 1076-1993*, (1994).

[5] ISO : "Information Processing System, Open Systems Interconnection, LOTOS - A Formal Description Technique Based on the Temporal Ordering of Observational Behaviour", *ISO 8807* (1989).

[6] ISO : "Final Committee Draft 15437 on Enhancements to LOTOS", *ISO/IEC JTC1/ SC21/ WG7*, (1998).

[7] Katagiri, H., Yasumoto, K., Kitajima, A., Higashino, T. and Taniguchi, K.: "Hardware Implementation of Communication Protocols Modeled by Concurrent EFSMs with Multi-Way Synchronization", *Proc. 37th Design Automation Conf. (DAC'2000)*, pp. 762–767 (2000).

[8] Kloos, C. D. and Moro, T. M. and Filho, G. R. and Lopez, A. M. : "VHDL Generation from a Timed Extension of the Formal Description Technique LOTOS within the FORMAT Project", *Microprocessing and Microprogramming*, 38, pp. 589-596 (1993).

[9] Michel, B., Jeroen, D. : "LP_SOLVE", *ftp://ftp.es.ele.tue.nl/pub/lp_solve.*

[10] Sisto, R. : "A Method to Build Symbolic Representations of LOTOS Specifications", *Proc. 15th IFIP Int. Symp. on Protocol Specification, Testing and Verification (PSTV-XV)*, pp. 331–346 (1995).

[11] Smith, J. and De Micheli, G.: "Automated Composition for Hardware Components", *Proc. 35th Design Automation Conf. (DAC'98)* (1998).

[12] Quemada, J., Larrabeiti, D. and Pavón, S. : "Compressing the State Space Representation of LOTOS Specifications", *Proc. 6th IFIP Int. Conf. on Formal Description Techniques (FORTE'93)*, pp. 19–34 (1993).

[13] Wytrebowicz, J. : "Hardware Specification Generated from Estelle", *Proc. 15th IFIP Int. Symp. on Protocol Specification, Testing and Verification (PSTV-XV)*, pp.435–450 (1996).

[14] Yasumoto, K. and Kitajima, A. and Higashino, T. and Taniguchi, K. : "Hardware Synthesis from Protocol Specifications in LOTOS", *Proc. Joint Int. Conf. on 11th Formal Description Techniques and 18th Protocol Specification, Testing, and Verification (FORTE/PSTV'98)*, pp. 405–420 (1998).

MODELING DISTRIBUTED EMBEDDED SYSTEMS IN *Multiclock* ESTEREL*

Basant Rajan
VERITAS Software India Pvt. Ltd.
S.No. 210 A/1, Range Hills,
Pune 411 020, India
basant@tcs.tifr.res.in

R.K. Shyamasundar
School of Technology and Computer Science
Tata Institute of Fundamental Research
Homi Bhabha Road, Bombay 400 005, India
shyam@tcs.tifr.res.in

Abstract In this paper, we show that the paradigm of *Multiclock* ESTEREL can be effectively used for the design of asynchronously communicating distributed systems. First we show that the protocol used in *Multiclock* ESTEREL for the modeling of VHDL can be used for the design of asynchronous interaction of processes, and an analysis can be made relative to speed or periodicity of the underlying processes for a safe implementation without missing any signals. The analysis also shows that one can arrive at a tradeoff between the periodicity and the buffer requirements on the average over a sequence of periods. Then, we illustrate the modeling of *communicating reactive processes* (which is essentially a network of ESTEREL nodes communicating via the rendezvous mechanism) as an instance of *Multiclock* ESTEREL.

Keywords: Embedded Systems, Multiclock Esterel, Synchronous languages.

1. INTRODUCTION

Programming languages (such as ESTEREL, Lustre, Signal, Statecharts etc.) based on the *perfect synchrony paradigm* [3] have proven useful for

*Partially supported by the Indo-French Centre (CEFIPRA) project 2202-1: *Formal Specification and Verification of Reactive Hybrid Systems.*

Tommaso Bolognesi and Diego Latella (Eds.), *Formal Methods for Distributed System Development.*

programming reactive systems where determinism is a must and logical concurrency is required for good programming style. One of the main reasons for its success is due to the fact that it permits the programmer to focus on the logic of reactions and makes it possible to use several automata-based verification systems for correctness proofs.

On the other hand, asynchronous languages (like Ada) have proven useful in distributed processing and implementing algorithms on a network of computer systems where nondeterminism is an appropriate abstraction at the logical and physical levels. Presently, each class of languages is unable to handle problems for which the other class is tailored.

However, complex systems do require the abilities of both types of languages. For instance, a robot driver could use a specific reactive program to control each articulation, but the global robot control may remain asynchronous because of limitations of networking capabilities.

In [4, 8, 2], a paradigm referred to as *Communicating Reactive Processes* (CRP) that provides a unification of perfect synchrony and asynchrony has been proposed. CRP explicitly distinguishes the parallelism at the network layer and ESTEREL layer. In a limited way, the concept of hierarchical network of synchronous nodes was introduced in *Hierarchical CRP* [12] without integrating the two notions of parallelism. In [1, 9, 10], the paradigm of *Multiclock* ESTEREL has been proposed as a unified framework for synchrony and asynchrony that overcomes the need to distinguish the various levels of concurrency. Some of the distinguishing features of *Multiclock* ESTEREL that set it apart from its predecessor ESTEREL are:

- *Multiclock* ESTEREL allows the specification of individual clocks for modules - *Multiclock* ESTEREL can now address a wider class of programs without sacrificing modularity.

- *Multiclock* ESTEREL is truly modular, even when using multiple clocks - By integrating the concept of asynchronous task execution along with the multiple clocks that *Multiclock* ESTEREL allows, *Multiclock* ESTEREL has managed to preserve modularity.

- *Multiclock* ESTEREL allows for asynchronous input events - Having allowed multiple clocks to govern the execution of different modules in the program and provided for interaction amongst them, *Multiclock* ESTEREL faced no difficulty dealing with asynchronous external events.

- *Multiclock* ESTEREL implements internal latching of all inputs - This feature serves to integrate seamlessly the domain of asynchronous events with the realm of synchronous execution semantics.

In this paper, we show that the framework of *Multiclock* ESTEREL and the protocols proposed for implementation can be effectively used for:

1 the specification and analysis of asynchronously interacting distributed systems.

2 the specification and analysis of networks of communicating processes with a unified view of synchrony and asynchrony. This in turn shows that classes of distributed systems such as quasi synchronous systems [5] can be effectively specified, implemented and verified in this framework.

The rest of the paper is organized as follows: An informal introduction to *Multiclock* ESTEREL is given in section 2 with an appendix in section 6 giving the informal semantics of *Multiclock* ESTEREL statements. Sections 3 and 4 describe the analysis and specification of asynchronous interaction among processes and the specification of CRP [4] in *Multiclock* ESTEREL respectively. These sections are followed by the concluding remarks.

2. DESCRIPTION OF *Multiclock* ESTEREL

Multiclock ESTEREL is based on the synchronous language ESTEREL which in turn is based on the *perfect synchrony hypothesis*. *Multiclock* ESTEREL is a language designed to support the explicit specification of clocks as is usual in the specification of hardware circuits. The perfect synchrony hypothesis has been adapted to the explicit specification of clocks in the sense that while in ESTEREL the reactions to external events were just controlled by the environment, in *Multiclock* ESTEREL reactions are determined by the clocks controlling the current execution by appropriately abstracting the input from the environment and reacting to it in the form of outputs.

The syntax and form of *Multiclock* ESTEREL closely resembles that of its precursor, ESTEREL. The general structure of a *Multiclock* ESTEREL program consists of a collection of named modules composed using the parallel or sequence operator. Each of these modules has an explicitly specified clock associated with it.

Each module in turn is built of either sub-modules or simple statements composed using either the parallel or sequence operator. *Multiclock* ESTEREL is for the most part block structured but supports explicit access control that *limits* the scope of objects. All variables and signals need to be declared before use as is the case in ESTEREL. Variables cannot be shared across modules composed in parallel.

Multiclock ESTEREL supports a simple complement of data types that includes *integers*, *floating point* numbers and *character strings* in addition to *signals*, *latches*, *clocks* and *sensors*. However, complex data structures like records are not supported.

While variables of simple types like *integers* are valid only in the modules defining them, signals can be shared across modules. Signals are used as an interface to communicate with the external world. Signals are distributed using an instantaneous broadcast mechanism just as in other synchronous programming languages like ESTEREL, SIGNAL *etc.*. This in turn implies that the statement emitting a signal will see the signal in its current execution environment.

Since every statement is bound to a clock, there is obviously a need to provide some amount of limited memory in the form of *latches* to interface the asynchronous environment to the reactive core. In the sequel, we shall describe latches[1] in section 2.1, expressions in section 2.2 and statements in section 6.

2.1. LATCHED SIGNALS

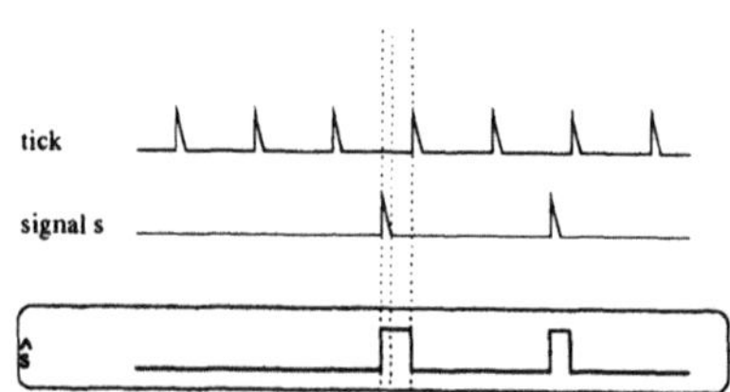

Figure 1 Clock, signal and latches

The capability of *Multiclock* ESTEREL to explicitly bind *clocks* to specific signals introduces certain problems. For instance, how does one treat events that occur *out of sync* with the current *tick* signal?

Figure 1 shows the relationship between a clock, a signal and the latched version of the signal as used in *Multiclock* ESTEREL .

- the waveform labeled *tick* is assumed to be the current clock. It is not necessary that the clock be periodic as shown.

- the waveform labeled *signal s* represents an arbitrary signal occurring in the environment of a *Multiclock* ESTEREL module.

- $\hat{s}$ denotes a signal that sustains the value of s until the *next* clock tick.

  The interval for which $\hat{s}$ sustains the signal s *includes* the next clock *tick*. In particular, $\hat{s}$ generated for any given event will never span *two* clock *ticks*. This in turn implies that if s occurs synchronously with a clock tick, its value will not remain latched until the next instant.

Latching serves to implement *limited* memory in asynchronous environments. *Multiclock* ESTEREL uses a technique similar to that used above to deal with signals that occur out of synchronization with the current clock. The use of *latched* signals makes interfacing modules/statements with different clocks easy. Signals may never be accessed using disparate clocks, but the *latched* versions of these signals can be read using any clock. Therefore parameter passing between modules with no common clock can be effectively accomplished using latched signals.

2.2. EXPRESSIONS

Multiclock ESTEREL supports two kinds of expressions.

1 The first kind ranges over signals and are referred to as *clock* expressions. Clock expressions are booleans functions over signals and cannot contain references to variables or sensors. They are used to specify bindings for the local clock or *tick*.

2 The second type of expression ranges over variables as well as *latched* version of signals and are referred to as signal expressions. These expressions are what *Multiclock* ESTEREL code can test for as part of its execution sequence. Note that clock expressions can only be used to bind clock signals. *Latches* of signals are derived entities.

3. MODELING INTERACTION BETWEEN ASYNCHRONOUS PROCESSES

In this section, we show that *Multiclock* ESTEREL and its implementation techniques can be used for modeling asynchronous interaction among processes. We begin with a method for modeling the VHDL feature of *scheduling events at later instants* in *Multiclock* ESTEREL as proposed in [1, 9]. The broad block diagram of the modeling is shown in Fig. 3.

Since the Esterel-module and the Signal-Driver are part of the Esterel code at the same node, we assume without loss of generality that the Esterel program can be denoted by a single node where the synchrony hypothesis holds. Under this assumption, the underlying protocol for scheduling of events is depicted in Fig. 2 where:

$S_1, \cdots, S_n$:	signals to be scheduled later
$latch\_S_1, \cdots, latch\_S_n$:	signals latched for the event scheduler
$latch\_\widehat{S}_1, \cdots, latch\_\widehat{S}_n$:	signals sent from the event-scheduler

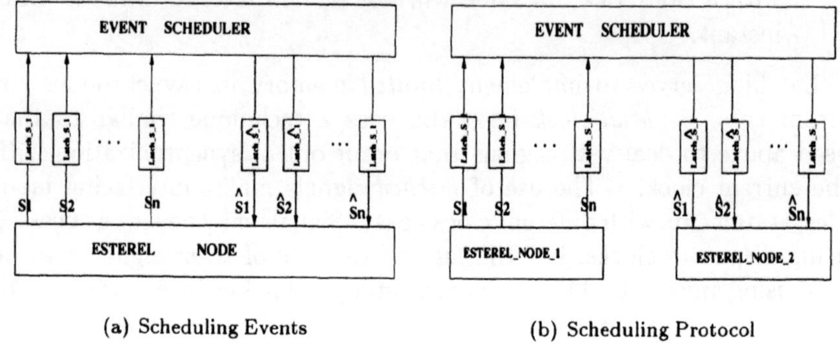

(a) Scheduling Events (b) Scheduling Protocol

Figure 2 Scheduling in Multiclock Esterel

The steps of the implementation as proposed in [1, 9] are given below:

1 On signal emission from Est-node (S_i) working on clock t_{est}, it sets the variable "$latch\_S_i$" to $value(S_i)$.

2 Variable "$latch\_S_i$" is read by the event-scheduler working on its clock t_{sch}.

3 The event-scheduler sets "$latch\_S_i$" to $\perp$ (denoting absence of signal) after reading it so that the read signal will not be mistaken as an unread signal in the next instant.

4 Similar reading/resetting takes place on the dual side.

The requirement of not missing any signal corresponds to the condition given below:

$$\forall_t \forall_i \ \widehat{S_i} = S_i \qquad (1)$$

Let t_{est} and t_{sch} be the clocks of the ESTEREL module and the scheduler-module respectively. Satisfaction of condition (1) implies the following condition:

$$t_{sch} \ \textit{"faster than"} \ t_{est} \cdots (2)$$

where the relation *"faster than"* is a relation that must ensure the following requirement:

> The clock tick of t_{sch} is frequent enough to consume and deliver it to the Esterel module on t_{est} before the ESTEREL module produces another item to be scheduled.

Notationally, we represent *"faster than"* by the symbol "$\sqsupseteq$". $\cdots$ (2a)

In the above protocol, the interaction was from the Esterel-module to itself via the event-scheduler. Now, let us see the extension of the same

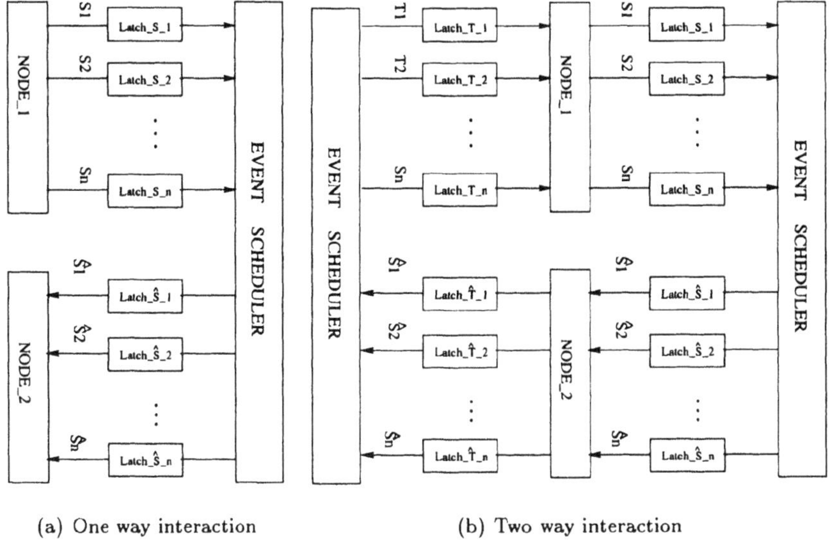

(a) One way interaction (b) Two way interaction

Figure 3 Communication in Multiclock Esterel

where one process interacts with another. The extended protocol with such an extension is depicted in Fig. 2.b where t_{N_1} and t_{N_2} are the clocks of nodes Node_1 and Node_2 respectively. The constraints required for not missing signals are given below:

$$t_{sch} \supseteq t_{N_1} \quad \cdots \quad (3)$$
$$t_{sch} \supseteq t_{N_2} \quad \cdots \quad (4)$$

Obviously, conditions (3)-(4) are not sufficient as we need to ensure relation (2a) between the two processes logically. The corresponding equation follows by considering Node_2 as the consumer and Node_1 as the producer and is given below:

$$t_{N_2} \supseteq t_{N_1} \quad \cdots (5)$$

Now let us consider the situation wherein the processes (Node_1 and Node_2) interact with each other. The scenario is depicted in Fig. 2 wherein for simplicity we have used the same scheduler instead of two schedulers. Now, in addition to conditions (3)-(5), we need the following condition as well (corresponding to the fact that Node_1 is the consumer and Node_2 is the producer):

$$t_{N_1} \supseteq t_{N_2} \quad \cdots (6)$$

From conditions (5) and (6), we can conclude the following condition:

Node_1 and Node_2 are *no faster than* each other ··· (7)

Condition (7) can be interpreted as follows:

Within any two clock ticks of one process the other process can have only one tick. ··· (8)

The above condition is exactly the condition used in [5] for the distributed set of processes called *quasi synchronous processes.* These are periodic processes asynchronously interacting with each other. In other words, it amounts to saying that the period of the processes are the same except that they can be out of phase. The advantage of the approach described above is that we can effectively use the framework of synchronous languages in terms of techniques and tools of verification of synchronous languages for the quasi synchronous systems.

The above conditions can be generalized relative to clocks and buffers. For instance, it is possible to consider the buffer requirements over a sequence of periods of certain processes so that it is possible to achieve the soundness requirement of not missing any signal. In fact, the semantic framework described in [13] can be used.

4. MODELING CRP

A new programming paradigm called Communicating Reactive Processes or *CRP*[4] that unifies the capabilities of asynchronous and synchronous concurrent programming languages is described in [3, 4] with a view to specifying complex reactive systems which usually have both synchronous and asynchronous features. A *CRP* program consists of independent locally reactive ESTEREL nodes that communicate with each other by *rendezvous* communication. Essentially, a *CRP* program consists of a network $M_1 \parallel M_2 \parallel M_n$ of ESTEREL reactive programs or *nodes*, each having its own input/output signals and its own notion of an instant.

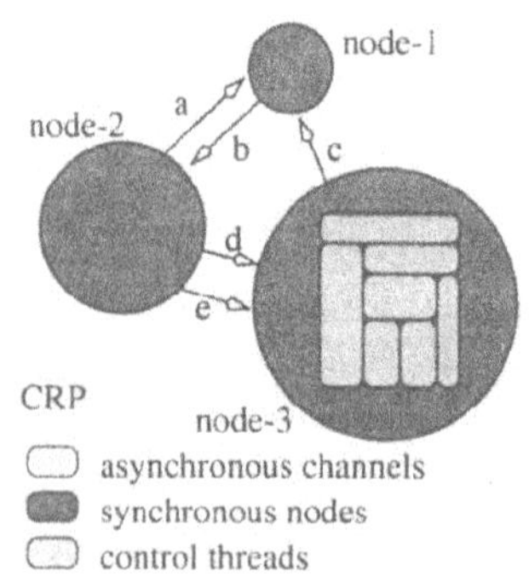

Figure 4 The *CRP* Environment

4.1. EMBEDDING *CRP*

Embedding CRP in *Multiclock* ESTEREL involves (i) controlling the visibility of the channels as required and (ii) the protocol for implementing the rendezvous and the guarded choice.

First, let us consider access control in *Multiclock* ESTEREL . In addition to normal block structured scoping, *Multiclock* ESTEREL permits the use of declarations to *restrict* the visibility of objects. This feature does not modify any of the underlying concepts involved in latching and clocking modules.

This is a generalization of the *CRP* scenario (Figure 4) where ESTEREL nodes had explicit names and channels declared between nodes were visible only to the nodes they connect.

Example 4.1 The ability to arbitrarily restrict the visibility of objects is a prerequisite if we are to correctly implement *CRP* programs within the framework of *Multiclock* ESTEREL because *CRP* programs involve the existence of named channels that are private to the nodes connected by them.

Now, let us turn to protocols required for implementation. The general framework for the implementation would involve implementing each *CRP* node as an independently clocked *Multiclock* ESTEREL module and using signals (which are automatically latched) to communicate amongst them. Private channels are implemented by restricting the scope of the signals constituting the channel.

Considering the configuration given in Figure 4 we could have the following partial implementation.

```
module node3 :
input d, e;
output c visible only to node1
   newtick clock3 in
      statement
   end
end
```

Note that the code for node *node-3* is free to execute on its own private clock *clock3*. This code segment however does not realize *rendezvous* as defined by the *CRP* paradigm. This example only provides the necessary framework to implement *rendezvous*.

We now extend Example 4.1 to implement *rendezvous* as defined in *CRP* on the lines of the implementation described in [2]. Note that we follow the synchronization protocol as given in [2].

Example 4.2 We give below code snippets that implement the *rendezvous* protocol of *CRP*. The code is divided into *sender* and *receiver* components.

```
01 module sender :
02 input rcpt, rabrt, labrt;
03 output data, conf, sabrt, success, fail;

05    present labrt else
06       emit data;
07       await case
08             labrt :  emit sabrt
09             emit fail
10             rcpt :  emit conf
11             emit success
12             rabrt :  emit fail
13          end
14       end
15 end
```

The sender module above works by first emitting the signal *data* and then waiting for either an *rabrt* or *rcpt* signal to arrive from the receiver. On receipt of the *rcpt* signal, the sender confirms the *rendezvous via* the *conf* signal. If on the other hand, the sender is locally preempted using the signal *labrt* it responds by emitting the *sabrt* signal and terminates. Receipt of signal *rabrt* also causes unsuccessful termination of the *rendezvous*. Note that the sender module has not defined an explicit clock and will run at the default one. Further note that preemption at the initial instance precludes the emission of signal *data*. A successful *rendezvous* is characterized by the presence of signal *success* at termination.

```
01 module receiver :
02 input data, sabrt, conf, labrt
03 output rcpt, rabrt, success, failure

05    present labrt else
06       await case
07          labrt :  present data then
08                      emit rabrt
09                      emit failure
10          end
11             data :  emit rcpt
12             await case
13                labrt :  emit failure
14                conf :  emit success
15                sabrt :  emit failure
16             end
17          end
18       end
19 end
```

The receiver module given above works by waiting for a *data* signal to arrive from a sender module and then acknowledging it by emitting a *rcpt* signal. It then awaits the arrival of a *conf* signal which will indicate successful termination. Note that preemption at the initial instant is transparent to the sender whereas preemption simultaneous with the receipt of the *data* signal causes the emission of the *rabrt* signal. Note also that preemption occurring after the *rcpt* signal is emitted will not prevent the sender from treating the *rendezvous* as complete. As in the case of the sender module, no explicit clock has been specified.

Example 4.2 above provides an overview of how *Multiclock* ESTEREL could be used to implement a *CRP* style *rendezvous*. However, this alone does not suffice to implement the generic *CRP* paradigm wherein *guarded choice* could be involved in *selecting* a particular *channel* from amongst many possibilities. The following example outlines how this could be achieved.

Example 4.3 We present an example of how *guarded choice* as in the *CRP* paradigm could be implemented in *Multiclock* ESTEREL. For the sake of brevity we consider only two *rendezvous* in each guard.

```
01 module receiver :
02 input dataA, sabrtA, confA,
      dataB, sabrtB, confB, labrt;
04 output rcptA, rabrtA,
      rcptB, rabrtB, success, failure;

07    present labrt else
08       await case
09          labrt :  present dataA then
10             emit rabrtA
11          end
12          present dataB then
13             emit rabrtB
14          end
15          dataA : emit rcptA
16          await case
17             labrt :  emit failure
18             confA : emit success
19             sabrtA : emit failure
20          end
21          dataB : emit rcptB
22          await case
23             labrt :  emit failure
24             confB : emit success
25             sabrtB : emit failure
26          end
27       end
28    end
29 end
```

The code for a receiver module above works by making a choice on which of signals *dataA* and *dataB* to respond to and then behaving like that was the only rendezvous being attempted. As a consequence of the deterministic nature of *Multiclock* ESTEREL, the implementation above is constrained to select signal *dataA* whenever both signal *dataA* and signal *dataB* occur together. Also note that the decision to implement choice in this manner precludes the possibility of allowing *mixed* guards (*i.e.* mixed input and output channels) as this could lead to implementation induced deadlocks.

```
01 module sender :
02 output dataA, sabrtA, confA, dataB,
03       sabrtB, confB, success, failure;
04 input rcptA, rabrtA,
05       rcptB, rabrtB, labrt;

06    present labrt else
07       emit dataA;
08       emit dataB;
09       await case
10          labrt :  emit sabrtA
11          emit sabrtB
12          rcptA : emit confA
13          emit sabrtB
14          emit success
15          rcptB : emit confB
16          emit sabrtA
17          emit success
18          rabrtA : present rabrtB then
19             emit failure
20          else
21             await case
22                labrt :  emit sabrtB
23                emit failure
24                rabrtB : emit failure
25                rcptB : emit confB
26                emit sabrtA
27                emit success
28             end
29          end
30          rabrtB : present rabrtA then
31             emit failure
32          else
33             await case
34                labrt :  emit sabrtA
35                emit failure
36                rabrtA : emit failure
37                rcptA : emit confA
38                emit sabrtB
39                emit success
40             end
41          end
42       end
43    end
44 end
```

The structure of the sender code above implementing *guarded rendezvous* makes it amply clear that it depends on the receiver to limit the choice. The sender emits both signal *dataA* and signal *dataB* and awaits a response of either signal *rcptA* or signal *rcptB.* Receipt of either one of these signals guarantees that the *rendezvous* can now successfully terminate or will be locally preempted, in which case, it is acceptable to issue an abort on the other channel. Note that receipt of an abort message from a receiver causes the sender to await a response from the remaining receiver unless there already is an abort message from it too.

It is obvious that in both the sender and receiver modules presented, only one *rendezvous* can terminate successfully. It is also easy to see that this implementation does not introduce any deadlock.

In the above, we have illustrated that the framework of *Multiclock* ESTEREL embed the CRP as an instance. Since the framework does not need to distinguish various levels of concurrencey and also the constraints of causality of ESTEREL (due to the underlying asynchronous interfaces), it can be generally used the specification and verification of a hierarchical network of processes.

5. CONCLUSIONS

We have shown how the paradigm of *Multiclock* ESTEREL and the protocols used for the implementation can be used for the specification and implementation of distributed systems and in particular, its advantages for the specification of globally asynchronous and locally synchronous systems (often called GALS in the hardware literature). In section 3, we illustrated the specification of quasi synchronous systems [5] in *Multiclock* ESTEREL and showed how such an analysis can be augmented for buffer requirements with the given underlying architectural constraints; in fact, such an analysis can be augmented with the analysis relative to various communication mechanisms as discussed in [13]. In section 4, we illustrated CRP [4] as an instance of *Multiclock* ESTEREL . Using these illustrations, it is possible [11] to effectively use *Multiclock* ESTEREL for the specification, design and implementation of time-triggered architectures [7] that are widely used for distributed embedded systems including web-based interactions. The study of the relationship of *Multiclock* ESTEREL with other clocked frameworks such as Lustre-scade [5], Mode automata [6] with the view of designing distributed systems is in progress.

References

[1] Basant Rajan. *Programming Languages: Specification & Design of Multiple-Clocked Systems.* PhD thesis, Tata Institute of Fundamental Research, 98.

[2] Basant Rajan and R.K. Shyamasundar. An Implementation of CRP. In *IASTED*, Singapore, 97. (Also TIFR/TCS-95/7 Bombay, India).

[3] G. Berry and G. Gonthier. The Esterel Synchronous Programming Language: Design semantics, Implementation. *SCP*, 19(2):87–152, Nov 92.

[4] G. Berry, S. Ramesh, and R.K. Shyamasundar. Communicating Reactive Processes. *20th ACM POPL*, pages 85–99, Jan 93.

[5] P. Caspi, C. Mazuet, R. Salem, and D. Weber. Formal design of distributed control systems with lustre. In *SafeCom99*, Sept 1999.

[6] Yann Remond F. Maraninchi. Mode-automata: About modes and states for reactive systems. volume ESOP of *LNCS*. Springer-Verlag, 98.

[7] H. Kopetz, M.Kucera, D. Millinger, C. Ebner, and I. Smaili. Interfacing time-triggered embedded systems. In *Proc. of the Int. Symposium on Internet Technology*, pages 180–186, Taipei, Taiwan, Apr 1998.

[8] B. Rajan and R.K. Shyamasundar. Networks of preemptible reactive processes: An implementation. In *Int. Conf. on VLSI Design*, New Delhi, India, Dec 1995.

[9] B. Rajan and R.K. Shyamasundar. *Multiclock* ESTEREL: A reactive framework for asynchronous design. In 13th *Intl. Conf. on VLSI Design*, pages 76–83, Calcutta, India, Jan 2000.

[10] B. Rajan and R.K. Shyamasundar. *Multiclock* ESTEREL: An asynchrnous framework for asynchronous design. In *Int. Parallel and Distributed Processing Symposium*, Cancun, Mexico, May 2000.

[11] Basant Rajan, S.A. Dahodwala, T.M. Topiwalla, and R.K Shyamasundar. Time triggered systems in *Multiclock* ESTEREL'Manuscript, 2000.

[12] S. Ramesh R.K. Shyamasundar. Semantics and verification of hierarchical programs. volume 999 of *LNCS*, pages 436–461. Springer-Verlag, 1995.

[13] T. Pitassi RK Shyamasundar, KT Narayana. Semantics of nondeterministic asynchronous broadcast networks. *Information and Computation*, 104:215–252, Jun 1993.

6. APPENDIX: INFORMAL SEMANTICS

Being a synchronous language, *Multiclock* ESTEREL statements follow a *zero delay* semantics and are therefore considered to execute (and terminate) instantaneously. The only exception to this being an explicit *pause* statement that terminates only on the next occurrence of the current clock.

Unlike ESTEREL, where there is an implicit *global* clock (the *tick*) at which the entire ESTEREL program makes transitions, in *Multiclock* ESTEREL modules are independently and explicitly clocked. This feature is a major deviation from ESTEREL and accounts for almost all the enhancements that *Multiclock* ESTEREL offers. Clocks can be specified as boolean expression over input signals.

We will now describe the individual constructs that *Multiclock* ESTEREL programs are built from. We shall let p, q range over programs and s over signals in the rules that follow.

Statement : ***nothing***
This statement does nothing and terminates instantaneously.

Statement : ***pause***
Waits until the next occurrence of the clock to which it is bound. In the classical ESTEREL it corresponds to waiting for the next input event.

Statement : ***emit*** *s*
This statement causes an instantaneous broadcast emission of the signal s and terminates instantaneously with the current clock.

Statement : p;q
The statement starts executing p immediately and behaves as it will as long as statement p (or its derivatives) remains active. Once p terminates control is passed to q and the behavior thereafter is defined by it. Due to the synchrony hypothesis, q starts in the same instant in which p terminates. For example, " `emit s1 ; emit s2` " results in the simultaneous broadcast of `s1` and `s2` and the statement terminates instantaneously.

Statement : ***loop*** *p* ***end***
The body of the loop p is started *instantaneously* the first time *and* whenever statement p or its derivatives terminates. For this reason, it is necessary that p does not terminate instantaneously. The `loop` statement never terminates but could be *exited* due to preemption.

Statement : ***present*** *s* ***then*** *p* ***else*** *q* ***end***
The statement tests for the presence of signal *s* and executes either statement *p* or statement *q* depending on whether signal *s* is present or absent. Both the test and the transfer of control to either statement *p* or statement *q* are done synchronously with the clock.

Statement : $p \| q$
Each of the composed statements share the same clock and make independent transitions of their own. Parallel components can communicate only through shared signals. Variables cannot be shared by parallel modules. Both the component statements are initiated synchronously with the initiation of the construct and the construct terminates synchronously with the termination of the component that terminates last.

Statement : ***signal*** *s* ***in*** *p* ***end***
The statement *p* is started immediately with a fresh signal *s* local to the block *p* overriding other bindings. It ensures that any reference to signal *s* within the statement block *p* is bound to the local signal *s* and this signal is not exported outside the block.

Statement : ***abort*** *p* ***when*** *s*
This constructs implements strong preemption in *Multiclock* ESTEREL. Statement *p* is started immediately and terminates either when statement *p* terminates normally or whenever the signal *s* occurs in the future, in which case the whole statement gets *preempted* instantaneously. That is, if signal *s* occurs during the execution of statement *p*, all processing scheduled for the current instant by statement *p* is abandoned.

Statement : ***weak abort*** *p* ***when*** *s*
Multiclock ESTEREL also provides for an alternate form of preemption termed weak preemption. The salient difference being that all processing scheduled for the current instant by statement when preemption is effected is allowed to execute to completion performing the *last wills*.

Statement : ***suspend*** *p* ***when immediate*** *s*
This construct executes statement *p* synchronously when it starts and terminates only when statement *p* does. However, statement *p* is prevented from executing whenever signal *s* is present. During this period however, changes to the environment are recorded. Execution of statement *p* resumes when signal *s* is no longer present.

*Statement : **trap** T **in** p **end***
The body **s** is started immediately and behaves as p until termination or an explicit exit. If statement p terminates, so does the **trap** construct. If p exits T, then the **trap** terminates in the same instant after completing the concurrent reactions. If p exits an enclosing trap U, then the exit is propagated upwards appropriately.

*Statement : **exit** T*
The statement exits the trap T instantaneously terminating the corresponding **trap** statement unless an enclosing trap is exited synchronously.

*Statement : **newtick** s **in** p **end***
This construct is what sets *Multiclock* ESTEREL apart from ESTEREL. This construct starts synchronously at the current clock and initiates execution of statement p on clock s. Statements in p do not start executing until clock s occurs. All statements in p see signals in their environment through values latched on the new clock s. The construct terminates synchronously with the first occurrence of the current clock after the termination of statement p which will in turn terminate synchronously with the new clock s. Further, statements in statement block p are insulated from any suspensions imposed on this construct. For all practical purposes, statement p can be considered an independent process.

*Statement : **weak newtick** s **in** p **end***
This is a variant of the newtick construct introduced above. This construct starts synchronously at the current clock and initiates execution of statement p on clock s. However, statements in statement block p are *not* insulated from any suspensions imposed on this construct. The keyword **weak** is used to signify that the change of clock is not sufficient to protect the enclosed statements from suspensions imposed on the construct.

Some of the derived constructs which are used widely are given in the following. The **halt** construct is given by "**loop** pause **end**". Similarly, the derived construct "**await** s" can be written as "**abort halt when** s". Similarly, "**sustain** s" can be rewritten as " **loop emit** s ; **pause end**. A module can be instantiated in another module through the **run** statement. For example " **run** DRAM" replaces the **run** statement by the body of the DRAM module.

Part Eight

FORMAL SEMANTICS

COMPACT NET SEMANTICS FOR PROCESS ALGEBRAS

Marco Bernardo, Marina Ribaudo
Università di Torino, Dipartimento di Informatica
Corso Svizzera 185, 10149 Torino, Italy
bernardo, marina@di.unito.it

Nadia Busi
Università di Bologna, Dipartimento di Scienze dell'Informazione
Mura Anteo Zamboni 7, 40127 Bologna, Italy
busi@cs.unibo.it

Abstract Process algebras and Petri nets are two well known formal methods. In the literature, several mappings from process algebras to Petri nets have been proposed to provide a truly concurrent semantic framework to concurrent programming languages. From an applicative point of view, such mappings facilitate the integration and the cross fertilization between the two formalisms, making it possible the development of a multiparadigm methodology for the modeling and analysis of concurrent systems since the early stages of their design. In this paper we define a new Petri net semantics for process algebras, we demonstrate that it improves the previous proposals with respect to the size of the resulting nets, and we illustrate the usefulness of the net semantics by showing how to reinterpret at the process algebra level results efficiently obtainable at the Petri net level.

1. INTRODUCTION

The complexity of modern computer and communication systems calls for the development of suitable description techniques for their modeling and analysis. The objective is to formally describe the relevant aspects of the systems under investigation since the early stages of their design, so that malfunctionings and inefficiency can be timely detected. Among such formal description techniques, in this paper we focus our attention on process algebras and Petri nets, which have been extensively investi-

Tommaso Bolognesi and Diego Latella (Eds.), *Formal Methods for Distributed System Development.*

gated in the last four decades (see, e.g., [9, 12]) not only in isolation but also jointly. Initially, the purpose was that of providing a truly concurrent semantic framework to concurrent programming languages [8]. This was accomplished by means of semantic mappings from process algebras to Petri nets. In this paper we concentrate on operational (as opposed to denotational) semantic mappings, which have been defined according to two different approaches.

In the *location oriented approach*, the Petri net corresponding to a process term contains a place for every position of the sequential subterms w.r.t. the static operators. As a consequence, the resulting nets turn out to be 1-safe. All the information about the syntactical structure of terms is encoded within places so that the relationships among sequential terms are smoothly preserved. This is exploited to define net transitions by means of inductive rules similar to those for the interleaving semantics for the process algebra, as shown in [7, 10].

In the *label oriented approach*, instead, each net place corresponds to a sequential subterm independently of its position w.r.t. static operators. As a consequence, instances of the same sequential term occurring in different positions w.r.t. static operators can be represented by multiple tokens within the same place. This is possible because the syntactical structure of the process term w.r.t. static operators is no longer fully retained within net places associated with sequential subterms. Therefore, the correspondence between the inductive rules for generating net transitions and the inductive rules of the interleaving semantics for the process algebra no longer holds. In this approach net transitions are defined by axioms. For instance, in [4] five axioms are employed to derive net transitions, with the resulting nets including places with outgoing inhibitor arcs to model scope extrusion and unguarded choice. As another example, in [1] only one axiom is employed, the price to be paid being the introduction of places with outgoing inhibitor and contextual arcs in which information about the syntactical structure is kept. The advantage of the label oriented approach is that the resulting nets are not necessarily 1-safe, so they can be more compact than those obtained with the location oriented approach. Furthermore some terms, for which an infinite net is generated in the location oriented approach, give rise to a finite net in the label oriented approach. This fact has been exploited e.g. in [5], where the finiteness of the generated nets is used to show an expressiveness gap bewteen two different semantics for a process algebra based on Linda coordination primitives.

More recently, it has been recognized that the net semantics for process algebras have an applicative relevance, as they facilitate the integration of the two formalisms making it possible the exchanging of modeling capabilities and analysis techniques between them. This is especially important since the two formalisms are characterized by complementary strengths and weaknesses. On the one hand, process algebras offer a compositional linguistic support to system modeling together with notions of equivalence and preorder. On the other hand, Petri nets provide a truly concurrent framework equipped with analysis techniques that avoid the construction of the underlying state space.

Given the applicative relevance of the net semantics for process algebras, the label oriented approach seems to be the approach of choice due to the compactness of the resulting nets. The weakness of this approach is that additional places with outgoing inhibitor and contextual arcs are employed to handle static operators. Actually, in [1] it has been proved that these places and arcs can be removed a posteriori; this makes the nets simpler but requires an additional computational step. In this paper we further elaborate on the label oriented net semantics of [1] by avoiding the introduction of additional places with inhibitor and contextual arcs altogether. This will be achieved by suitably decorating actions within the sequential terms associated with net places in order to keep track of the occurrences of static operators, without falling in the excess of information of the location oriented approach.

This paper is organized as follows. In Sect. 2 and 3 we present process algebras and Petri nets, respectively. In Sect. 4 we define the new label oriented net semantics for our example process algebra and we prove its correctness by showing that the interleaving semantics for the process algebra can be retrieved from it. In Sect. 5 we discuss the finiteness and the compactness of the generated nets, the ability to detect some kinds of symmetry, and the reinterpretation at the process algebraic level of results efficiently obtainable at the net level. Finally, Sect. 6 concludes the paper with ongoing and future work. Due to lack of space, the reader is referred to [2] for proofs and additional examples.

2. PROCESS ALGEBRA

The process algebra we shall use throughout the paper is composed of a set *Act* of actions, ranged over by a, b, c, d and including the invisible action τ, and the standard operators. Let *Const* be a set of constants ranged over by A, B, C and let $Rel = \{\varphi : Act \longrightarrow Act \mid \varphi^{-1}(\tau) = \{\tau\}\}$ be a set of action relabeling functions preserving action observability.

Definition 1 The set $\mathcal{L}$ of process terms is generated by the grammar

$$E ::= \sum_{i \in I} a_i.E_i \mid E/L \mid E[\varphi] \mid E \parallel_S E \mid A$$

where I is a finite set of indices and $L, S \subseteq Act - \{\tau\}$. We denote by

$sort(E)$ the set of actions occurring in E and by $\mathcal{G}$ the set of closed and guarded terms of $\mathcal{L}$. ■

Thus we have standard operators such as the guarded alternative composition operator (denoted by $\underline{0}$ whenever $I = \emptyset$), the hiding operator, the relabeling operator, the parallel composition operator with CSP like synchronization, and the constant invocation. We recall that the hiding, relabeling and parallel composition operators are said static operators.

$\sum_{i \in I} a_i.E_i$ expresses a nondeterministic choice among several alternatives. If action a_k, $k \in I$, is executed first, then the derivative term is E_k and the other summands are discarded. E/L behaves as term E except that each executed action is turned into τ whenever it belongs to L. $E[\varphi]$ behaves as term E except that each executed action a becomes $\varphi(a)$. $E_1 \parallel_S E_2$ asynchronously executes actions of E_1 or E_2 which do not belong to S and synchronously executes identical actions of E_1 and E_2 which belong to S. Finally, each constant A comes equipped with an equation of the form $A \triangleq E$ defining its behavior.

$$\sum_{i \in I} a_i.E_i \xrightarrow{a_i} E_i \quad i \in I$$

$$\frac{E \xrightarrow{a} E'}{E/L \xrightarrow{a} E'/L} \text{ if } a \notin L \qquad \frac{E \xrightarrow{a} E'}{E/L \xrightarrow{\tau} E'/L} \text{ if } a \in L$$

$$\frac{E \xrightarrow{a} E'}{E[\varphi] \xrightarrow{\varphi(a)} E'[\varphi]} \qquad \frac{E \xrightarrow{a} E'}{A \xrightarrow{a} E'} \text{ if } A \triangleq E$$

$$\frac{E_1 \xrightarrow{a} E_1'}{E_1 \parallel_S E_2 \xrightarrow{a} E_1' \parallel_S E_2 \quad E_2 \parallel_S E_1 \xrightarrow{a} E_2 \parallel_S E_1'} \text{ if } a \notin S$$

$$\frac{E_1 \xrightarrow{a} E_1' \quad E_2 \xrightarrow{a} E_2'}{E_1 \parallel_S E_2 \xrightarrow{a} E_1' \parallel_S E_2'} \text{ if } a \in S$$

Table 1 Operational interleaving semantics for $\mathcal{G}$

The interleaving semantics for $\mathcal{G}$ is represented by a labeled transition system (LTS for short) where states are in correspondence with terms and labels are actions. Such a LTS is generated according to the inductive rules shown in Table 1.

Definition 2 The interleaving semantics of $E \in \mathcal{G}$ is the LTS $\mathcal{I}[\![E]\!] = (S_{E,\mathcal{I}}, Act, \longrightarrow_{E,\mathcal{I}}, E)$ where $S_{E,\mathcal{I}}$ is the least subset of $\mathcal{G}$ reachable from E and $\longrightarrow_{E,\mathcal{I}}$ is $\longrightarrow$ restricted to $S_{E,\mathcal{I}} \times Act \times S_{E,\mathcal{I}}$. ■

3. PETRI NETS

A Petri net [12] is a bipartite graph whose classes of nodes are called places (representing system conditions and resources) and transitions (representing system activities), respectively. Unlike LTSs, Petri nets support a distributed notion of state given by the marking of places with tokens, which is formalized through a multiset. We use m as metavariable for multisets, "$\{\!|$" and "$|\!\}$" as brackets for multisets, "$\_ \oplus \_$" ("$\_ \ominus \_$") to denote multiset union (difference), and $\mathcal{M}(S)$ ($\mathcal{P}(S)$) to denote the collection of multisets over (subsets of) set S.

Definition 3 A *Petri net* is a tuple (P, Act, T, M_0) where P is a set of places, $T \subseteq \mathcal{M}(P) \times Act \times \mathcal{M}(P)$ is a set of transitions, and $M_0 \in \mathcal{M}(P)$ is the initial marking. The set $\mathcal{M}(P)$ of possible markings will be ranged over by M. ■

In the graphical representation of a Petri net, places are drawn as circles containing black dots called tokens (which describe the current marking) while transitions are drawn as boxes with the appropriate labels. Each transition t can be written as $({}^\bullet t, a, t^\bullet)$ where ${}^\bullet t$ is the weighted preset of t (places where tokens are consumed) and $t^\bullet$ is the weighted postset of t (places where tokens are produced). Given a transition t, we draw an arc from each place in ${}^\bullet t$ to t as well as from t to each place in $t^\bullet$, where each arc is labeled with the multiplicity of the related place.

Definition 4 Let $N = (P, Act, T, M_0)$ be a Petri net.

- $t \in T$ is enabled at marking $M \in \mathcal{M}(P)$ if and only if ${}^\bullet t \subseteq M$. The firing of t produces marking $M' = (M \ominus {}^\bullet t) \oplus t^\bullet$, written $M\,[a\rangle\, M'$.
- The reachability graph (RG for short) of N is the LTS $\mathcal{RG}[\![N]\!] = (RS(M_0), Act, [\rangle, M_0)$ where $RS(M_0)$ is the least subset of $\mathcal{M}(P)$ reachable from M_0. ■

4. NET SEMANTICS

4.1. NET PLACES

We present in three steps our improved net semantics for $\mathcal{G}$. The first step consists of introducing a suitable set of places as well as defining a function which decomposes terms into their sequential components and maps them onto places.

Since sequential terms are represented by guarded alternative compositions, these are used to formalize places. In order to keep track of the static operators, we suitably decorate actions within net places thus avoiding the introduction of inhibitor and contextual arcs and the excess of information of the location oriented approach. Each action a becomes $a^{\theta,R,\sigma}$. Decoration $\theta \in \{\tau, \varepsilon\}$ is used to remember to hide certain actions. Decoration R is a set of conflicts employed to avoid clashes within the scope of a relabeling operator. Its presence is required by the fact that actions will be directly relabeled within net places. We define *Conf* as an infinite set of conflicts (ranged over by r) on which the operation $^{-} : \mathit{Conf} \longrightarrow \mathit{Conf}, \bar{\bar{r}} = r$ is defined. Finally, decoration σ is a string of combinators used to correctly handle synchronizations. We define *Comb* as an infinite set of combinators (ranged over by k) and we consider the set Comb^* of combinator strings (ranged over by σ), on which the following two operations are defined (i) $^{-} : \mathit{Comb} \longrightarrow \mathit{Comb}$, $\bar{\bar{k}} = k$ and (ii) $\odot : \mathit{Comb}^* \times \mathit{Comb}^* \nrightarrow \mathit{Comb}^*$, $\sigma k \odot \sigma \bar{k} = \sigma$. The latter is extended to multisets over Comb^* as follows:

$$\bigodot m = \begin{cases} \bigodot(\{\!|\, \sigma \,|\!\} \oplus m') & \text{if } \exists \sigma, k, m'. \, m = \{\!|\, \sigma k, \sigma \bar{k} \,|\!\} \oplus m' \\ m & \text{otherwise} \end{cases}$$

Definition 5 The set of places is defined by $\mathcal{V} = \{\sum_{i \in I} a_i^{\theta_i, R_i, \sigma_i}.E_i \mid \theta_i \in \{\tau, \varepsilon\} \wedge R_i \in \mathcal{P}(\mathit{Conf}) \wedge \sigma_i \in \mathit{Comb}^* \wedge E_i \in \mathcal{G}'\}$ where the summation is associative and commutative and $\mathcal{G}'$ is the set of closed and guarded terms defined on the same syntax as $\mathcal{G}$ with *Act* replaced by $\mathit{Act}' = \mathit{Act} \times \{\tau, \varepsilon\} \times \mathcal{P}(\mathit{Conf}) \times \mathit{Comb}^*$, (a, θ, R, σ) denoted $a^{\theta,R,\sigma}$, and $a^{\varepsilon,\emptyset,\varepsilon}$ denoted a. $\mathcal{V}$ will be ranged over by V while $\mathcal{M}(\mathcal{V})$ will be ranged over by Q. ■

Decorations will be introduced by the decomposition function through substitutions. In order to avoid undesired substitutions, we identify binders arising from the static operators. In E/L, all the actions in L are bound in E. In $E[\varphi]$, all the actions in $\mathit{Dom}(\varphi) = \{c \in \mathit{Act} \mid \varphi(c) \neq c\}$ are bound in E. For operational convenience, we rewrite the synchronization set S of $E_1 \parallel_S E_2$ as the function $s = \{(a, a) \mid a \in S\}$. In $E_1 \parallel_s E_2$, all the actions in $\pi_1(s) = \{a \mid \exists b. \, (a, b) \in s\}$ are bound in E_1 and E_2. When dealing with substitutions, we make explicit the actions occurring in the body E of each constant definition $A \stackrel{\Delta}{=} E$ by rewriting the definition itself as $A(a_1, \ldots, a_n) \stackrel{\Delta}{=} E$, where $\{a_1, \ldots, a_n\} = sort(E)$. The semantic rule for constants in Table 1 is accordingly modified.

Definition 6 Let us assume a total ordering $\preceq$ be defined over Act. The decomposition function $dec : \mathcal{G}' \longrightarrow \mathcal{M}(\mathcal{V})$ is defined by structural induction as follows (assuming $A(a_1, \ldots, a_n) \stackrel{\Delta}{=} E$):

$$
\begin{array}{l}
dec(\sum_{i \in I} a_i^{\theta_i, R_i, \sigma_i}.E_i) = \{\!| \sum_{i \in I} a_i^{\theta_i, R_i, \sigma_i}.E_i |\!\} \\
dec(E/L) = dec(E\{a^{\tau, \emptyset, \varepsilon}/a \mid a \in L\}) \\
dec(E[\varphi]) = dec(E\{d^{\theta, R \cup R_a, \sigma}/a \mid a \in sort(E) \wedge (a, d^{\theta, R, \sigma}) \in \varphi \wedge \\
\qquad\qquad\qquad\qquad R_a = \{r_{ca} \mid \varphi(c) = \varphi(a) \wedge c \prec a \wedge r_{ca} \text{ fresh}\} \cup \\
\qquad\qquad\qquad\qquad\qquad \{\bar{r}_{ac} \mid \varphi(c) = \varphi(a) \wedge a \prec c \wedge r_{ac} \text{ fresh}\}\}) \\
dec(E_1 \|_s E_2) = dec(E_1\{d^{\theta, R, \sigma k_a}/a \mid (a, d^{\theta, R, \sigma}) \in s \wedge k_a \text{ fresh}\}) \oplus \\
\qquad\qquad\qquad\qquad dec(E_2\{d^{\theta, R, \sigma \bar{k}_a}/a \mid (a, d^{\theta, R, \sigma}) \in s \wedge k_a \text{ fresh}\}) \\
dec(A(b_1^{\theta_1, R_1, \sigma_1}, \ldots, b_n^{\theta_n, R_n, \sigma_n})) = dec(E\{b_i^{\theta_i, R_i, \sigma_i}/a_i \mid 1 \leq i \leq n\})
\end{array}
$$

with the syntactical substitutions on $\mathcal{G}'$ defined by structural induction as follows:

$$
\begin{array}{l}
(\sum_{i \in I} a_i^{\theta_i, R_i, \sigma_i}.E_i)\{b^{\theta, R, \sigma}/a\} = \sum_{i \in I} c_i^{\theta'_i, R'_i, \sigma'_i}.E_i\{b^{\theta, R, \sigma}/a\} \\
E/L\{b^{\theta, R, \sigma}/a\} = \begin{cases} E\{b^{\theta, R, \sigma}/a\}/L & \text{if } a \notin L \\ E/L & \text{if } a \in L \end{cases} \\
E[\varphi]\{b^{\theta, R, \sigma}/a\} = \begin{cases} E\{b^{\theta, R, \sigma}/a\}[\varphi'] & \text{if } a \notin Dom(\varphi) \\ E[\varphi'] & \text{if } a \in Dom(\varphi) \end{cases} \\
(E_1 \|_s E_2)\{b^{\theta, R, \sigma}/a\} = \begin{cases} E_1\{b^{\theta, R, \sigma}/a\} \|_s E_2\{b^{\theta, R, \sigma}/a\} & \text{if } a \notin \pi_1(s) \\ E_1 \|_{s'} E_2 & \text{if } a \in \pi_1(s) \end{cases} \\
A(a_1^{\theta_1, R_1, \sigma_1}, \ldots, a_n^{\theta_n, R_n, \sigma_n})\{b^{\theta, R, \sigma}/a\} = A(c_1^{\theta'_1, R'_1, \sigma'_1}, \ldots, c_n^{\theta'_n, R'_n, \sigma'_n})
\end{array}
$$

where:

$$
\begin{array}{rcl}
\varphi' & = & \{(d_i, c_i^{\theta'_i, R'_i, \sigma'_i}) \mid \exists a_i^{\theta_i, R_i, \sigma_i}.\, (d_i, a_i^{\theta_i, R_i, \sigma_i}) \in \varphi\} \\
s' & = & \{(d_i, c_i^{\theta'_i, R'_i, \sigma'_i}) \mid \exists a_i^{\theta_i, R_i, \sigma_i}.\, (d_i, a_i^{\theta_i, R_i, \sigma_i}) \in s\}
\end{array}
$$

and:

$$
c_i^{\theta'_i, R'_i, \sigma'_i} = \begin{cases} b^{\theta_i \theta, R_i \cup R, \sigma_i \sigma} & \text{if } a_i = a \\ a_i^{\theta_i, R_i, \sigma_i} & \text{if } a_i \neq a \end{cases}
$$

■

It is worth recalling that, in the clause for the parallel composition of the location oriented approach, an explicit track of the operator would be kept by decomposing $E_1 \|_S E_2$ into $dec_{\text{loc}}(E_1) \|_S id \oplus id \|_S dec_{\text{loc}}(E_2)$. Here, instead, according to the label oriented approach, we do not explicitly keep track of the parallel composition operator. As a consequence, instances of the same sequential term occurring within the scope of a parallel composition operator can be mapped onto the same place; e.g., term $a.\underline{0} \parallel_\emptyset a.\underline{0}$ is mapped onto a single place $a.\underline{0}$ with multiplicity two.

Let us see some examples showing how the decomposition works. In the case of hiding, the decomposition function maps term $(a.\underline{0})/\{a\}$ onto place $a^{\tau,\emptyset,\varepsilon}.\underline{0}$. This decoration provides the information that action a is hidden and this information will be subsequently used when generating net transitions.

In the case of relabeling, term $(a.\underline{0} + b.\underline{0})[\varphi]$, where $\varphi(a) = \varphi(b) = b$, is mapped onto place $b^{\varepsilon,\{r\},\emptyset}.\underline{0} + b^{\varepsilon,\{\bar{r}\},\emptyset}.\underline{0}$. In this way, the two actions are kept distinct within the scope of the relabeling operator by their complementary conflicts r and $\bar{r}$. Such conflicts are necessary whenever parallel composition operators occur within the scope of the relabeling operator, as we shall see shortly. Again, this information will be used when generating net transitions.

In the case of parallel composition, term $a.\underline{0} \parallel_{\{a\}} a.\underline{0}$ is mapped onto the set of places $\{a^{\varepsilon,\emptyset,k}.\underline{0}, a^{\varepsilon,\emptyset,\bar{k}}.\underline{0}\}$ which can synchronize because they have complementary combinators k and $\bar{k}$.

As far as the interplay of parallel composition and hiding is concerned, binders are exploited to avoid the synchronization of hidden actions. As an example, in the term $(a.\underline{0})/\{a\} \parallel_{\{a\}} a.\underline{0}$ the synchronization of the two a actions is not possible because the left hand one is hidden. This is reflected at the net level by the fact that the generated places $a^{\tau,\emptyset,\varepsilon}.\underline{0}$ and $a^{\varepsilon,\emptyset,\bar{k}}.\underline{0}$ cannot synchronize because their combinators do not match. In particular, the combinator of the left hand a action is ε instead of k because that a action occurs within the scope of binder $/\{a\}$.

As far as the interplay of parallel composition and relabeling is concerned, the related decorations are used in a combined way in order to avoid the synchronization between different actions that collapse to the same action after their relabeling. As an example, in the term $(a.\underline{0} \parallel_{\{b\}} b.\underline{0})[\varphi]$, where $\varphi(a) = \varphi(b) = b$, no synchronization on b is possible because the parallel composition is within the scope of the relabeling and not the other way around. Again, this is reflected at the net level by the fact that the generated places $b^{\varepsilon,\{r\},k}.\underline{0}$ and $b^{\varepsilon,\{\bar{r}\},\bar{k}}.\underline{0}$ cannot synchronize, despite the fact that they have complementary combinators k and $\bar{k}$, because of the presence of complementary conflicts r and $\bar{r}$.

We observe that the identification of binders to avoid undesired substitutions is necessary mainly because, in the case of the relabeling operator, actions are directly relabeled within net places. As an example, consider term $(a.\underline{0} + b.\underline{0})/\{a\}[\varphi]$, where $\varphi(a) = \varphi(b) = b$, which can execute action τ or action b. When decomposing this term, there are two possible translations causing errors if $/\{a\}$ is not correctly taken into account. In the former case, one may think of relabeling actions occurring in $/\{a\}$ as well, thus erroneously obtaining term $(b^{\varepsilon,\{r\},\varepsilon}.\underline{0}+b^{\varepsilon,\{\bar{r}\},\varepsilon}.\underline{0})/\{b\}$ which gives rise to place $b^{\tau,\{r\},\varepsilon}.\underline{0}+b^{\tau,\{\bar{r}\},\varepsilon}.\underline{0}$ from which only a transition labeled with τ can be generated. In the latter case, one may think of not considering $/\{a\}$ at all, thereby erroneously obtaining term $(b^{\varepsilon,\{r\},\varepsilon}.\underline{0} + b^{\varepsilon,\{\bar{r}\},\varepsilon}.\underline{0})/\{a\}$ which gives rise to place $b^{\varepsilon,\{r\},\varepsilon}.\underline{0} + b^{\varepsilon,\{\bar{r}\},\varepsilon}.\underline{0}$ from which only a transition labeled with b can be generated. In conclusion, actions occurring in binders can be neither relabeled nor ignored. For this reason, we decompose the term above into place $a^{\tau,\emptyset,\varepsilon}.\underline{0} + b^{\varepsilon,\{\bar{r}\},\varepsilon}.\underline{0}$, where we note that the a action is not relabeled to b because it occurs in $/\{a\}$, from which two transitions labeled with τ resp. b can be generated, as expected. As another example, consider term $(a.\underline{0}+b.\underline{0}) \parallel_{\{a\}} (a.\underline{0}+b.\underline{0})[\varphi]$, where $\varphi(a) = \varphi(b) = b$, which can synchronously execute an a action which is turned into b or asynchronously execute two b actions. Similarly to the previous example, if we relabel $\parallel_{\{a\}}$ as $\parallel_{\{b\}}$ we erroneously introduce an additional synchronization between the two original b actions, while if we do not consider $\parallel_{\{a\}}$ at all we get rid of the synchronization between the two original a actions. In order to avoid the two cases above, we preliminarily rewrite the term as $(a.\underline{0}+b.\underline{0}) \parallel_{\{(a,a)\}} (a.\underline{0}+b.\underline{0})[\varphi]$. This gives rise to $(a.\underline{0} + b^{\varepsilon,\{\bar{r}\},\varepsilon}.\underline{0}) \parallel_{\{(a,b^{\varepsilon,\{r\},\varepsilon})\}} (a.\underline{0} + b^{\varepsilon,\{\bar{r}\},\varepsilon}.\underline{0})$ which results in the set of places $\{b^{\varepsilon,\{r\},k}.\underline{0} + b^{\varepsilon,\{\bar{r}\},\varepsilon}.\underline{0}, b^{\varepsilon,\{r\},\bar{k}}.\underline{0} + b^{\varepsilon,\{\bar{r}\},\varepsilon}.\underline{0}\}$ from which three transitions labeled with b can be generated, as expected. As usual, the application of each syntactical substitution may be preceded by alpha conversion to avoid binding free names. This is necessary e.g. when decomposing term $(a.\underline{0})/\{b\}[\varphi]$ where $\varphi(a) = \varphi(b) = b$.

Example 7 Let us a consider a simple system composed of two identical processes accessing the same memory module in mutual exclusion. This system can be described in $\mathcal{G}$ as follows:

$$\begin{aligned} System &\triangleq ((Proc \parallel_{\emptyset} Proc) \parallel_{\{acq,rel\}} Mem)/\{comp\} \\ Proc &\triangleq comp.acq.use.rel.Proc \\ Mem &\triangleq acq.rel.Mem \end{aligned}$$

where *comp* denotes a local computation (which is therefore hidden) while *acq*, *use*, and *rel* denote the acquisition, use, and release of the memory module, respectively. The decomposition of *System* computed by applying the rules in Def. 6 is $\{\!|\ comp^{\tau,\emptyset,\varepsilon}.acq^{\varepsilon,\emptyset,k_1}.use.rel^{\varepsilon,\emptyset,k_2}.Proc$, $comp^{\tau,\emptyset,\varepsilon}.acq^{\varepsilon,\emptyset,k_1}.use.rel^{\varepsilon,\emptyset,k_2}.Proc, acq^{\varepsilon,\emptyset,\bar{k}_1}.rel^{\varepsilon,\emptyset,\bar{k}_2}.Mem\ |\!\}$. Note that one place appears twice in the multiset, correctly capturing the presence of two independent replicas of *Proc*. ∎

4.2. NET TRANSITIONS

The second step consists of connecting net places by means of net transitions. Unlike the location oriented approach, we will not have a net transition generating rule for each algebraic operator because static operators are not fully retained within places. This lack of information is however compensated for by action decorations introduced within places.

Let us reconsider the examples at the end of Sect. 4.1. Term $(a.\underline{0})/\{a\}$ is mapped onto place $a^{\tau,\emptyset,\varepsilon}.\underline{0}$ which becomes the preset of a transition labeled with τ by exploiting the information stored in the decoration. Term $(a.\underline{0} + b.\underline{0})[\varphi]$, where $\varphi(a) = \varphi(b) = b$, is mapped onto place $b^{\varepsilon,\{r\},\varepsilon}.\underline{0} + b^{\varepsilon,\{\bar{r}\},\varepsilon}.\underline{0}$ which becomes the preset of two (collapsing) transitions both labeled with b. Term $a.\underline{0} \parallel_{\{a\}} a.\underline{0}$ is mapped onto the set of places $\{a^{\varepsilon,\emptyset,k}.\underline{0}, a^{\varepsilon,\emptyset,\bar{k}}.\underline{0}\}$ which becomes the preset of a transition labeled with a using the combinators k and $\bar{k}$ stored in the decorations.

Formally, the set of transitions $\longrightarrow_{\mathcal{N}}$ is defined as the least subset of $\mathcal{P}(\mathcal{V}) \times Act \times \mathcal{M}(\mathcal{V})$ generated by the axiom:

$$\{\!|\sum_{h=1}^{n_i} a_{i,h}^{\theta_{i,h},R_{i,h},\sigma_{i,h}}.E_{i,h} \mid 1 \le i \le n \,|\!\} \xrightarrow{b}_{\mathcal{N}} \bigoplus_{i=1}^{n} \{\!|dec(E_{i,h_i}) \mid 1 \le h_i \le n_i\} \,|\!\}$$

where $n \in \mathbf{N}_+$ and $n_i \in \mathbf{N}_+$ for all $i = 1, \ldots, n$. The preset of the transition is a finite nonempty multiset of places, from each of which a summand is selected (the one with index i, h_i). If the preset contains more than one place, then the generated transition results from the synchronization of the actions of the selected summands. The postset of the transition is the multiset composed of the places representing the decomposition of the derivative terms of the selected summands in the preset. The axiom is subject to the following conditions which are explained below:

1 For all $i, i' = 1, \ldots, n$, $a_{i,h_i} = a_{i',h_{i'}}$ and $\theta_{i,h_i} = \theta_{i',h_{i'}}$.

2 For all $i, i' = 1, \ldots, n$, it is not the case that $\{r, \bar{r}\} \subseteq R_{i,h_i} \cap R_{i',h_{i'}}$ whenever $i \neq i'$.

3 $\bigodot_{i=1}^{n} \sigma_i = \{\!|\epsilon|\!\}$.

4 $b = \begin{cases} a & \text{if } \theta_{1,h_1} = \epsilon \\ \tau & \text{if } \theta_{1,h_1} = \tau \end{cases}$

Condition 1 guarantees that the selected summands in the preset have the same action at the same visibility level. Conditions 2 and 3 ensure that the selected summands in the preset correctly synchronize by considering the decorations of the related actions. More precisely, condition 2 avoids the generation of undesired synchronization transitions which are due to a relabeling operator. Condition 3, instead, checks for the correct reduction of combinators to the empty string, which means that a synchronization is possible. Finally, condition 4 determines the action labeling the transition according to the hiding related decoration of one of the selected summands.

Example 8 Let us a consider again the simple system of Ex. 7. Starting from the two places $V_1 = comp^{\tau,\emptyset,\varepsilon}.acq^{\varepsilon,\emptyset,k_1}.use.rel^{\varepsilon,\emptyset,k_2}.Proc$ and $V_2 = acq^{\varepsilon,\emptyset,\bar{k}_1}.rel^{\varepsilon,\emptyset,\bar{k}_2}.Mem$ forming its decomposition, a single transition labeled with τ can be generated whose preset is V_1 and whose postset is given by V_3 obtained from $dec(acq^{\varepsilon,\emptyset,k_1}.use.rel^{\varepsilon,\emptyset,k_2}.Proc) = \{\!| acq^{\varepsilon,\emptyset,k_1}.use.rel^{\varepsilon,\emptyset,k_2}.Proc |\!\}$. On the contrary, no transition can be generated having as preset the place V_2 because combinator k_1 does not reduce to the empty string. Such a reduction is instead possible when considering V_3 and V_2 together. They constitute the preset of a transition labeled with acq, whose postset is composed of places V_4 obtained from $dec(use.rel^{\varepsilon,\emptyset,k_2}.Proc) = \{\!| use.rel^{\varepsilon,\emptyset,k_2}.Proc |\!\}$ and V_5 obtained from $dec(rel^{\varepsilon,\emptyset,\bar{k}_2}.Mem) = \{\!| rel^{\varepsilon,\emptyset,\bar{k}_2}.Mem |\!\}$. By proceeding in this way, we can generate all the transitions in the net corresponding to *System*. ■

4.3. NETS ASSOCIATED WITH TERMS

The third step consists of associating with each term an appropriate Petri net by exploiting the previous two steps.

Definition 9 The net semantics of $E \in \mathcal{G}$ is the net $\mathcal{N}[\![E]\!] = (P_{E,\mathcal{N}}, Act, \longrightarrow_{E,\mathcal{N}}, dec(E))$ where $P_{E,\mathcal{N}}$ is the least subset of $\mathcal{V}$ such that:

- $dec(E) \subseteq P_{E,\mathcal{N}}$;
- if $Q_1 \subseteq P_{E,\mathcal{N}}$ and $Q_1 \xrightarrow{a}_{\mathcal{N}} Q_2$, then $Q_2 \subseteq P_{E,\mathcal{N}}$,

and $\longrightarrow_{E,\mathcal{N}}$ is $\longrightarrow_{\mathcal{N}}$ restricted to $\mathcal{P}(P_{E,\mathcal{N}}) \times Act \times \mathcal{M}(P_{E,\mathcal{N}})$. ■

Example 10 Let us a consider again the simple system of Ex. 7 and 8. The corresponding Petri net is shown in Fig. 1. ■

4.4. RETRIEVABILITY RESULT

We now show that the net semantics for $\mathcal{G}$ is correct w.r.t. the interleaving semantics, which means that the net semantics and the interleaving semantics of a given term represent the same system. Following [10], this is achieved by proving for all $E \in \mathcal{G}$ that $\mathcal{RG}[\![\mathcal{N}[\![E]\!]]\!]$ is strongly bisimilar [9] to $\mathcal{I}[\![E]\!]$. Note that these two LTSs are not isomorphic in general.

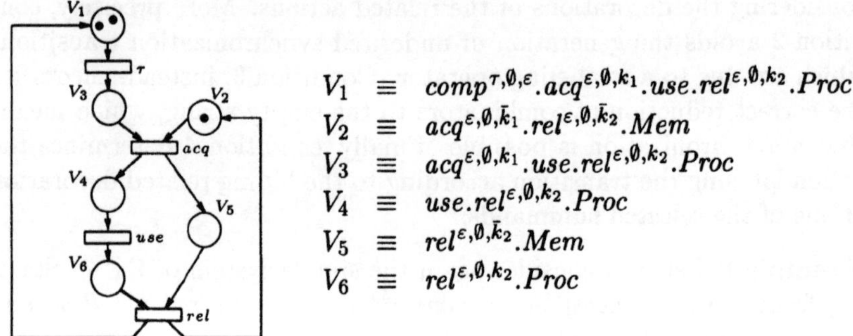

Figure 1 $\mathcal{N}[\![System]\!]$

For instance, Fig. 2 shows both $\mathcal{RG}[\![\mathcal{N}[\![System]\!]]\!]$ and $\mathcal{I}[\![System]\!]$. As it can be noted, there is no 1-1 correspondence between the states in the two LTSs, but a strong bisimulation can be recognized in which several states in $\mathcal{I}[\![System]\!]$ correspond to a single state in $\mathcal{RG}[\![\mathcal{N}[\![System]\!]]\!]$. This correspondence is represented through the subscripts in the name of the states; e.g., s_1 and s'_1 are bisimilar to M_1.

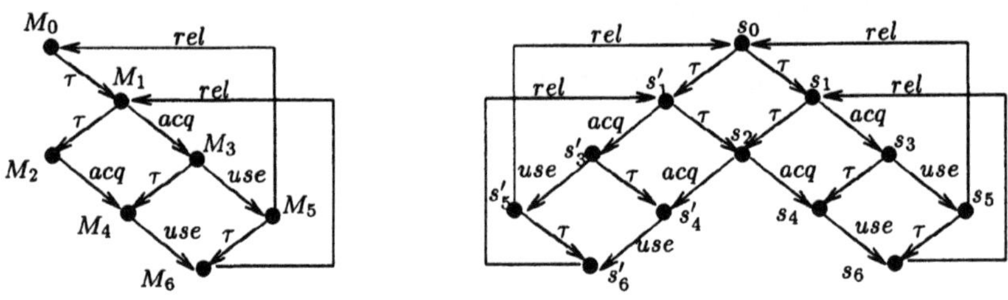

Figure 2 $\mathcal{RG}[\![\mathcal{N}[\![System]\!]]\!]$ and $\mathcal{I}[\![System]\!]$

Theorem 11 *Let* $\mathcal{B} = \{(E, Q) \in \mathcal{G}' \times \mathcal{M}(\mathcal{V}) \mid Q = dec(E)\}$. *Then* $\mathcal{B}$ *is a strong bisimulation.* ■

5. PROPERTIES OF THE SEMANTICS

5.1. COMPARISON AT THE NET LEVEL

In the location oriented approach of [7, 10] the resulting nets are 1-safe. As explained in Sect. 4.1, this is a consequence of the decomposition clause $dec_{\mathrm{loc}}(E_1 \parallel_S E_2) = dec_{\mathrm{loc}}(E_1) \parallel_S id \oplus id \parallel_S dec_{\mathrm{loc}}(E_2)$ for the parallel composition operator, which always keeps distinct the two subnets corresponding to the two operand terms of the parallel composition operator. In our approach, instead, the two subnets can collapse into a single one depending on E_1, E_2, and S.

Theorem 12 *Let $E_1, E_2 \in \mathcal{G}$ be two sequential terms. Then $|dec(E_1 \,\|_S\, E_2)| \leq 2 = |dec_{loc}(E_1 \,\|_S\, E_2)|$. Moreover, if $E_1 \equiv E_2$ and $(sort(E_1) \cup sort(E_2)) \cap S = \emptyset$, then $|dec(E_1 \,\|_S\, E_2)| = 1$.* ∎

The theorem above straightforwardly generalizes to an arbitrary number of sequential terms and gives an idea of the compactness that can be gained using the proposed label oriented net semantics. This justifies why the label oriented approach should be preferred to the location oriented approach from an applicative point of view, especially when dealing with large systems in which some parts are replicated.

5.2. COMPARISON AT THE STATE LEVEL

As shown in Fig. 2, $\mathcal{RG}[\![\mathcal{N}[\![System]\!]]\!]$ is smaller than $\mathcal{I}[\![System]\!]$. Actually, with an easy adaptation of the proof of the retrievability result, it can be shown that the reverse never happens.

Theorem 13 *Let $E \in \mathcal{G}$ be a finite state term and let $state(Z)$ be the set of states of LTS Z. Then $|state(\mathcal{RG}[\![\mathcal{N}[\![E]\!]]\!])| \leq |state(\mathcal{I}[\![E]\!])|$. In particular, if $E \equiv E_1 \,\|_S\, E_2$, then $|state(\mathcal{RG}[\![\mathcal{N}[\![E]\!]]\!])| < |state(\mathcal{I}[\![E]\!])|$ whenever $E_1 \equiv E_2$ and $(sort(E_1) \cup sort(E_2)) \cap S = \emptyset$.* ∎

The second part of the theorem above straightforwardly generalizes to an arbitrary number of terms. Again, this is particularly important when dealing with large systems in which some parts are replicated since aggregated (w.r.t. the strong bisimulation equivalence) state spaces are obtained without applying minimization algorithms such as [11].

5.3. FINITENESS RESULT

Another nice property of our net semantics is that, in certain cases, it yields finite Petri nets for infinite state terms. As an example, consider $A \stackrel{\Delta}{=} d.(a.\underline{0} \,\|_{\emptyset}\, b.A)$. Then $\mathcal{N}[\![A]\!]$ is the finite net depicted in Fig. 3, while $\mathcal{N}_{loc}[\![A]\!]$ has infinitely many places and $\mathcal{I}[\![A]\!]$ has infinitely many states.

Theorem 14 *Let $E \in \mathcal{G}$ be an infinite state term. If the three following conditions hold:*

- *For each hiding operator $\_/L$ occurring within a recursive constant definition, $sort(E) \cap L = \emptyset$.*
- *For each relabeling operator $\_[\varphi]$ occurring within a recursive constant definition, $sort(E) \cap Dom(\varphi) = \emptyset$.*

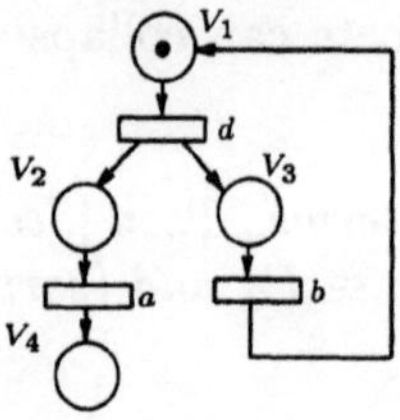

$$V_1 \equiv d.(a.\underline{0} \,||_\emptyset\, b.A)$$
$$V_2 \equiv a.\underline{0}$$
$$V_3 \equiv b.A$$
$$V_4 \equiv \underline{0}$$

Figure 3 **Finite net semantics of the infinite state term $A \triangleq d.(a.\underline{0} \,||_\emptyset\, b.A)$**

- *For each parallel composition operator $\_ \,||_S\, \_$ occurring within a recursive constant definition, $sort(E) \cap S = \emptyset$.*

then $\mathcal{N}[\![E]\!]$ is finite while $\mathcal{N}_{\mathrm{loc}}[\![E]\!]$ has infinitely many places and $\mathcal{I}[\![E]\!]$ has infinitely many states. ■

The theorem above shows that, for the considered class of infinite state terms, a finite hence analyzable representation can be obtained only with our label oriented net semantics.

5.4. STRUCTURAL REINTERPRETATION

For Petri nets there are several structural analysis techniques which can be used to derive system properties avoiding the generation of the underlying state space. The most important structural technique is the one based on the computation of P-invariants and T-invariants [12].

Definition 15 Let $N = (P, Act, T, M_0)$ be a Petri net such that $|P| = m$ and $|T| = n$. The *incidence matrix* of N is the matrix $\mathbf{A} = [a_{i,j}] \in \mathbf{Z}^{\,n \times m}$ where $a_{i,j} = i^\bullet(j) - {}^\bullet i(j)$. A *P-invariant* is an integer solution of $\mathbf{A} \cdot \mathbf{y} = \mathbf{0}$. A *T-invariant* is an integer solution of $\mathbf{A}^{\mathrm{T}} \cdot \mathbf{x} = \mathbf{0}$. ■

P-invariants single out places that do not change their token count during transition firings, whereas T-invariants indicate how often each transition has to fire in order to reproduce a given marking. P-invariants and T-invariants are computed starting from the incidence matrix and are independent from any initial marking, which is only instrumental for their interpretation. As an example of system properties that can be proved by exploiting invariants, it can be shown that a Petri net is bounded if for each of its places there exists a P-invariant associating a positive value with that place, while a necessary condition in order for a Petri net to be live and bounded is that for each of its transitions there exists a T-invariant associating a positive value with that transition.

A desirable property of a net semantics is that of allowing system properties derived from a Petri net to be reinterpreted at the level of the algebraic specification mapped onto that net. In this way one can take advantage of the efficiency of the structural techniques developed for Petri nets, thus importing such techniques in a formalism that does not directly support them. As far as invariants are concerned, our net semantics possesses this property because net places are in correspondence with (decorated) sequential subterms of the algebraic specification and net transitions are labeled with actions occurring in the algebraic specification.

Example 16 Let us consider again the system introduced in Ex. 7, whose net semantics is depicted in Fig. 1. Its incidence matrix $\mathbf{A}$ is

	V_1	V_2	V_3	V_4	V_5	V_6
τ	-1	0	1	0	0	0
acq	0	-1	-1	1	1	0
use	0	0	0	-1	0	1
rel	1	1	0	0	-1	-1

The net has four P-invariants $\mathbf{y}_1^{\mathrm{T}} = [\,1\;0\;1\;1\;0\;1\,]$, $\mathbf{y}_2^{\mathrm{T}} = [\,1\;0\;1\;0\;1\;0\,]$, $\mathbf{y}_3^{\mathrm{T}} = [\,0\;1\;0\;0\;1\;0\,]$, and $\mathbf{y}_4^{\mathrm{T}} = [\,0\;1\;0\;1\;0\;1\,]$ and one T-invariant $\mathbf{x}^{\mathrm{T}} = [\,1\;1\;1\;1\,]$.

By exploiting the P-invariants above, we are able to prove that the memory is used in a mutually exclusive way. Since each P-invariant identifies a subset of the places of the net whose token count is invariant under transition firing, we know that

$$\begin{aligned} M(V_1) + M(V_3) + M(V_4) + M(V_6) &= 2 \\ M(V_1) + M(V_3) + M(V_5) &= 2 \\ M(V_2) + M(V_5) &= 1 \\ M(V_2) + M(V_4) + M(V_6) &= 1 \end{aligned}$$

where M denotes the current marking while the constant occurring in each equation is determined by the initial marking. To prove mutual exclusion, we have to demonstrate that there is no state in which both processes can execute action *use*. At the net level, $V_4 \equiv use._{k_2}rel.Proc$ is the place representing a process ready to perform action *use*, so we have to demonstrate that there is no marking in which place V_4 contains more than one token. This can obviously be done by constructing the reachability graph of the net, but it can be also done more efficiently by exploiting the information provided by the fourth P-invariant, as it establishes that, for each reachable marking, the sum of the number of tokens in places V_2, V_4, and V_6 is one.

By exploiting the only T-invariant, instead, we are able to prove that the initial state of the algebraic specification is a home state, because the initial marking of the net enables the transition sequence τ, acq, use, rel whose transition count vector coincides with the T-invariant. ■

6. CONCLUSION

In this paper we have presented (and proved correct) a label oriented net semantics for a process algebra, which improves previous net semantics w.r.t. the size of the generated nets and allows system properties efficiently derived at the net level to be reinterpreted at the algebraic level. Because of its practical advantages, such a label oriented net semantics has been recently implemented to integrate the stochastically timed process algebra EMPA based software tool TwoTowers [3] with the generalized stochastic Petri net based software tool GreatSPN [6], in order to create a multiparadigm tool for the analysis of concurrent and distributed systems which benefits from the complementary strengths of process algebras and Petri nets.

As far as future work is concerned, we plan to investigate the possibility of extending our approach to a symbolic framework where value passing process algebras and colored Petri nets are considered.

References

[1] M. Bernardo, N. Busi, R. Gorrieri, *"A Distributed Semantics for EMPA Based on Stochastic Contextual Nets"*, in Computer Journal 38:492-509, 1995

[2] M. Bernardo, N. Busi, M. Ribaudo, *"Compact Net Semantics for Process Algebras"*, Tech. Rep. UBLCS-2000-02, University of Bologna (Italy), 2000

[3] M. Bernardo, W.R. Cleaveland, S.T. Sims, W.J. Stewart, *"TwoTowers: A Tool Integrating Functional and Performance Analysis of Concurrent Systems"*, in Proc. of FORTE/PSTV '98, Kluwer, pp. 457-467, 1998

[4] N. Busi, R. Gorrieri, *"A Petri Net Semantics for π-Calculus"*, in Proc. of CONCUR '95, LNCS 962:145-159, 1995

[5] N. Busi, R. Gorrieri, G. Zavattaro, *"On the Expressiveness of Linda Coordination Primitives"*, in Information and Computation 156:90-121, 2000

[6] G. Chiola, G. Franceschinis, R. Gaeta, M. Ribaudo, *"GreatSPN 1.7: Graphical Editor and Analyzer for Timed and Stochastic Petri Nets"*, in Performance Evaluation 24:47-68, 1995

[7] P. Degano, R. De Nicola, U. Montanari, *"A Distributed Operational Semantics for CCS Based on Condition/Event Systems"*, in Acta Informatica 26:59-91, 1988

[8] U. Goltz, *"On Representing CCS Programs by Finite Petri Nets"*, in Proc. of MFCS '88, LNCS 324:339-350, 1988

[9] R. Milner, *"Communication and Concurrency"*, Prentice Hall, 1989

[10] E.-R. Olderog, *"Nets, Terms and Formulas"*, Cambridge University Press, 1991

[11] R. Paige, R.E. Tarjan, *"Three Partition Refinement Algorithms"*, in SIAM Journal of Computing 16:973-989, 1987

[12] W. Reisig, *"Petri Nets: An Introduction"*, Springer-Verlag, 1985

A CONCISE COMPOSITIONAL STATECHARTS SEMANTICS DEFINITION

Michael von der Beeck
Department of Computer Science
Munich University of Technology
Arcisstr. 21, D-80290 München, Germany
beeck@in.tum.de

Abstract Statecharts is a well-known specification language for modeling system behavior. Its intuitive graphical appeal for specifying hierarchical, concurrent state machines conceals the difficulties in formalizing its semantics. The causes are in Statecharts' two-level semantics when demanding compositionality, as is necessary in order to obtain easy comprehensibility and modularity of system specifications. This paper suggests a compositional approach for formalizing the Statecharts semantics directly on sequences of micro steps using labeled transition systems as semantical domain. Though we consider causality, global consistency, as well as the synchrony hypothesis, our approach results in a very short and concise semantics definition of Statecharts.

Keywords: Statecharts, operational semantics, compositionality

1. INTRODUCTION

Statecharts is a visual language widely accepted for specifying the behavior of *reactive* and *embedded systems* [4]. It constitutes an essential part of many well-known object-oriented specification notations like UML [2] and ROOM [19]. The Statecharts language extends traditional *finite-state machines* by concepts of *hierarchy* and *concurrency*. Hierarchy is achieved by embedding one Statechart in a state of another Statechart – resulting in an Or-state. Concurrency is supported by a complex Statechart (an And-state) composed of several simultaneously active sub-Statecharts communicating with each other via a sort of broadcasting mechanism. Furthermore, in addition to traditional finite-state machines the Statecharts language allows transitions labeled by pairs of events, where the first component is referred to as *trigger* and may include *negated events*, and the second component is referred to as

Tommaso Bolognesi and Diego Latella (Eds.), *Formal Methods for Distributed System Development.*

action. Intuitively, if the environment offers the positive events in the trigger, but not the negated ones, then the transition is triggered and can be executed, thereby performing a so-called *micro step* and generating the events contained in the action which in turn can trigger new transitions while disabling others. When this chain reaction comes to a halt, one execution step – also referred to as a *macro step* – is complete.

On the one hand the Statecharts language has been proven very successful for specification purposes due to its intuitive syntax and semantics, on the other hand precisely defining its semantics has been a long-term challenge. Many proposals for defining the Statecharts semantics [3, 5, 6, 9, 10, 11, 12, 14, 17, 18, 20, 21] have emerged [22, 7]. Semantical intricacies arise if the following wishful properties – Statecharts specific as well as quite general semantic ones – shall be simultaneously achieved by the Statecharts semantics to be defined:

- synchrony hypothesis
  This hypothesis [1] requires that the reaction time of a system is short in comparison with the environment, i.e. the system reacts to an input, before the environment sends the next input.

- causality principle
  Causality holds if there exists a (causal) ordering among the transitions of a macro step such that each transition t of the macro step is triggered by the events which are offered by the environment or which are generated by transitions occuring before t in that ordering.

- global consistency
  Global consistency requires that every transition of a macro step should not only be enabled by the events generated in previous micro steps of this macro step, but by the events generated in all micro steps of this macro step.

- compositionality
  Compositionality ensures that the semantics of a Statechart is only defined in terms of the semantics of its components. This general semantic property provides essential means supporting development and verification in a structural way.

Several Statecharts semantics definitions circumvent difficulties by ignoring some of these properties or by only having defined the semantics in an informal way. Other approaches combine these properties, but at the expense of resulting in quite a lengthy and complex Statecharts semantics definition.

Therefore it is the aim of this paper to define a Statecharts semantics which offers all of the aforementioned features, but which nevertheless is as concise and easily comprehensible as possible.

We present an operational Statecharts semantics of minimal length directly defined on sequences of micro steps. We use labelled transition systems as the semantic domain and SOS rules (Structured Operational Semantics [16]) in order to achieve a compositional semantics.

The remainder of this paper is organized as follows. The next section offers a brief introduction to Statecharts and the way their semantics are traditionally defined. In Section 3 we present our definition of Statecharts semantics. Related work is considered in Section 4, while Section 5 provides our conclusions.

2. TRADITIONAL STATECHARTS

In this section we present the syntax and traditional semantics of Statecharts. As the starting point for our new semantics definition we define Statecharts terms, which constitute a textual notation providing the abstract syntax of (visual) Statecharts. Then we briefly describe our reference semantics. The presentation of syntax and semantics is supported by an example, which we will later consider again for illustrating our new semantics definition.

2.1. SYNTAX

Example. Consider the Statechart in Fig. 1. It consists of an *And-state*, labeled by n_1, which denotes the parallel composition of the two Statecharts labeled by n_2 and n_3, both of which are *Or-states*, i.e. both describe a sequential state machine. Or-state n_2 is further refined by Or-state n_4 and basic state n_5 which are connected via transitions t_1 and t_2. Label c of t_1 specifies that t_1 is triggered by c, i.e., by the occurence of event c. Or-state n_4 is further refined by basic states n_8 and n_9, connected by transition t_4 with label $a \wedge \neg c$ specifying that t_4 is triggered if event a occurs, but not event c. Or-state n_3 consists of two basic states n_6 and n_7 connected via transition t_3 with label a/c, so that upon occurence of event a transition t_3 can execute thereby generating event c.

Statecharts Terms. Statecharts is a visual language. However, for our purposes it is convenient to represent Statecharts not visually but by terms. This is also done in related work [9, 12, 20], and our approach closely follows the one described by Maggiolo-Schettini et al. [12]. For-

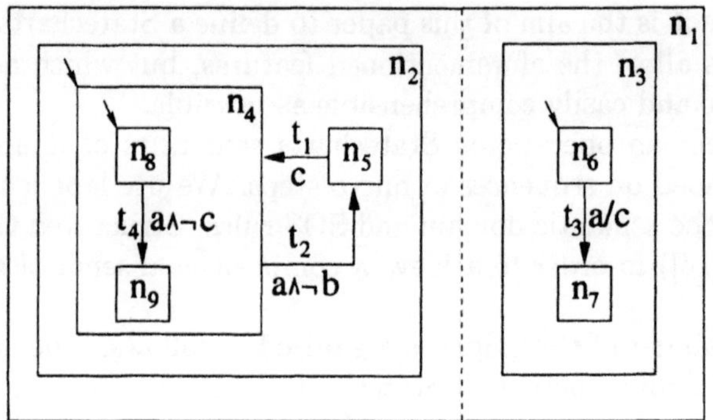

Figure 1 Statechart Example

mally, let $\mathcal{N}$ be a countable set of names for Statecharts states, $\mathcal{T}$ be a countable set of names for Statecharts transitions, and Π be a countable set of Statecharts events. Moreover, we associate with every event $e \in \Pi$ its negated counterpart $\neg e$. We also lift negation to negated events by defining $\neg\neg e =_{\text{df}} e$. Furthermore, we write $\neg I$ for $\{\neg e \,|\, e \in I\}$. Then, the set SC of Statecharts terms is defined to be the least set satisfying the following rules.

1 **Basic state:** If $n \in \mathcal{N}$, then $s = [n]$ is a Statecharts term.

2 **Or-state:** If $n \in \mathcal{N}$, $s_1, \ldots, s_k$ are Statecharts terms for $k > 0$, $\rho = \{1, \ldots, k\}$, $l \in \rho$, and $T \subseteq \mathsf{TR} =_{\text{df}} \mathcal{T} \times \rho \times 2^{\Pi \cup \neg \Pi} \times 2^{\Pi} \times \rho$, then $s = [n : (s_1, \ldots, s_k); l; T]$ is a Statecharts term. Here, $s_1, \ldots, s_k$ are the sub-states of s, T is the set of transitions between these states, s_1 is the default state of s, and s_l is the currently active sub-state of s.

3 **And-state:** If $n \in \mathcal{N}$ and $s_1, \ldots, s_k$ are Statecharts terms for $k > 0$, then $s = [n : (s_1, \ldots, s_k)]$ is a Statecharts term. Here, $s_1, \ldots, s_k$ are the (parallel) sub-states of s.

We define $\mathsf{root}(s) =_{\text{df}} n$. If $\hat{t} = (t, i, I, O, j) \in T$ is a transition of Or-state $[n : (s_1, \ldots, s_k); l; T]$, then we define $\mathsf{name}(\hat{t}) =_{\text{df}} t$, $\mathsf{source}(\hat{t}) =_{\text{df}} s_i$, $\mathsf{trg}(\hat{t}) =_{\text{df}} I$, $\mathsf{act}(\hat{t}) =_{\text{df}} O$, and $\mathsf{target}(\hat{t}) =_{\text{df}} s_j$. $\mathsf{trg}(\hat{t})$ is called the *trigger part* and $\mathsf{act}(\hat{t})$ is called the *action part* of $\hat{t}$. Furthermore, we define $\mathsf{trg}^+(\hat{t}) =_{\text{df}} \mathsf{trg}(\hat{t}) \cap \Pi$ and $\mathsf{trg}^-(\hat{t}) =_{\text{df}} \mathsf{trg}(\hat{t}) \cap \neg\Pi$.

We assume the following for SC: (i) all state names and transition names are mutually disjoint, (ii) no transition t produces an event that contradicts its trigger, i.e., $\mathsf{trg}(t) \cap \neg\mathsf{act}(t) = \emptyset$, and (iii) no transition t produces an event that is included in its trigger, i.e., $\mathsf{trg}(t) \cap \mathsf{act}(t) =$

$\emptyset$. As a consequence of (i), states and transitions in Statecharts terms are uniquely referred to by their name. For convenience, we sometimes identify a Statecharts state s and a transition t with its name $\mathsf{root}(s)$ and $\mathsf{name}(t)$, respectively. Furthermore, let $\mathsf{trans}(s)$ be the set of transitions within a Statecharts term s.

The Statecharts term corresponding to the Statechart example presented in Fig. 1 is term s_1, which is defined as follows.

$$
\begin{array}{ll}
s_1 = [n_1 : (s_2, s_3)] & s_2 = [n_2 : (s_4, [n_5]); 1; \{\hat{t}_1, \hat{t}_2\}] \\
s_3 = [n_3 : ([n_6], [n_7]); 1; \{\hat{t}_3\}] & s_4 = [n_4 : ([n_8], [n_9]); 1; \{\hat{t}_4\}] \\
\hat{t}_1 = (t_1, 5, \{c\}, \emptyset, 4) & \hat{t}_2 = (t_2, 4, \{a, \neg b\}, \emptyset, 5) \\
\hat{t}_3 = (t_3, 6, \{a\}, \{c\}, 7) & \hat{t}_4 = (t_4, 8, \{a, \neg c\}, \emptyset, 9)
\end{array}
$$

The Statecharts variant considered here does not include all "features" present in some other variants. In particular, we prohibit *interlevel transitions* and – only for space limitations – do not treat the *history mechanism*.

2.2. TRADITIONAL SEMANTICS

In this section, we sketch the semantics of Statecharts terms adopted from Maggiolo-Schettini, Peron, and Tini [12] which will serve for us as a reference semantics and which is a slight variant of the "traditional" Statecharts semantics, as proposed by Pnueli and Shalev [17].[1]

As mentioned before, a Statechart s reacts to the arrival of some external events by triggering enabled micro steps, possibly in a chain-reaction–like manner, thereby performing a macro step. More precisely, a macro step comprises a maximal set of micro steps, or transitions, that

- are *triggered* by events which are either offered by the environment or generated by other micro steps of the same macro step,
- are mutually *consistent*,
- are mutually *compatible*,
- are *relevant*, and
- obey the principle of *causality*.

The Statecharts principle of *global consistency*, which prohibits an event to be present and absent in the same macro step, is subsumed by the notions of *triggered* and *compatible*.

[1]The variation was motivated by the fact that Pnueli and Shalev's step-construction procedure – which will be presented in the following – can fail.

Figure 2 Step-construction algorithm

```
procedure step-construction(s, E);
var T := ∅;
while T ⊂ enabled(s, E, T) do
   choose t ∈ enabled(s, E, T) \ T;
   T := T ∪ {t}
od;
return T
```

Now, we briefly introduce these notions. Let $s \in \mathsf{SC}$, $t \in \mathsf{trans}(s)$, $T \subseteq \mathsf{trans}(s)$, and $E \subseteq \Pi$.

Transition t is *consistent* with all transitions in T, denoted by $t \in \mathsf{consistent}(s, T)$, if t is not in the same parallel component as any transition in T. Transition t is *compatible* to all transitions in T, denoted by $t \in \mathsf{compatible}(s, T)$, if no event produced by t appears negated in the trigger of a transition in T. Transition t is *relevant* for Statechart term s, denoted by $t \in \mathsf{relevant}(s)$, if the source state of t is currently active. Transition t is *triggered* by event set E, denoted by $t \in \mathsf{triggered}(s, E)$, if the positive, but not the negative trigger events of t are in E. Finally, we say that transition t is *enabled* in s with respect to event set E and transition set T, if $t \in \mathsf{enabled}(s, E, T)$, where $\mathsf{enabled}(s, E, T) =_{\mathrm{df}} \mathsf{relevant}(s) \cap \mathsf{consistent}(s, T) \cap \mathsf{triggered}(s, E \cup \bigcup_{t \in T} \mathsf{act}(t)) \cap \mathsf{compatible}(s, T)$.

Observe that the maximality of each macro step implements the synchrony hypothesis of Statecharts.

Macro Step. Function enabled constitutes the essential means to define a macro step. Unfortunately, for given Statechart s and event set E a solution T^* of equation $T = \mathsf{enabled}(s, E, T)$ need not constitute a macro step, since causality is neglected here. However, Maggiolo-Schettini et al. [12] provide an operational approach for causally justifying the triggering of each transition of a macro step using the nondeterministic *step-construction* algorithm presented in Fig. 2.

Given Statecharts term s and set E of events, $\mathsf{step\text{-}construction}(s, E)$ nondeterministically computes one macro step, i.e. a set of transitions, out of the set of all possible macros steps. For event set I offered by the environment and set T of transitions computed by $\mathsf{step\text{-}construction}(s, I)$, the Statecharts term s may evolve in a single *macro step* $s \overset{I}{\underset{O}{\Longrightarrow}} s'$ to Statecharts term s', thereby executing the transitions in T and producing the events $O =_{\mathrm{df}} \bigcup_{t \in T} \mathsf{act}(t)$.

In the following we consider some macro steps of our Statechart example from Fig. 1. For convenience, we abbreviate a Statechart term by its active basic states - e.g. writing (n_8, n_6) for term s_1 defined before. Let $\Pi = \{a, b, c\}$ and assume that the environment only offers event a. Then transition t_4 can execute resulting in macro step $(n_8, n_6) \overset{\{a\}}{\underset{\emptyset}{\Longrightarrow}} (n_9, n_6)$. Note that though transition t_3 is also triggered, it can not be executed together with t_4 in the same macro step, because this would violate global consistency, since t_3 would generate event c whose negative conterpart $\neg c$ is contained in the trigger of t_4. But t_2 and t_3 can execute in the same macro step, since both are triggered by event a and since global consistency is fulfilled resulting in macro step $(n_8, n_6) \overset{\{a\}}{\underset{\{c\}}{\Longrightarrow}} (n_5, n_7)$. Note, that the execution of t_3 alone does not constitute a valid macro step in the considered situation, because a macro step has to be maximal. However, if the environment offers the event set $\{a, b\}$ or $\{a, b, c\}$, then t_3 constitutes a complete macro step.

3. NEW SEMANTICS DEFINITION

3.1. AIM

We aim at the development of the definition of a Statecharts semantics which satisfies the following properties:

- general properties (i.e. not specific for the Statecharts language): The semantics shall be formal, as concise as possible, comprehensible, modular, and easily modifiable. These properties suggest to define a compositional semantics.

- Statecharts specific properties: The semantics shall fulfill causality, the synchrony hypothesis, and global consistency, as already provided by our (non-compositional) reference semantics.

3.2. APPROACH

We directly define the semantics on sequences of micro steps. This is in contrast to our other approaches [10, 11], where we defined a fine-grained semantics on single micro steps. Due to the direct approach a very concise Statecharts semantics results. We use labelled transition systems (LTS) as the semantic domain, where LTS-states model Statechart terms and LTS-transitions model sequences of Statechart micro steps. Considering micro step sequences – which may not be of maximal length – instead of macro steps is a prerequisite to achieve a compositional semantics, because a non-maximal micro step sequence of a Statecharts term s constitutes a potential macro step (or parts thereof) of an And-state with sub-state s.

Our semantics combines causality and synchrony on the macro step level, and meets compositionality on the level of micro step sequences. In contrast to the work of Maggiolo-Schettini et al. [12] and Uselton and Smolka [20] our semantics gets along with quite simple labels.

Note that due to a result of Huizing and Gerth [7] it is impossible to define a single-level LTS semantics that combines causality, synchrony, and compositionality on the level of macro steps if the labels are simply pairs of event sets in the style "trigger/action".

3.3. FORMAL SEMANTICS DEFINITION

Let LTS be the set of *labeled transition systems.* Then the semantics $[\![s]\!]$ of a Statecharts term $s \in \mathsf{SC}$ is given by the labeled transition system $(\mathsf{SC}, 2^{\Pi} \times 2^{\neg\Pi} \times 2^{\Pi} \times 2^{\mathcal{T}}, \longrightarrow, s) \in \mathsf{LTS}$, where

- SC is the set of states,
- $2^{\Pi} \times 2^{\neg\Pi} \times 2^{\Pi} \times 2^{\mathcal{T}}$ the set of labels,
- $\longrightarrow \subseteq \mathsf{SC} \times (2^{\Pi} \times 2^{\neg\Pi} \times 2^{\Pi} \times 2^{\mathcal{T}}) \times \mathsf{SC}$ the transition relation, and
- s the start state.

For the sake of simplicity, we abbreviate $(s, (I^+, I^-, O, L), s') \in \longrightarrow$ by $s \xrightarrow[O]{(I^+,I^-)}_L s'$. We say that *Statechart term s may change with label (I^+, I^-, O) to Statechart term s'*, where s and s' are called the *source* and the *target states* of this transition, respectively, and I^+ and I^- are called the *positive* and *negative input part* of the label, respectively, whereas O is called the *output part* of the label.[2] If we are not interested in the target state, we write $s \xrightarrow[O]{(I^+,I^-)}_L$ instead of $\exists s'.\ s \xrightarrow[O]{(I^+,I^-)}_L s'$ and say that Statechart s *may engage in a transition labeled with* (I^+, I^-, O). We define $(s_1, \ldots, s_k)_{[l \mapsto s']} =_{\mathrm{df}} (s_1, \ldots, s_{l-1}, s', s_{l+1}, \ldots, s_k)$ for $1 \leq l \leq k$ and $s' \in \mathsf{SC}$, and define $\vec{s} =_{\mathrm{df}} (s_1, \ldots, s_k)$. We write $s - t \rightarrow s'$ instead of $(\mathsf{source}(t) = s \ \wedge \ \mathsf{target}(t) = s')$ for $t \in \mathsf{TR}$ and $s, s' \in \mathsf{SC}$.

Furthermore, we need function default : SC $\longrightarrow$ SC which for a given Statecharts term s sets the default state of every Or-state s' within s to the currently active sub-state of s'.

$$
\begin{aligned}
\mathsf{default}([n]) &=_{\mathrm{df}} [n] \\
\mathsf{default}([n : (s_1, \ldots, s_k); l; T]) &=_{\mathrm{df}} [n : (\mathsf{default}(s_1), s_2, \ldots, s_k); 1; T] \\
\mathsf{default}([n : (s_1, \ldots, s_k)]) &=_{\mathrm{df}} [n : (\mathsf{default}(s_1), \ldots, \mathsf{default}(s_k))]
\end{aligned}
$$

[2] Here we ignore the set L of transition names, because L is not necessary for the semantics definition itself. L is only used to compute macro steps out of the semantics.

Transition relation $\longrightarrow$ is now defined by Table 1 using SOS rules. Due to space limitations we use the rule formats

$$name\frac{premise}{conclusion}\ (side\ condition) \quad \text{as well as} \quad name\frac{premise}{conclusion}\ (side\ condition).$$

Table 1 SOS rules

$$\text{OR-1}\ \frac{-}{[n:\vec{s};i;T] \xrightarrow[\mathsf{act}(t)]{(\mathsf{trg}^+(t),\mathsf{trg}^-(t))}_{\{\mathsf{name}(t)\}} [n:\vec{s}_{[l\mapsto \mathsf{default}(s_l)]};l;T]}\ \left(\begin{array}{c} t\in T, \\ s_i - t - s_l \end{array}\right)$$

$$\text{OR-2}\ \frac{s_l \xrightarrow[O]{(I^+,I^-)}_L s'_l}{[n:\vec{s};l;T] \xrightarrow[O]{(I^+,I^-)}_L [n:\vec{s}_{[l\mapsto s'_l]};l;T]}$$

$$\text{AND}\ \frac{\left(\forall m\in M: s_m \xrightarrow[O_m]{(I^+_m,I^-_m)}_{L_m} s'_m\right) \wedge \left(\forall i,j\in M: (\neg I^-_i)\cap O_j = \emptyset\right)}{[n:\vec{s}] \xrightarrow[\bigcup_{m\in M} O_m]{\left(\bigcup_{l\in N}\left(I^+_{f(l)}\setminus\left(\bigcup_{j=1}^{l-1} O_{f(j)}\right)\right),\ \bigcup_{m\in M} I^-_m\right)}_{\bigcup_{m\in M} L_m} [n:\vec{s'}]}$$

$$\left(\begin{array}{c} M\subseteq K=\{1,\ldots,k\},\ N=\{1,\ldots,|M|\}, \\ f:N\to M \text{ bijection },\ \forall p\in K\setminus M:\ s'_p=s_p \end{array}\right)$$

Explanation of the SOS rules:

- OR-1 (Statechart transition execution)
  This rule describes the execution of a Statechart transition $t \in T$ of an Or-state $[n : \vec{s}; i; T]$. It defines that the Or-state with currently active sub-state s_i may change with label $(\mathsf{trg}^+(t), \mathsf{trg}^-(t), \mathsf{act}(t))$ to Or-state $[n : \vec{s}_{[l\mapsto \mathsf{default}(s_l)]}; l; T]$ with currently active sub-state s_l, if $t \in T$, $\mathsf{source}(t) = s_i$, and $\mathsf{target}(t) = s_l$. The definition of

the transition label is motivated as follows: In order to execute the trigger, all positive events $\mathsf{trg}^+(t)$ in $\mathsf{trg}(t)$ must be offered by the environment, and all negative events $\mathsf{trg}^-(t)$ in $\mathsf{trg}(t)$ must be absent. Then t can execute by offering the output $\mathsf{act}(t)$.

- OR-2 (Propagation of a change of a sub-state)
  If Statechart term s_l (with $l \in \{1, \ldots, k\}$) may change with label (I^+, I^-, O) to Statechart term s'_l, then Or-state $[n : \vec{s}; l; T]$ with currently active sub-state s_l may change with the same label to Or-state $[n : \vec{s}_{[l \mapsto s'_l]}; l; T]$ with currently active sub-state s'_l.

- AND (Synchronization of parallel states)
  This rule defines a sort of synchronization between (parallel) sub-states of an And-state. More concretely, AND defines for all such sub-states s_m of And-state $[n : \vec{s}]$, which may engage in a transition, the interaction between s_m and the other sub-states s_j of $[n : \vec{s}]$, where $1 \leq j \leq k$, $j \neq m$. (Remember that $\vec{s} =_{\mathrm{df}} (s_1, \ldots, s_k)$.) If $M \subseteq \{1, \ldots, k\}$, such that

  1 for all $m \in M$ sub-state s_m may change with label (I^+_m, I^-_m, O_m) to sub-state s'_m and
  2 the global consistency condition $\forall i, j \in M : (\neg I_i^-) \cap O_j = \emptyset$ is fulfilled,

  then And-state $[n : \vec{s}]$ may change to And-state $[n : \vec{s'}]$, where $\forall p \in K \setminus M : s'_p = s_p$.
  The label of this state change is justified as follows:

  – To explain the definition of the positive input part, we consider every parallel component $s_{f(l)}$ of $[n : \vec{s}]$ separately for $l \in N$ (or equivalently for every $f(l) \in M$).[3] $I^+_{f(l)}$ is the set of positive events which have to be offered by the environment as a precondition for the state change from $s_{f(l)}$ to $s'_{f(l)}$ if no events have been generated within the corresponding sequence of micro steps. But since we can assume that the state change from $s_{f(l)}$ to $s'_{f(l)}$ is the l-th state change, i.e. the l-th micro step, within the micro step sequence from $[n : \vec{s}]$ to $[n : \vec{s'}]$, we can further assume that exactly all the state changes from $s_{f(j)}$ to $s'_{f(j)}$ for $j \in \{1, \ldots, l-1\}$ have

[3]Note, that we use the bijection $f : \{1, \ldots, |M|\} \rightarrow M$ in order to express an ordering of micro step executions within a sequence of micro steps which is independent of the order of substates s_m in $[n : \vec{s}]$ for $(1 \leq m \leq k)$.

already taken place in this micro step sequence. Due to these state changes the set $\bigcup_{j=1}^{l-1} O_{f(j)}$ of events has been generated, before a possible state change from $s_{f(l)}$ to $s'_{f(l)}$ occurs. Taking these generated events into account, the set of positive events which still have to be provided by the environment, so that the state change from $s_{f(l)}$ to $s'_{f(l)}$ can occur, is the set $I^+_{f(l)} \setminus \left(\bigcup_{j=1}^{l-1} O_{f(j)}\right)$. Thus the positive input part for the micro step sequence, i.e. for the state change from $[n : \vec{s}]$ to $[n : \vec{s'}]$, is given by the set $\bigcup_{l \in N} \left(I^+_{f(l)} \setminus \left(\bigcup_{j=1}^{l-1} O_{f(j)}\right)\right)$.

- The negative input part and the output part are simply given by the union of all I^-_m and O_m for $m \in M$, respectively.

Rule AND reveals that our semantics is defined on the granularity of micro step sequences instead of on a single micro step.

We define a notion of *macro step* for our semantics as follows:

Macro Step. Let $s, s' \in \mathsf{SC}$, $I, O \subseteq \Pi$. Then: $s \overset{I}{\underset{O}{\Rightarrow}} s'$ *if and only if* $(\exists I^+ \subseteq \Pi, I^- \subseteq \neg\Pi, L \subseteq \mathcal{T}.(s \xrightarrow[O]{(I^+,I^-)}_L s' \wedge I^+ \subseteq I \wedge (\neg I^-) \cap I = \emptyset)) \wedge$ $(\nexists \hat{I}^+, \hat{O} \subseteq \Pi, \hat{I}^- \subseteq \neg\Pi, \hat{L} \subseteq \mathcal{T}.(s \xrightarrow[\hat{O}]{(\hat{I}^+,\hat{I}^-)}_{\hat{L}} \wedge \hat{I}^+ \subseteq I \wedge (\neg \hat{I}^-) \cap I = \emptyset \wedge$ $L \subsetneq \hat{L}))$
In this case we say that Statechart term s may perform a *macro step* with input I and output O to Statechart term s'.

This macro step definition employs the information about the set L of Statechart transitions executed if a (semantical) transition $s \xrightarrow[O]{(I^+,I^-)}_L s'$ takes place by requiring that L must be maximal. Thus a macro step consists of a maximal number of micro steps, whereas the AND-rule of our Statecharts semantics allows micro step sequences of non-maximal length. Therefore the semantics considers all potential macro steps which can arise if the considered Statechart is put in parallel with another Statechart context. This approach realizes a compositional Statecharts semantics on sequences of micro steps. But note that as soon as macro steps are computed, i.e. when maximality of micro step sequences is required, compositionality is lost.

To sum up, our semantics approaches the combination of causality, synchrony, and compositionality by satisfying the first two properties on the macro step level and by approaching macro step compositionality due to the semantics' compositionality on micro step sequences.

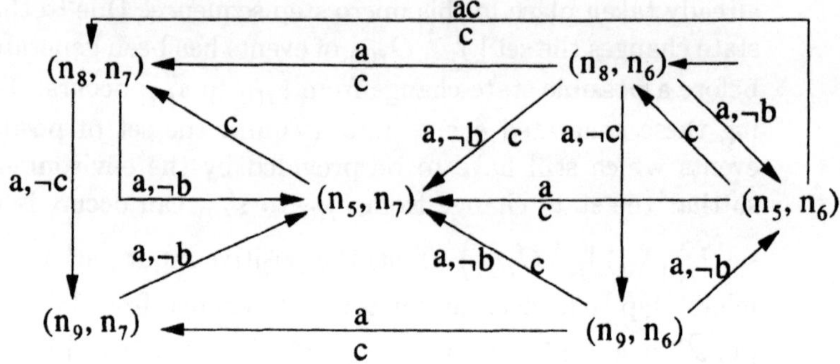

Figure 3 Semantics for the Statechart Example from Fig. 1

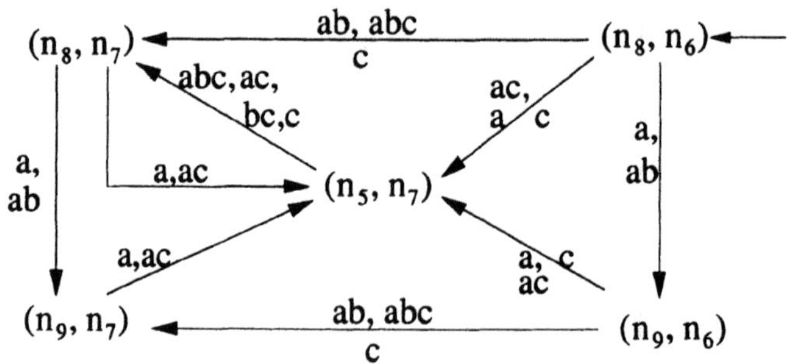

Figure 4 Macro Steps for the Statechart Example from Fig. 1

3.4. EXAMPLE

Consider again our Statecharts example from Fig. 1. Its semantics, i.e. a labeled transition system, and the possible macro steps are graphically represented as state transition diagrams in Fig. 3 and Fig. 4.[4] For example, the macro steps $(n_8, n_6) \overset{\{a\}}{\underset{\emptyset}{\Longrightarrow}} (n_9, n_6)$ and $(n_8, n_6) \overset{\{a\}}{\underset{\{c\}}{\Longrightarrow}} (n_5, n_7)$ already considered before can be found in Fig. 4, where we represent a macro step $\ldots \overset{I}{\underset{O}{\Longrightarrow}} \ldots$ by writing I to the left of or above, whereas O to the right of or below the arrow. Furthermore, we write e.g. ab instead of $\{a, b\}$. Different possibilities for I at the same arrow are separated by ",". Therefore, the arrow from (n_8, n_6) to (n_5, n_7) also represents macro step $(n_8, n_6) \overset{\{ac\}}{\underset{\{c\}}{\Longrightarrow}} (n_5, n_7)$. Note, that we have skipped empty output sets.

[4] A Statecharts term is again abbreviated by its active basic states.

The above-mentioned macro steps from (n_8, n_6) to (n_5, n_7) both result from the semantical transition $(n_8, n_6) \xrightarrow[\{c\}]{(\{a\},\{\neg b\})} (n_5, n_7)$, which can be found in Fig. 3. A semantical transition is represented analogously to a macro step, but both – positive and negative – input parts occur to the left of or above the arrow, where e.g. an input $(\{a\}, \{\neg b\})$ is abbreviated by $a, \neg b$ and an input $(\{a\}, \emptyset)$ is abbreviated by a. The semantical transition considered before results from one of the following two sequences of micros steps: $(n_8, n_6) \xrightarrow[\{c\}]{(\{a\},\emptyset)} (n_8, n_7) \xrightarrow[\emptyset]{(\{a\},\{\neg b\})} (n_5, n_7)$ and $(n_8, n_6) \xrightarrow[\emptyset]{(\{a\},\{\neg b\})} (n_5, n_6) \xrightarrow[\{c\}]{(\{a\},\emptyset)} (n_5, n_7)$.

3.5. EQUIVALENCE OF SEMANTICS

We can now formalize our intuition of the semantic relation between the traditional Statecharts semantics from Section 2.2 and our Statecharts semantics from Section 3.3 by relating their corresponding macro step notions as follows:

$$\forall s, s' \in \mathsf{SC},\ \forall I, O \subseteq \Pi : \quad s \overset{I}{\underset{O}{\Longrightarrow}} s' \text{ if and only if } s \overset{I}{\underset{O}{\Rrightarrow}} s'.$$

4. RELATED WORK

We compare work concerning the formal semantics of Statecharts and quite similiar languages along the *three semantic dimensions of Statecharts*: causality, synchrony (hypothesis), and compositionality. Huizing and Gerth [7] have been the first considering these three properties simultaneously.

The starting point is provided by Harel et al. [6] developing a formal Statecharts semantics satisfying causality and synchrony, but ignoring compositionality. Subsequent work of Pnueli and Shalev [17] follows the same approach, but also considers global consistency.

In order to achieve synchrony and compositionality, some work neglects causality. So Scholz [18] uses streams as semantic domain resulting in a very concise, but non-causal fixed point semantics. Maraninchi [13] presents *Argos*, a language with a visual appeal very similar to Statecharts and a solid algebraic foundation. However, its compositional semantics – defined via SOS-rules as labelled transition systems – significantly differs from classical Statecharts semantics. For example, Argos is deterministic, abstracts from "non-causal" specifications, and allows a sequential component to fire more than once within a macro step.

Combining causality and compositionality, but at the expense of synchrony, has been the motivation for Mikk et al. [14] as well as Damm et al. [3]. However, in these cases one can not argue that they ignored synchrony, because they started from the Statecharts variant used in the STATEMATE tool [5]. In this variant an event generated within a macro step will not be sensed in the same, but the next macro step. The work of Latella et al. [8] as well as Paltor and Lilius [15] formalizes the semantics of UML-Statecharts [2] – the latter ignoring compositionality. Also UML-Statecharts do not fulfill the synchrony hypothesis.

From our point of view, the most interesting work concerns formalizing Statecharts semantics combining all three dimensions – causality, synchrony, and compositionality. Here, two classes can be distinguished:

The first class is based on a process algebra approach. At first the Statecharts language is embedded in a process algebra, for which then a structured operational semantics based on labelled transition systems is defined. Uselton and Smolka [21] and Levi [9] follow this approach by explicitly defining a causality relation on events and event sets, respectively. In our earlier work [11] we also present a process algebra approach, but do so without an explicitly defined causal ordering.

The second class is characterized by a "direct" approach defining an operational Statecharts semantics using labelled transition systems. Uselton and Smolka [20] again require a causality relation on events, whereas Maggiolo-Schettini et al. [12] use labels consisting of four-tuples which include complex and intricate information about causal orderings, global consistency, and negated events. Our present work also fits into this class. But in contrast to the previous work [20] and [12], our semantics does without an explicit causality relation and, moreover, is very concise.

5. CONCLUSIONS

The definition of a formal semantics for the Statecharts language has been a long-term challenge. A considerable number of proposals has been developed which differ in the Statecharts syntax to be supported and the set of semantical properties to be fulfilled.

We have presented a formal operational Statecharts semantics definition which combines causality and synchrony on the macro step level with compositionality on the level of micro step sequences. The essential additional property of this semantics definition is provided by its very conciseness.

References

[1] G. Berry and G. Gonthier. The ESTEREL synchronous programming language: Design, semantics, implementation. *Science of Computer Programming*, 19(2):87–152, 1992.

[2] G. Booch, J. Rumbaugh, and I. Jacobson. *The Unified Modeling Language User Guide.* Addison-Wesley, 1999.

[3] W. Damm, B. Josko, H. Hungar, and A. Pnueli. A compositional real-time semantics of STATEMATE designs. In W.-P. de Roever, H. Langmaack, and A. Pnueli, editors, *Compositionality: The Significant Difference*, volume 1536 of *Lecture Notes in Computer Science*, pages 186–238. Springer-Verlag, September 1997.

[4] D. Harel. Statecharts: A visual formalism for complex systems. *Science of Computer Programming*, 8:231–274, 1987.

[5] D. Harel and A. Naamad. The STATEMATE semantics of Statecharts. *ACM Transactions on Software Engineering*, 5(4):293–333, October 1996.

[6] D. Harel, A. Pnueli, J. Schmidt, and R. Sherman. On the formal semantics of Statecharts. In *Symposium on Logic in Computer Science (LICS '87)*, pages 56–64, Ithaca, NY, USA, June 1987. IEEE Computer Society Press.

[7] C. Huizing and R. Gerth. On the semantics of reactive systems. Technical report, Eindhoven University of Technology, 1990.

[8] D. Latella, I. Majzik, and M. Massink. Towards a formal operational semantics of UML Statechart diagrams. In *Formal Methods for Open Object-based Distributed Systems.* Chapman & Hall, 1999.

[9] F. Levi. *Verification of Temporal and Real-Time Properties of Statecharts.* PhD thesis, University of Pisa-Genova-Udine, Pisa, Italy, February 1997.

[10] G. Lüttgen, M. von der Beeck, and R. Cleaveland. A Compositional Approach to Statecharts Semantics. submitted for publication.

[11] G. Lüttgen, M. von der Beeck, and R. Cleaveland. Statecharts via process algebra. In J. C. M. Baeten and S. Mauw, editors, *Concurrency Theory (CONCUR '99)*, volume 1664 of *Lecture Notes in Computer Science*, Eindhoven, The Netherlands, August 1999. Springer-Verlag.

[12] A. Maggiolo-Schettini, A. Peron, and S. Tini. Equivalences of Statecharts. In U. Montanari and V. Sassone, editors, *CONCUR '96 (Concurrency Theory)*, volume 1119 of *Lecture Notes in Computer Science*, pages 687–702, Pisa, Italy, August 1996. Springer-Verlag.

[13] F. Maraninchi. Operational and compositional semantics of synchronous automaton compositions. In R. Cleaveland, editor, *CONCUR '92 (Concurrency Theory)*, volume 630 of *Lecture Notes in Computer Science*, pages 550–564, Stony Brook, NY, USA, August 1992. Springer-Verlag.

[14] E. Mikk, Y. Lakhnech, and M. Siegel. Hierarchical automata as model for Statecharts. In *Proceedings of Asian Computing Science Conference (ASIAN '97)*, volume 1345 of *Lecture Notes in Computer Science*. Springer-Verlag, December 1997.

[15] I. Paltor and J. Lilius. Formalising UML state machines for model checking. In R. France and B. Rumpe, editors, *UML'99 - The Unified Modeling Language. Beyond the Standard.*, volume 1723 of *LNCS*. Springer, 1999.

[16] G. Plotkin. A structural approach to operational semantics. Technical Report DAIMI-FN-19, Computer Science Department, Aarhus University, Denmark, 1981.

[17] A. Pnueli and M. Shalev. What is in a step: On the semantics of Statecharts. In T. Ito and A. Meyer, editors, *Theoretical Aspects of Computer Software (TACS '91)*, volume 526 of *Lecture Notes in Computer Science*, pages 244–264, Sendai, Japan, September 1991. Springer-Verlag.

[18] P. Scholz. *Design of Reactive Systems and their Distributed Implementation with Statecharts.* PhD thesis, Munich University of Technology, Munich, Germany, August 1998.

[19] B. Selic, G. Gullekson, and P. Ward. *Real-time Object Oriented Modeling and Design.* J. Wiley, 1994.

[20] A. Uselton and S. Smolka. A compositional semantics for Statecharts using labeled transition systems. In B. Jonsson and J. Parrow, editors, *CONCUR '94 (Concurrency Theory)*, volume 836 of *Lecture Notes in Computer Science*, pages 2–17, Uppsala, Sweden, August 1994. Springer-Verlag.

[21] A. Uselton and S. Smolka. A process-algebraic semantics for Statecharts via state refinement. In *Proceedings of PROCOMET '94.* North Holland/Elsevier, 1994.

[22] M. von der Beeck. A comparison of Statecharts variants. In H. Langmaack, W. Roever, and J. Vytopil, editors, *Formal Techniques in Real-Time and Fault-Tolerant Systems (FTRTFT '94)*, volume 863 of *Lecture Notes in Computer Science*, pages 128–148, Lübeck, Germany, September 1994. Springer-Verlag.

IMPLEMENTING CCS IN MAUDE

Alberto Verdejo and Narciso Martí-Oliet
Depto. de Sistemas Informáticos y Programación
Universidad Complutense de Madrid, Spain
{alberto,narciso}@sip.ucm.es

Abstract We explore the features of rewriting logic and the language Maude as a logical and semantic framework for representing both the semantics of CCS, and a modal logic for describing local capabilities of CCS processes. Although a rewriting logic representation of the CCS semantics was given previously, it cannot be directly executed in the default interpreter of Maude. Moreover, it cannot be used to answer questions such as which are the successors of a process after performing an action, which is used to define the semantics of the modal logic. Basically, the problems are the existence of new variables in the righthand side of the rewrite rules and the nondeterministic application of the semantic rules, inherent to CCS. We show how these problems can be solved by exploiting the reflective properties of rewriting logic, which allow controlling the rewriting process. This executable specification plus the reflective control of the rewriting process can be used to analyze CCS processes.

1. INTRODUCTION

Rewriting logic was introduced by Meseguer [8] as a unified model of concurrency in which several well-known models of concurrent systems can be represented in a common framework. This goal was further extended by Martí-Oliet and Meseguer [7] to the idea of rewriting logic as a *logical and semantic framework*. It was shown that many other logics, widely different in nature, can be represented inside rewriting logic in a natural and direct way. The general way in which such representations are achieved is by:

- Representing formulas or proof-theoretic structures such as sequents, as terms in an order-sorted equational data type whose equations express structural axioms natural to the logic in question.
- Representing the rules of deduction of a logic as rewrite rules that transform certain patterns of formulas into other patterns modulo the given structural axioms.

Tommaso Bolognesi and Diego Latella (Eds.), *Formal Methods for Distributed System Development.*

Similar techniques can be used to naturally specify and prototype many languages and systems in rewriting logic. In particular, the similarities between rewriting logic and structural operational semantics were noted by Meseguer [8] and further explored by Martí-Oliet and Meseguer [7]. As an illustrative example, Martí-Oliet and Meseguer [7] completely developed a representation of Milner's CCS [9] in rewriting logic. However, this representation cannot be directly executed in the default interpreter of Maude [3], a high-performance language and system supporting both equational and rewriting logic computation.

Basically, the problems are the existence of new variables in the right-hand side of the rewrite rules and the nondeterministic application of the semantic rules, inherent to CCS. We show how these problems can be solved by exploiting the reflective capabilities of rewriting logic and of Maude, that allow representing rewriting logic inside itself [2], and in particular controlling the rewriting process. This executable specification plus the reflective control of the rewriting process can be used to analyze CCS processes and to answer different questions such as which are the successors of a process after performing an action. In summary, we have managed to make the representation of CCS *executable* by using reflective techniques in such a way that it can be used to define in Maude the semantics of Hennessy-Milner modal logic [6].

In the rest of this section we review the representation of the CCS semantics in Maude, in order to see in detail how the problems we mentioned above arise.[1] First, we show the representation of the CCS syntax in two functional modules defining actions and processes.

```
(fmod ACTION is  protecting QID .
  sorts Label Act .  subsorts Qid < Label < Act .
  op tau : -> Act .  *** silent action
  op ~_ : Label -> Label .
  var N : Label .  eq ~ ~ N = N .
endfm)
(fmod PROCESS is  protecting ACTION .
  sorts ProcessId Process .  subsort Qid < ProcessId < Process .
  op 0 : -> Process .                              *** inaction
  op _._ : Act Process -> Process [prec 25] .      *** prefix
  op _+_ : Process Process -> Process [prec 35] .  *** summation
  op _|_ : Process Process -> Process [prec 30] .  *** composition
  op _'[_/_'] : Process Label Label -> Process [prec 20] .
                      *** relabelling: [b/a] relabels "a" to "b"
  op _\_ : Process Label -> Process [prec 20] .    *** restriction
endfm)
```

[1] For lack of space, the reader is assumed to be familiar with the Maude syntax [3], as well as CCS and its operational semantics [9].

Full CCS is represented, including (possibly recursive) process definitions by means of *contexts*.

```
(fmod CCS-CONTEXT is  protecting PROCESS .
  sorts BadProcess Context BadContext .
  subsort Process < BadProcess .  subsort Context < BadContext .
  op _=def_ : ProcessId Process -> Context .
  op nil : -> Context .
  op _&_ : BadContext BadContext -> BadContext [assoc comm id: nil] .
  op _definedIn_ : ProcessId Context -> Bool .
  op def : ProcessId Context -> BadProcess .
  op not-defined : -> BadProcess .
  op context : -> Context .
  vars X X' : ProcessId .  var P : Process .  var C : Context .
  cmb (X =def P) & C : Context if not(X definedIn C) .
  eq X definedIn nil = false .
  eq X definedIn ((X' =def P) & C) = (X == X') or (X definedIn C) .
  eq def(X, nil) = not-defined .
  eq def(X, ((X' =def P) & C)) = if X == X' then P else def(X, C) fi .
endfm)
```

The general idea for implementing in rewriting logic the operational semantics of CCS, is to translate each semantic rule into a rewrite rule where either the premises are rewritten to the conclusion, or the conclusion is rewritten to the premises. In previous work [7] the first approach was followed; here we adopt the second one, because we want to be able to prove in a bottom-up way that a given transition is valid in CCS.

The CCS transition $P \xrightarrow{a} P'$ is represented in Maude by the term P `--` a `->` P', of sort `Judgement`, built by means of the operator

```
op _--_->_ : Process Act Process -> Judgement [prec 50] .
```

In general, a semantic rule has a conclusion and a set of premises, each one represented by a judgement. So we need a sort to represent sets of judgements:

```
sort JudgementSet .  subsort Judgement < JudgementSet .
op emptyJS : -> JudgementSet .
op __ : JudgementSet JudgementSet -> JudgementSet
        [assoc comm id: emptyJS prec 60] .
var J : Judgement .  eq J J = J .
```

The union constructor is written with empty syntax (`__`), and declared associative (`assoc`), commutative (`comm`), and with identity element the empty set (`id: emptyJS`). Matching and rewriting take place *modulo* such properties. Idempotency is specified by means of an explicit equation.

A semantic rule is implemented as a rewrite rule where the singleton set consisting of the judgement representing the conclusion is rewritten to the set consisting of the judgements representing the premises.

For example, for the restriction operator of CCS, we have the rule[2]

```
crl [res] :  P \ L -- A -> P' \ L
          => -----------------------
                   P -- A -> P'        if (A =/= L) and (A =/= ~ L) .
```

and for the axiom schema defining the prefix operator we have

```
rl [pref] :    A . P -- A -> P
          => -------------------
                   emptyJS .
```

Thus, a transition $P \xrightarrow{a} P'$ is possible in CCS if and only if the judgement representing it can be rewritten to the empty set of judgements by rewrite rules of the form described above that define the operational semantics of CCS in a backwards search fashion.

However, we have found problems while working with this approach in the current version of Maude. The first one is that sometimes new variables appear in the premises which are not in the conclusion. For example, in one of the semantic rules for the parallel operator we have

```
rl [par] :      P | Q -- tau -> P' | Q'
          => --------------------------------
               P -- L -> P'   Q -- ~ L -> Q' .
```

where L is a new variable in the righthand side of the rewrite rule. Rules of this kind cannot be directly used by the Maude default interpreter; they can only be used at the metalevel using a strategy to instantiate the extra variables.

Another problem is that sometimes several rules can be applied to rewrite a judgement. For example, for the summation operator we have, because of its intrinsic nondeterminism,[3]

```
rl [sum] :  P + Q -- A -> P'        rl [sum] :  P + Q -- A -> Q'
         => -------------------              => -------------------
                P -- A -> P' .                      Q -- A -> Q' .
```

In general, not all of these possibilities lead to an empty set of judgements. So we have to deal with the whole tree of possible rewritings of a judgement, searching if one of the branches leads to `emptyJS`.

In Section 2, we show how these problems can be solved in the current version of the Maude system by using reflection, obtaining an executable semantics, where we can prove if a transition $P \xrightarrow{a} P'$ is possible.

---

[2]Using the fact that text beginning with --- is a comment in Maude, rules are displayed in such a way as to emphasize the correspondence with the usual presentation in textbooks, although in this case the conclusion is above the horizontal line.

[3]Only one rule is enough by declaring _+_ to be commutative. In this case, nondeterminism appears because of possible multiple matches (modulo commutativity) against pattern P + Q.

In Section 3, we extend this representation in order to be able to answer different kinds of questions, such as if process P can perform action a (and we do not care about the process it becomes), or which are the *successors* of a process P after performing actions in a given set As, that is,

$$succ(P, As) = \{P' \mid P \xrightarrow{a} P' \wedge a \in As\}.$$

In Section 4 we show how we can define in Maude the semantics of the Hennessy-Milner modal logic for describing local capabilities of CCS processes.

2. EXECUTABLE CCS SEMANTICS

In this section we show how the problem of new variables in the righthand side of a rewrite rule is solved by using the concept of *explicit metavariables* presented by Stehr and Meseguer [11], and how nondeterministic rewriting is controlled by using a *search strategy* [1, 4].

New variables in the righthand side of a rule represent "unknown" values when we are rewriting; by using metavariables we make explicit this lack of knowledge. The semantics with explicit metavariables has to bind them to concrete values when these values become known.

For the moment, metavariables are only needed as actions in the judgements, so we declare a new sort for metavariables as actions:

```
sort MetaVarAct .
op ?'(_')A : Qid -> MetaVarAct .
var NEW1 : Qid .
```

We also introduce a new sort `Act?` of "possible actions," and modify the operator for building judgements in order to deal with it:

```
sort Act? .  subsorts Act MetaVarAct < Act? .
var ?A : MetaVarAct .  var A? : Act? .
op _--_->_ : Process Act? Process -> Judgement [prec 50] .
```

As mentioned above, a metavariable will be bound when its concrete value becomes known, so we need a new judgement stating that a metavariable is bound to a concrete value

```
op '[_:=_'] : MetaVarAct Act -> Judgement .
```

and a way to propagate this binding to the rest of judgements where the bound metavariable may be present. Since this propagation has to reach all the judgements in the current state of the inference process, we introduce an operation to enclose the set of judgements, and a rule to propagate a binding

```
op '{'{_'}'} : JudgementSet -> Configuration .
var JS : JudgementSet .
```

```
rl [bind] : {{ [?A := A] JS }} => {{ <act ?A := A > JS }} .
```

where we use several overloaded, auxiliary functions `<act_:=_>_` to perform the substitutions (see complete specification in the full version [13]).

Now we are able to redefine the rewrite rules implementing the CCS semantics, taking care of metavariables. For the prefix operator we maintain the previous axiom schema and add a new rule for the case when a metavariable appears in the judgement

```
rl [pref] :  A . P -- A -> P          rl [pref] :  A . P -- ?A -> P
         => --------------------               => --------------------
                   emptyJS  .                           [?A := A]  .
```

Note how the metavariable `?A` present in the lefthand side judgement is bound to the concrete action `A` taken from the process `A.P`. This binding will be propagated to any other judgement in the set of judgements containing `A.P -- ?A -> P`.

For the summation operator, we generalize the rules allowing a more general variable `A?` of sort `Act?`, since the behavior is the same independently of whether a metavariable or an action appears in the judgement:

```
rl [sum] :  P + Q -- A? -> P'          rl [sum] :  P + Q -- A? -> Q'
        => -------------------                 => -------------------
             P -- A? -> P' .                        Q -- A? -> Q' .
```

Nondeterminism is again present; we deal with it in Section 2.1.

For the parallel operator, there are two rules for the cases when one of the composed processes performs an action on its own,

```
rl [par] : P | Q -- A? -> P' | Q      rl [par] : P | Q -- A? -> P | Q'
        => -----------------------            => -----------------------
             P -- A? -> P' .                        Q -- A? -> Q' .
```

and two additional rules dealing with the case when communication happens between both processes,

```
rl [par] :                P | Q -- tau -> P' | Q'
        => ----------------------------------------------
           P -- ?(NEW1)A -> P'    Q -- ~ ?(NEW1)A -> Q' .

rl [par] :                     P | Q -- ?A -> P' | Q'
        => ------------------------------------------------------------
           P -- ?(NEW1)A -> P'    Q -- ~ ?(NEW1)A -> Q'   [?A := tau] .
```

where we have overloaded the ~ operator `op ~_ : Act? -> Act? .`

Note how the term `?(NEW1)A` is used to represent a *new metavariable*. Rewriting has to be controlled by a *strategy* that instantiates the variable `NEW1` with a new (quoted) identifier each time one of the above rules is applied, in order to build *new* metavariables. The strategy presented in Section 2.1 does this as well as implementing the search in the tree of possible rewritings.

There are two rules dealing with the restriction operator of CCS, depending on whether an action or a metavariable occurs in the lefthand side judgement

```
crl [res] :  P \ L -- A -> P' \ L
        => -----------------------
                P -- A -> P'        if (A =/= L) and (A =/= ~ L) .

rl [res] :          P \ L -- ?A -> P' \ L
       => ------------------------------------------
             P -- ?A -> P'   [?A =/= L]   [?A =/= ~ L] .
```

In the latter case, we cannot use a conditional rewrite rule as in the former case, because the condition (?A =/= L) and (?A =/= ~ L) cannot be checked until we know the concrete value of the metavariable ?A. Hence, we have to add a new kind of judgement used to state constraints between metavariables, which is eliminated when it is fulfilled,

```
op '[_=/=_'] : Act? Act? -> Judgement .
crl [dist] : [A =/= A'] => emptyJS  if A =/= A' .
```

where (normal) actions are used.

For the relabelling operator of CCS we have similar rewrite rules, and for process identifiers we only need the generalization of the original rule by means of a more general variable A?.

```
crl [def] :           X -- A? -> P'?
        => ------------------------------
             def(X, context) -- A? -> P'?   if (X definedIn context) .
```

Using these rules, we can begin to pose some questions about the capability of a process to perform an action. For example, we can ask if the process 'a.'b.0 can perform action 'a (becoming 'b.0) by rewriting the configuration composed of a judgement representing that transition:

```
Maude> (rew {{ 'a . 'b . 0 -- 'a -> 'b . 0 }} .)
Result Configuration : {{ emptyJS }}
```

Since a configuration consisting of the empty set of judgements is reached, we can conclude that the transition is possible.

However, if we ask if the process 'a.0 + 'b.0 can perform action 'b becoming process 0, we get as result

```
Maude> (rew {{ 'a . 0 + 'b . 0 -- 'b -> 0 }} .)
Result Configuration : {{ 'a . 0 -- 'b -> 0 }}
```

representing that the given transition is not possible, which is not the case. The problem is that {{'a.0 + 'b.0 -- 'b -> 0}} can be rewritten in two different ways, and only one of them leads to a configuration consisting of the empty set of judgements.

Therefore, we need a strategy to search the tree of all possible rewrites.

2.1. SEARCHING IN THE TREE OF REWRITINGS

Rewriting logic is reflective [2], that is, there is a finitely presented rewrite theory $\mathcal{U}$ that is *universal* in the sense that we can represent any finitely presented rewrite theory $\mathcal{R}$ (including $\mathcal{U}$ itself) and any terms t, t' in $\mathcal{R}$ as terms $\overline{\mathcal{R}}$ and $\overline{t}, \overline{t'}$ in $\mathcal{U}$, and we then have the following equivalence: $\mathcal{R} \vdash t \longrightarrow t' \Leftrightarrow \mathcal{U} \vdash \langle \overline{\mathcal{R}}, \overline{t} \rangle \longrightarrow \langle \overline{\mathcal{R}}, \overline{t'} \rangle$.

In Maude, key functionality of the universal theory $\mathcal{U}$ has been efficiently implemented in the functional module `META-LEVEL`, where Maude terms are reified as elements of a data type `Term`, Maude modules are reified as terms in a data type `Module`, the process of reducing a term to normal form is reified by a function `meta-reduce`, and the process of applying a rule of a system module to a subject term is reified by a function `meta-apply` [3].

In this section we show how the reflective properties of Maude [2] can be used to control the rewriting of a term and the search in the tree of possible rewritings of a term. The depth-first strategy is based on previous work [1, 4], although modified to deal with the substitution of metavariables explained in the previous section.

The module implementing the search strategy is parameterized with respect to a constant equal to the metarepresentation of the Maude module which we want to work with. Hence, we define a parameter theory with a constant `MOD` representing the module, and a constant `labels` representing the list of labels of rewrite rules to be applied:

```
(fth AMODULE is  including META-LEVEL .
  op MOD : -> Module .
  op labels : -> QidList .
endfth)
```

The module containing the strategy, extending `META-LEVEL`, is then the parameterized module `SEARCH[M :: AMODULE]`.

Since we are defining a strategy to search a tree of possible rewritings, we need a notion of search goal. For the strategy to be general enough, we assume that the module `MOD` has an operation `ok` (defined at the object level), which returns a value of sort `Answer` such that

- `ok(T) = solution` means that the term `T` is one of the terms we are looking for, that is, `T` denotes a solution;
- `ok(T) = no-solution` means that the term `T` is not a solution and no solution can be found below `T` in the search tree;
- `ok(T) = maybe-sol` means that `T` is not a solution, but we do not know if there are solutions below it.

The strategy controls the possible rewritings of a term by means of `meta-apply`. `meta-apply(M,T,L,S,N)` applies (discarding the first `N` successful matches) a rule of module `M` with label `L`, partially instantiated with substitution `S`, to the term `T`. It returns the resulting fully reduced term and the representation of the match used in the reduction.

We saw before the necessity of instantiating the new variables in the righthand side of a rewrite rule in order to create new metavariables. We have to provide a substitution in such a way that the rules are always applied without new variables in the righthand side. For simplicity we will assume that a rule has at most three new variables called `NEW1`, `NEW2`, and `NEW3`. These variables are then substituted by new identifiers, which are quoted numbers. Hence, we define a new operation `meta-apply'` which receives the greatest number used to substitute variables in T and uses three new numbers to create three new (metarepresented) identifiers.

```
vars N M : MachineInt .  var L : Qid .  var T : Term .
op subst : MachineInt -> Substitution .
eq subst(M) = (('NEW1@Qid <- {conc('' , index(' , M + 1))}'Qid);
               ('NEW2@Qid <- {conc('' , index(' , M + 2))}'Qid);
               ('NEW3@Qid <- {conc('' , index(' , M + 3))}'Qid)) .
op meta-apply' : Term Qid MachineInt MachineInt -> Term .
eq meta-apply'(T,L,N,M) = extTerm(meta-apply(MOD,T,L,subst(M),N)) .
```

`meta-apply'` returns *one* of the possible one-step rewritings at the top level of a given term. Our first step is to define an operation `allRew` that returns *all* the possible *one-step sequential* rewritings [8] of a given term by using rewrite rules with labels in the list `labels`. The third argument of `allRew` represents the greatest number `M` used to substitute new variables in `T`. There is a `TermList` sort in module `META-LEVEL`, but it does not have an identity element, which we need to represent the case when no rule can be applied. So we extend it as follows:

```
op ~ : -> TermList .  var TL : TermList .
eq ~, TL = TL .  eq TL, ~ = TL .
```

The operations needed to find all the possible rewritings, and their definitions, are as follows:

```
op allRew : Term QidList MachineInt  -> TermList .
op topRew : Term Qid MachineInt MachineInt -> TermList .
op lowerRew : Term Qid MachineInt -> TermList .
op rewArguments : Qid TermList TermList Qid MachineInt -> TermList .
op rebuild : Qid TermList TermList TermList -> TermList .
var LS : QidList . vars C S OP : Qid . vars Before After : TermList .
eq allRew(T, nil, M) = ~ .
eq allRew(T, L LS, M) = topRew(T, L, 0, M),  *** rew at the top of T
                        lowerRew(T, L, M),   *** rew of subterms
                        allRew(T, LS, M) .   *** rew with labels LS
```

```
eq topRew(T, L, N, M) =
    if meta-apply'(T, L, N, M) == error* then ~
    else (meta-apply'(T, L, N, M) , topRew(T, L, N + 1, M)) fi .
eq lowerRew({C}S, L, M) = ~ .
eq lowerRew(OP[TL], L, M) = rewArguments(OP, ~, TL, L, M) .
eq rewArguments(OP, Before, T, L, M) =
    rebuild(OP, Before, allRew(T, L, M), ~) .
eq rewArguments(OP, Before, (T, After), L, M) =
    rebuild(OP, Before, allRew(T, L, M), After) ,
    rewArguments(OP, (Before, T), After, L, M) .
eq rebuild(OP,Before,~,After) = ~ .
eq rebuild(OP,Before,T,After) = meta-reduce(MOD,OP[Before,T,After]) .
eq rebuild(OP,Before,(T, TL),After) =
   meta-reduce(MOD, OP[Before,T,After]), rebuild(OP,Before,TL,After) .
```

Now we can define a strategy to search in the (conceptual) tree of all possible rewritings of a term T for a term that satisfies the ok predicate. Each node of the search tree is a pair, whose first component is a term and whose second component is a number representing the greatest number used as identifier for new variables in the process of rewriting the term. The tree nodes that have been generated but not yet been checked are maintained in a sequence.

```
sorts Pair PairSeq .  subsort Pair < PairSeq .
op <_',_> : Term MachineInt -> Pair .
op nil : -> PairSeq .
op _|_ : PairSeq PairSeq -> PairSeq [assoc id: nil] .
var PS : PairSeq .
```

We need an operation to build these pairs from the list of terms produced by allRew:

```
op buildPairs : TermList MachineInt -> PairSeq .
eq buildPairs(~, N) = nil .  eq buildPairs(T, N) = < T , N > .
eq buildPairs((T, TL), N) = < T , N > | buildPairs(TL, N) .
```

The operation rewDepth starts the search by calling the operation rewDepth' with the root of the search tree. rewDepth' returns the first solution found in a depth-first way. If there is no solution, the error* term is returned.

```
op rewDepth : Term -> Term .
op rewDepth' : PairSeq -> Term .
eq rewDepth(T) = rewDepth'(< meta-reduce(MOD, T), 0 >) .
eq rewDepth'(nil) = error* .
eq rewDepth'(< T , N > | PS) =
   if meta-reduce(MOD, 'ok[T]) == {'solution}'Answer then T
   else (if meta-reduce(MOD, 'ok[T]) == {'no-solution}'Answer then
              rewDepth'(PS)
         else rewDepth'(buildPairs(allRew(T,labels,N),N + 3) | PS)
         fi) fi .
```

Now we can test the CCS semantics with some examples using different judgements. First, we define a module `CCS-OK` extending the CCS syntax and semantic rules by defining some process constants to be used in the examples, and the predicate `ok` that states when a configuration is a solution. In this case a configuration denotes a solution when it is the empty set of judgements, representing that the set of judgements at the beginning is provable by means of the semantic rules.

```
(mod CCS-OK is  including CCS-SEMANTICS .
  ops p1 p2 : -> Process .
  eq p1 = ('a . 0) + ('b . 0 | ('c . 0 + 'd . 0)) .
  eq p2 = ('a . 'b . 0 | (~ 'c . 0) ['a / 'c]) \ 'a .
  sort Answer .
  ops solution  no-solution  maybe-sol : -> Answer .
  op ok : Configuration -> Answer .
  var JS : JudgementSet .
  eq ok({{ emptyJS }}) = solution .
  ceq ok({{ JS }}) = maybe-sol if JS =/= emptyJS .
endm)
```

In order to instantiate the parameterized generic module `SEARCH`, we use the Full Maude `up` function to obtain the metarepresentation of module `CCS-OK`, and then we declare a view [3]:

```
(mod META-CCS is  including META-LEVEL .
  op METACCS : -> Module .
  eq METACCS = up(CCS-OK) .
endm)
(view ModuleCCS from AMODULE to META-CCS is
  op MOD to METACCS .
  op labels to ('bind 'pref 'sum 'par 'res 'dist 'rel 'def) .
endv)
(mod SEARCH-CCS is
  protecting SEARCH[ModuleCCS] .
endm)
```

Now we can test the examples. First we can prove that process `p1` can perform action `'c` becoming `'b.0 | 0`.[4]

```
Maude> (red rewDepth({{ p1 -- 'c -> 'b . 0 | 0 }})). )
Result Term : {{ emptyJS }}
```

We can also prove that `p2` cannot perform action `'a` (but see later).

```
Maude> (red rewDepth({{p2 -- 'a -> ('b.0 | (~'c.0) ['a/'c])\'a}})).)
Result Term : error*
```

[4]We refer to the Maude manual [3] for indications about how to introduce metarepresented terms. Here we use $\overline{t}$ for the metarepresentation of term t.

In these examples, we have had to provide the resulting process. In the positive proof there is no problem, but in the negative proof, that is, that p2 cannot perform action 'a, the given proof is not completely correct: We have proved that process p2 cannot perform action 'a *becoming* `('b.0 | (~'c.0) ['a/'c])\'a`, but we have not proved that there is no way in which p2 can execute action 'a. We will see in the next section how this can be proved.

3. HOW TO OBTAIN NEW RESULTS

We are now interested in answering questions such as: Can process P perform action a (without caring about the process it becomes)? That is, we want to know if $P \xrightarrow{a} P'$ is possible, but P' is *unknown.* This is the same problem we found when new variables appear in the premises of a semantic rule. The solution, as we did with actions, is to define metavariables as processes, adding a new sort of possible processes, and modifying the operator used to build the basic transition judgements.

```
sort MetaVarProc .  op ?'(_')P : Qid -> MetaVarProc .
sort Process? .  subsorts Process MetaVarProc < Process? .
var ?P : MetaVarProc .
op _--_->_ : Process? Act? Process? -> Judgement [prec 50] .
```

We also have to define a new kind of judgements that binds metavariables with processes, a rule to propagate these bindings, and operations that perform the substitution (see full version [13]). For the CCS operators, new rules have to be added to deal with metavariables in the second process of the transition judgement. For example, for the prefix operator we have to add two new rules:

```
rl [pref] :  A . P -- A -> ?P        rl [pref] :  A . P -- ?A -> ?P
        => ------------------                 => ----------------------
               [?P := P] .                        [?A := A] [?P := P] .
```

Now, we can prove that process p2 *cannot* perform action 'a, by rewriting the judgement `p2 -- 'a -> ?('any)P`, where the metavariable `?('any)P` means that we do not care about the resulting process.

```
Maude> (red rewDepth({{ p2 -- 'a -> ?('any)P }}) .)
Result Term : error*
```

Another interesting question is which are the *successors* of a process P after performing actions in a given set As, that is,

$$succ(P, As) = \{P' \mid P \xrightarrow{a} P' \wedge a \in As\}.$$

Since we can use metavariables as processes, we have to rewrite a judgement like `P -- A -> ?('proc)P` instantiating the variable `A` with actions in the given set. Those rewritings will bind the metavariable

`?('proc)P` with the successors of `P`, but we find two problems. The first one is that we lose the bindings between metavariables and processes when they are substituted by applying the rewrite rule `bind`. To solve this, we have to modify the operator to build configurations, by also keeping a set of bindings already produced, which will be saved by the `bind` rule:

```
op '{'{_|_'}'} : JudgementSet JudgementSet -> Configuration  .
rl [bind] : {{ [?P := P] JS | JS' }} =>
            {{ (<proc ?P := P > JS) | [?P := P] JS' }} .
```

We have to change also the function `ok`, adding this new argument.

Another problem is that `rewDepth` only returns one solution, but we can modify it in order to get all the solutions, that is, in order to explore the whole tree of rewritings finding *all* the nodes that satisfy the function `ok`. The operation `allSol` returns a set with all the solutions.

```
sort TermSet .  subsort Term < TermSet .
op '{'} : -> TermSet .
op _U_ : TermSet TermSet -> TermSet [assoc comm id: {}] .
ceq T U T' = T if meta-reduce(MOD, '_==_[T, T']) == {'true}'Bool .
op allSol : Term -> TermSet .
eq allSol(T) = allSolDepth(< meta-reduce(MOD,T), 0 >) .
op allSolDepth : PairSeq -> TermSet .
eq allSolDepth(nil) = {} .
eq allSolDepth(< T , N > | PS) =
  if meta-reduce(MOD, 'ok[T]) == {'solution}'Answer then
             (T U allSolDepth(PS))
  else (if meta-reduce(MOD, 'ok[T]) == {'no-solution}'Answer then
            allSolDepth(PS)
       else allSolDepth(buildPairs(allRew(T,labels,N),N + 3) | PS)
       fi)
  fi .
```

Now we can define (in an extension of module `SEARCH-CCS`) a function `succ` which, given the metarepresentation of a process and a set of metarepresentations of actions, returns the set of metarepresentations of the process successors.

```
op succ : Term TermSet -> TermSet .
eq succ(T, {}) = {} .
eq succ(T, A U AS) = filter(allSol(
        ''{'{_|_'}'} [ '_--_->_ [ T, A, '?'(_')P [{''proc}'Qid]] ,
        {'emptyJS}'JudgementSet ]), '?'(_')P [{''proc}'Qid])
     U succ(T,AS) .
```

where `filter` (see full version [13]) is used to remove all the bindings involving metavariables different from `?('proc)P`.

4. MODAL LOGIC FOR CCS PROCESSES

In this section we show how we can define in Maude the semantics of a modal logic for CCS processes by using the functions of previous sections. We introduce a *modal* logic for describing local capabilities of CCS processes. This logic, which is a version of Hennessy-Milner logic [6], and its semantics are presented by Stirling [12]. Formulas are as follows:

$$\Phi ::= \mathtt{tt} \mid \mathtt{ff} \mid \Phi_1 \wedge \Phi_2 \mid \Phi_1 \vee \Phi_2 \mid [K]\Phi \mid \langle K \rangle \Phi$$

where K is a set of actions. The satisfaction relation describing when a process P satisfies a property Φ, $P \models \Phi$, is defined as follows:

$$\begin{array}{ll} P \models \mathtt{tt} & \\ P \models \Phi_1 \wedge \Phi_2 & \text{iff } P \models \Phi_1 \text{ and } P \models \Phi_2 \\ P \models \Phi_1 \vee \Phi_2 & \text{iff } P \models \Phi_1 \text{ or } P \models \Phi_2 \\ P \models [K]\Phi & \text{iff } \forall Q \in \{P' \mid P \stackrel{a}{\longrightarrow} P' \wedge a \in K\} \,.\, Q \models \Phi \\ P \models \langle K \rangle \Phi & \text{iff } \exists Q \in \{P' \mid P \stackrel{a}{\longrightarrow} P' \wedge a \in K\} \,.\, Q \models \Phi \end{array}$$

First, we define a sort `HMFormula` of modal logic formulas and operations to build these formulas:

```
(mod MODAL-LOGIC is  protecting CCS-SUCC .
  sort HMFormula .
  ops tt ff : -> HMFormula .
  ops _/\_  _\/_ : HMFormula HMFormula -> HMFormula .
  ops `[_`]_  <_>_ : TermSet HMFormula -> HMFormula .
```

We define the modal logic semantics in the same way as we did with the CCS semantics, that is, by defining rewrite rules that rewrite a judgement $P \models \Phi$ into the set of judgements which have to be fulfilled:

```
  op _|=_ : Term HMFormula -> Judgement .
  op forall : TermSet HMFormula -> JudgementSet .
  op exists : TermSet HMFormula -> JudgementSet .
  var P : Term .  vars K PS : TermSet .  vars Phi Psi : HMFormula .
  rl [true] : P |= tt            =>  emptyJS .
  rl [and]  : P |= Phi /\ Psi  =>  (P |= Phi) (P |= Psi) .
  rl [or]   : P |= Phi \/ Psi  =>  P |= Phi .
  rl [or]   : P |= Phi \/ Psi  =>  P |= Psi .
  rl [box]  : P |= [ K ] Phi   =>  forall(succ(P, K), Phi) .
  rl [diam] : P |= < K > Phi   =>  exists(succ(P, K), Phi) .
  eq forall({}, Phi)       = emptyJS .
  eq forall(P U PS, Phi) = (P |= Phi) forall(PS, Phi) .
  rl [ex] : exists(P U PS, Phi) => P |= Phi .
```

These rules are also nondeterministic. For example, the application of the two rules `or` is nondeterministic because they have the same left-hand side, and the rule `ex` is also nondeterministic because of multiple matchings modulo associativity and commutativity. We can instantiate the module **SEARCH** in Section 2.1 with the metarepresentation of the module containing the definition of the modal logic semantics.

As an example, we show some modal formulas satisfied by a vending machine **'Ven** [12] defined in a CCS context as

```
eq context = (('Ven =def ('2p . 'VenB + '1p . 'VenL))  &
             ('VenB =def 'big . 'collectB . 'Ven)      &
             ('VenL =def 'little . 'collectL . 'Ven)) .
```

and how they can be proved in Maude.

One of the properties that the vending machine fulfills is that a button cannot be depressed initially, that is, before money is deposited.

```
Maude> (red rewDepth('Ven |= [ 'big U 'little ] ff) .)
Result Term : emptyJS
```

Another interesting property of **'Ven** is that after a **'2p** coin is inserted, the little button cannot be depressed whereas the big one can:

```
Maude> (red rewDepth('Ven |= ['2p](((['little] ff) /\ (<'big> tt))) .)
Result Term : emptyJS
```

5. CONCLUSION

We have represented the CCS structural operational semantics in rewriting logic in a general way, solving the problems of new variables in the righthand side of the rules and nondeterminism by means of the reflective features of rewriting logic and its realization in the Maude module **META-LEVEL**.

We have seen how the semantics can be extended to answer questions about the capability of a process to perform an action, by considering metavariables as processes in the same way as we had done with actions. Having metavariables as processes allows also answering which are the successors of a process after performing an action, and this allows a representation of the semantics of the Hennessy-Milner modal logic in a similar way to its mathematical definition.

In the full version of this paper [13], in addition to more details, we have extended (using the same techniques) the semantics representation to sequences of actions or *traces*. We have also implemented *weak transition semantics*, $P \stackrel{a}{\Longrightarrow} P'$, which does not observe τ transitions. Finally, we have extended the modal logic representation introducing new modalities $[\![K]\!]$ and $\langle\!\langle K \rangle\!\rangle$ defined by means of the weak transition relation.

Other specific-purpose tools, such as the Concurrency Workbench [5], are more efficient and expressive than our tool, which is much in the style of a prototype, where we can play not only with processes and their capabilities, but with the semantics, represented at a very high mathematical level, adding or modifying some rules.

Although our implementation is much in the style of logic programming, one advantage is the possibility of working with algebraic specifications modulo equational axioms. Moreover, other strategies could be employed rather than depth-first, still keeping the same underlying specification.

Work in progress applies all these techniques to the specification language E-LOTOS, developed within ISO for the formal specification of open distributed concurrent systems [10].

Acknowledgements We are grateful to J. Meseguer and the referees for their helpful comments on earlier versions of this paper.

References

[1] R. Bruni. *Tile Logic for Synchronized Rewriting of Concurrent Systems.* PhD thesis, Dipartimento di Informatica, Università di Pisa, 1999.

[2] M. Clavel. *Reflection in General Logics and in Rewriting Logic with Applications to the Maude Language.* PhD thesis, University of Navarre, 1998.

[3] M. Clavel, F. Durán, S. Eker, P. Lincoln, N. Martí-Oliet, J. Meseguer, and J. Quesada. *Maude: Specification and Programming in Rewriting Logic.* SRI International, Jan. 1999, revised Aug. 1999.

[4] M. Clavel, F. Durán, S. Eker, P. Lincoln, N. Martí-Oliet, J. Meseguer, and J. Quesada. Using Maude. In *Proc. FASE 2000*, LNCS 1783. Springer, 2000.

[5] R. Cleaveland, J. Parrow, and B. Steffen. The Concurrency Workbench: A semantics-based tool for the verification of finite-state systems. *ACM Transactions on Programming Languages and Systems*, 15(1):36–72, Jan. 1993.

[6] M. Hennessy and R. Milner. Algebraic laws for nondeterminism and concurrency. *Journal of the ACM*, 32(1):137–161, Jan. 1985.

[7] N. Martí-Oliet and J. Meseguer. Rewriting logic as a logical and semantic framework. Technical Report SRI-CSL-93-05, SRI International, 1993. To appear in *Handbook of Philosophical Logic*, Kluwer Academic Publishers.

[8] J. Meseguer. Conditional rewriting logic as a unified model of concurrency. *Theoretical Computer Science*, 96:73–155, 1992.

[9] R. Milner. *Communication and Concurrency.* Prentice-Hall, 1989.

[10] J. Quemada, editor. Final committee draft on Enhancements to LOTOS. ISO/IEC JTC1/SC21/WG7 Project 1.21.20.2.3., May 1998.

[11] M.-O. Stehr and J. Meseguer. Pure type systems in rewriting logic. In *Proc. of LFM'99: Workshop on Logical Frameworks and Meta-Languages*, France, 1999.

[12] C. Stirling. Modal and temporal logics for processes. In *Logics for Concurrency: Structure vs Automata*, LNCS 1043, pages 149–237. Springer, 1996.

[13] A. Verdejo and N. Martí-Oliet. Executing and verifying CCS in Maude. Technical Report 99-00, Dpto. Sistemas Informáticos y Programación, Universidad Complutense de Madrid, Feb. 2000.

Part Nine

INVITED PAPERS ON VERIFICATION AND SECURITY

FROM REFUTATION TO VERIFICATION

John Rushby*

Computer Science Laboratory, SRI International, Menlo Park CA 94025 USA

Rushby@csl.sri.com

Abstract Model checking has won some industrial acceptance in debugging designs. Theorem proving and formal verification are less popular. An approach built around automated abstractions could integrate theorem proving with model checking in an acceptable way and provide a bridge between refutation and verification.

Keywords: Model checking, theorem proving, formal models, assurance, debugging.

1. INTRODUCTION

Formal methods have achieved a modest degree of acceptance in some industries (e.g., communications protocols and processor design), but mainly for "refutation"—that is, for purposes associated with debugging and testing. Formal verification—that is, establishing some some notion of correctness—is much less widely practiced, and is largely restricted to regulated industries.

In this short note, I examine some reasons for the relative success of formal methods in refutation, and their relative failure in verification, and propose that formal verification methods should adopt some of the characteristics of those used for refutation. I further propose that refutation and verification should be seen as points on a continuum and that by adopting techniques that allow moving smoothly along this continuum we may be able to increase both the utility and acceptance of formal methods.

2. FORMAL MODELING AND ANALYSIS

The rôle of formal methods in computer science is analogous to that of mathematical modeling and calculation in traditional engineering dis-

*This work was partially supported by DARPA through USAF Rome Laboratory contract F30602-96-C-0204 and by NASA Langley Research Center contract NAS1-20334.

Tommaso Bolognesi and Diego Latella (Eds.), *Formal Methods for Distributed System Development.*

ciplines. In those disciplines, engineers build mathematical models of some aspects of their design (e.g., the shape of an airfoil) and of its environment (e.g., the properties of fluids in motion) and use calculation to predict the behavior of the design (e.g., its lift and drag). These calculations are used in the design loop, and also to verify the behavior and properties of the final design. The utility of such verification is contingent on how faithfully the finished artifact implements its design, on the faithfulness of the environment models employed, and on the accuracy of the calculations. Validating the faithfulness of the models and the accuracy of the calculations are separate problems from verification, though usually some "end-to-end" checks are performed by testing the final artifact.

Formal methods in computer science follow a similar pattern, except that formal logic provides the mathematical underpinnings for computer science, as opposed to the calculus and differential equations that underpin most traditional engineering disciplines. Hence, the calculations performed in formal methods are based on automated deduction: theorem proving, model checking, and related techniques. These techniques, and the problems they address, are computationally far more demanding than those for differential equations: many problems in automated deduction are undecidable, and those that are decidable have awesome computational complexity. This computational intractability presents practitioners of formal methods with difficult choices: analysis of highly detailed and accurate formal models generally requires the power of human-guided theorem proving, while fully automated procedures such as model checking generally require the formal model to be simplified and approximated to bring it within their capabilities (they may also restrict the analysis to relatively simple properties such as those that can be expressed in a temporal logic).

The difficulty with approximate models is that they may not be faithful the the real system; hence verification performed using such models is of dubious validity. Refutation is also vulnerable to approximation in modeling—a reported bug may be a consequence of over-aggressive approximation and may not be present in the real design—but it is usually straightforward to distinguish whether a bug is real or not: we simply transform the scenario that manifests the bug in the model into a scenario for the real system (or for a simulator, or for a more accurate mathematical model of the system) and test it. This is the basis for the industrial acceptance of model-checking: examining *all* the behaviors of an approximate model often reveals more bugs, or higher-value bugs (those that would otherwise be caught only much later in the lifecycle), than examining *some* of the behaviors of the real system (as in traditional testing or simulation). The model-checking process itself is largely automated, and putative bugs are furnished with a counterexample trace that can be used (often automatically) to construct a scenario for the real system that will determine whether the bug is present there also, or is merely a consequence of the approximate modeling employed.

The difficulty with faithful models is that their analysis is usually outside the scope of fully automated procedures, and must instead rely on heuristic or human-guided theorem proving. Heuristic theorem proving is fine when it succeeds; in the more usual case that it fails, however, its user must try to distinguish whether this is due to a flawed design (i.e., a bug) or to an inadequate heuristic. Theorem proving methods seldom provide a counterexample trace that might help in making this distinction and instead depend on the user having a strong understanding of the operation of the heuristic. If the failure is due to an inadequate heuristic, then strong understanding of its operation is again required to modify either the heuristic or the model in an effective manner.

Human guidance is an alternative to heuristic theorem proving. Human-guided theorem proving is usually directly interactive: the human directs the strategy of the proof, while automated proof procedures take care of the tactical details. Again, it requires strong understanding of the operation of the prover to guide it successfully, and to diagnose and recover from tactical failures.

I believe that the chief reason for the poor industrial acceptance of methods based on theorem proving is that considerable skill is required to operate a theorem prover, and that skill concerns the operation of the prover more than the properties of the model being analyzed. Hence, designers are often uninterested in acquiring the necessary skill: their interests focus on their designs, not on the vagaries of adjuncts such as theorem provers. In contrast, the interactions required to use a model checker chiefly concern construction of an approximate model that is adequately faithful to the original design, yet within the scope of the model checker. This process concerns the design, and is similar to a design activity, and is acceptable to practitioners.

The choices described above present a rather unattractive dilemma: use automated techniques with approximate models that are adequate for refutation but do not extend to verification; or use accurate models that can be used to verify strong properties, but whose analysis depends on theorem proving methods that are not well-accepted in industry. Fortunately, there is an escape through the horns of this dilemma, and this is described in the next section.

3. AUTOMATED ABSTRACTION

The approximations that render a model suitable for automated analysis such as model checking do not necessarily have to sacrifice the ability to establish correctness. The approximations that are typically performed are Draconian: large or unbounded data structures are chopped down to small finite sizes—a queue may be restricted to length three, or a datapath to two bits. But more subtle approximations may be equally effective in rendering the model tractable to automated analysis, while preserving certain correctness properties: for example, a queue could be modeled as being in one of three states—empty, full, or in-between. Use of such *abstractions* in model checking has been understood for some time [5,10]. Typically, a number of verification conditions (putative theorems) must be proved to establish that the proposed abstraction is a conservative approximation to the original model and preserves the property of interest, and then model checking or some other automated procedure can be applied to the abstracted model. As usual, reported bugs must be checked to see if they are real or the consequence of too aggressive an approximation, but if analysis of the abstracted model reveals no bugs, then we have succeeded in verifying the property for the original model. The problem with this approach is that is is often almost as hard to prove the verification conditions as it would be to establish the property directly for the original model.

An alternative to post-hoc justification of an approximate model constructed by hand is to use automated methods to *construct* the approximate model. One method for doing this performs abstract interpretation over the data structures appearing in the model (e.g., [6]); another, generally more powerful, method replaces selected predicates in the model by Boolean variables (e.g., [3,12]). Calculation of the abstracted model in the latter case requires checking a large number of verification conditions; compared to those for post-hoc justification of an approximate model constructed by hand, these verification conditions tend to be more numerous, but simpler. Heuristic theorem proving techniques combined with powerful decision procedures are often able to discharge most of these conditions automatically. Furthermore, it is not a catastrophe if theorem proving fails to discharge a true verification condition: the approximate model will be more conservative than necessary, but it may still be good enough for model checking to succeed. If it does not, then the counterexample can help refine the construction, or can suggest additional predicates on which to abstract [11].

4. ITERATED, INTEGRATED, ANALYSIS

Construction of accurate abstractions often requires knowledge of some invariants of the original model—and the combination of automated abstraction and model checking can sometimes be used in a recursive fashion to help develop these. The idea is to construct an even simpler abstraction than the one desired, and then use a model checker to calculate the reachable states of that abstraction (most model checkers can do this). The reachable states characterize the strongest invariant of that abstraction; their concretization (the inverse of abstraction) is an invariant—and plausibly a strong one—of the original model. An invariant discovered in this way can often be strengthened further by a few iterations of a strongest postcondition computation (performed by theorem proving).

This approach suggests iterated application of these techniques (one abstraction is used to calculate an invariant that helps calculate another abstraction, and so on), which in turn suggests a blackboard architecture in which a common intermediate representation is used to accumulate useful properties and abstractions of a design in a form that many different tools can both use and contribute to. A common intermediate representation also provides a way to support many source notations without inordinate cost. Verification systems that explore this approach include SAL [2], which is under development at SRI, and InVeSt [4], which is a collaboration between Verimag and SRI (InVeSt is built on PVS and is being integrated with SAL). Other experiments using related methods have been reported recently [1,7].

The great benefit of this approach is that human guidance to the process of iterated abstraction and model checking is conducted in terms of properties related to the design: the user suggests predicates to be abstracted, examines counterexample traces or derived invariants, and suggests new predicates. Theorem proving is a central element in the mechanization, but is fully automated and conducted behind the scenes in circumstances that can tolerate the occasional failure to prove a true theorem.

Furthermore, by providing additional methods of abstraction that are *not* property-preserving (such as those that simply chop data structures down to a small size), this architecture and approach can be used for refutation as well as for verification. By increasing the sophistication of the abstractions employed, and by accumulating invariants, we can proceed smoothly from fairly crude approximate models that may be adequate for finding bugs early in the design loop, to more faithful models that reveal deeper bugs, and eventually to property-preserving abstractions that are able to verify aspects of the final design.

Even in environments where refutation is all that is required, this approach can increase the power and convenience of the refutation process by automating the construction of a graduated series of approximate models. In regulated environments, or others where true verification is required, this approach may make it possible to involve the design engineers more fully in the process (rather than having a separate formal verification group that operates largely outside the design loop) and may also bring to these environments the cost-benefits of employing formal methods for refutation as well as for verification.

References

[1] Parosh Aziz Abdulla, Aurore Annichini, Saddek Bensalem, Ahmed Bouajjani, Peter Habermehl, and Yassine Lakhnech. Verification of infinite-state systems by combining abstraction and reachability analysis. In Halbwachs and Peled [8], pages 146–159.

[2] Saddek Bensalem, Vijay Ganesh, Yassine Lakhnech, César Muñoz, Sam Owre, Harald Rueß, John Rushby, Vlad Rusu, Hassen Saïdi, N. Shankar, Eli Singerman, and Ashish Tiwari. An overview of SAL. In C. Michael Holloway, editor, *LFM 2000: Fifth NASA Langley Formal Methods Workshop*, NASA Langley Research Center, Hampton, VA, June 2000. Available at `http://shemesh.larc.nasa.gov/fm/Lfm2000/Proc/`.

[3] Saddek Bensalem, Yassine Lakhnech, and Sam Owre. Computing abstractions of infinite state systems compositionally and automatically. In Hu and Vardi [9], pages 319–331.

[4] Saddek Bensalem, Yassine Lakhnech, and Sam Owre. InVeSt: A tool for the verification of invariants. In Hu and Vardi [9], pages 505–510.

[5] Edmund M. Clarke, Orna Grumberg, and David E. Long. Model checking and abstraction. *ACM Transactions on Programming Languages and Systems*, 16(5):1512–1542, September 1994.

[6] James Corbett, Matthew Dwyer, John Hatcliff, Corina Pasareanu, Robby, Shawn Laubach, and Hongjun Zheng. Bandera: Extracting finite-state models from Java source code. In *22nd International Conference on Software Engineering*, pages 439–448, IEEE Computer Society, Limerick, Ireland, June 2000.

[7] Satyaki Das, David L. Dill, and Seungjoon Park. Experience with predicate abstraction. In Halbwachs and Peled [8], pages 160–171.

[8] Nicolas Halbwachs and Doron Peled, editors. *Computer-Aided Verification, CAV '99*, Volume 1633 of Springer-Verlag *Lecture Notes in Computer Science*, Trento, Italy, July 1999.

[9] Alan J. Hu and Moshe Y. Vardi, editors. *Computer-Aided Verification, CAV '98*, Volume 1427 of Springer-Verlag *Lecture Notes in Computer Science*, Vancouver, Canada, June 1998.

[10] C. Loiseaux, S. Graf, J. Sifakis, A. Bouajjani, and S. Bensalem. Property preserving abstractions for the verification of concurrent systems. *Formal Methods in System Design*, 6:11–44, 1995.

[11] Hassen Saïdi. Model checking guided abstraction and analysis. In *Seventh International Static Analysis Symposium (SAS'00)*, Santa Barbara CA, June 2000. To appear.

[12] Hassen Saïdi and N. Shankar. Abstract and model check while you prove. In Halbwachs and Peled [8], pages 443–454.

PROCESS ALGEBRAIC ANALYSIS OF CRYPTOGRAPHIC PROTOCOLS

Michele Boreale   Rocco De Nicola   Rosario Pugliese
Dipartimento di Sistemi e Informatica, Università di Firenze
{boreale,denicola,pugliese}@dsi.unifi.it

Abstract Recent approaches to the analysis of crypto-protocols build on concepts which are well-established in the field of process algebras, such as labelled transition systems (lts) and observational semantics. We outline some recent work in this direction that stems from using cryptographic versions of the pi-calculus – most notably Abadi and Gordon's *spi-calculus* – as protocol description languages. We show the impact of these approaches on a specific example, a simplified version of the Kerberos protocol.

Keywords: Process calculi, Reasoning about security, Protocol verification, Semantics, Formal methods

1. INTRODUCTION

Recent approaches to the analysis of crypto-protocols build on a few concepts which are well-established in the field of process algebras, such as labelled transition systems (lts) and observational semantics. We outline here some recent work in this direction that stems from the use of cryptographic versions of the pi-calculus [10] – most notably Abadi and Gordon's *spi-calculus* [2, 1] – as protocol description languages. We show the impact of these approaches on a specific example, a simplified version of the Kerberos protocol [8].

A major line of research is centered around the notion of observational equivalence, which permits making such notions as 'attacker' and 'secrecy' rigorous. For instance, according to [2], a way of asserting that a protocol, represented by a process term $P(d)$, keeps datum d secret is requiring that $P(d)$ be *equivalent* to $P(d')$, for every other d'. Observational equivalences based on context-closure, like *may-testing* [6, 3, 2] and *barbed equivalence* [11], appear to be appropriate in this setting. The intuition behind them is precisely that no external context (which

Tommaso Bolognesi and Diego Latella (Eds.), *Formal Methods for Distributed System Development.*

in the present setting can be read as 'attacker') may notice any difference when running in parallel with $P(d')$ or $P(d)$. The definitions of these equivalences, while rigorous and intuitive, suffer from universal quantification over contexts (attackers), that makes equivalence checking very hard. We have shown in previous work [4] how to avoid such quantification and obtain more tractable characterizations. The latters are based on an 'environment-sensitive' lts (as opposed to the 'standard' lts, which only explains process intentions). The states of this lts are *configurations* $\sigma \triangleright P$, where P is a process and σ represents the current environment's knowledge about names and keys. The moves of the process are constrained by this knowledge. Trace equivalence and weak bisimilarity are defined on top of this lts, and proven to be in agreement with may-testing and barbed equivalence, respectively. A major advantage of the new equivalences is represented by the congruence laws that make compositional reasoning on protocols possible. In this paper, we use bisimilarity and related compositional laws to formalize and verify a few security properties of a key-exchange protocol, inspired by the simplified version of Kerberos in [5]. The verification implies establishing a few process equalities that, we believe, would be hard to prove if one relied on the original, contextual definitions.

Another line of research based on environment-sensitive lts' relies on the analysis of the traces a protocol can execute, to find out whether any 'insecure' state (one, e.g., in which some confidential information has been passed to the environment) can ever be reached. The resulting method is less amenable to compositional reasoning, but seems well-suited for mechanization. We hint at some ongoing work (by one of us in collaboration with Martin Abadi and Marcelo Fiore), indicating that a full automation of the method can be achieved by considering a *symbolic* version of the environment-sensitive lts.

The rest of the paper is organized as follows. Section 2 introduces the language, the environment-sensitive lts and bisimilarity. Section 3 presents a number of properties of bisimilarity, mainly compositional proof rules. These rules are exploited in Section 4, to analyze a simplified version of Kerberos. The trace analysis of the protocol, based on a symbolic version of the lts, is touched upon in Section 5. A few concluding remarks are reported in Section 6.

2. THE LANGUAGE

In this section, we introduce syntax, operational semantics and bisimilarity for our language.

Syntax

The syntax of the calculus is summarized in Table 1. *Names* can be used as communication channels, primitive data or encryption keys: we do not distinguish between these three kinds of objects (notationally, we prefer letters $h, k, \ldots$, when we want to stress the use of a name as a key). *Messages* are obtained from names either via pairing or via shared-key encryption. In particular, message $\{M\}_k$ represents the ciphertext obtained by encrypting message M under key k, using a shared-key encryption system. Encryptions can be arbitrarily nested. *Expressions* are obtained applying encryption, decryption, pairing and projection operators to names and ciphertexts. For example, the result of evaluating $\mathtt{dec}_\eta(\zeta)$ is the text obtained by decrypting the ciphertext ζ using the value of η as a key. Logical *formulae* generalize the usual equality operator of the π-calculus with a predicate $name(\cdot)$, which tests for the format of the argument (plain name or a compound ciphertext), and with a 'let' construct that binds the value of some expression ζ to a name z. *Processes* are built using a set of operators which include those from the standard π-calculus, plus two new operators: boolean guard and encryption/decryption.

An informal explanation of the operators might be the following: $\mathbf{0}$ is the process that does nothing; $\eta(x).P$ represents input of a generic message x along η: the only useful case is when η is a name, otherwise the whole process is stuck; $\overline{\eta}\zeta.P$ represents output of ζ along η: the only useful case is when η is a name and ζ is a message, otherwise the whole process is stuck; $P + Q$ can behave either as P or Q: the choice might either be triggered by the environment, or by internal computations of P or Q; $P \,|\, Q$ is the parallel execution of P and Q; $(\nu\, a)\, P$ creates a new name a which is only known to P; $!\, P$ behaves like unboudedly many copies of P running in parallel, i.e. $P \mid P \mid P \mid \cdots$; ϕP behaves like P if the formula ϕ is logically true, otherwise it is stuck; $\mathtt{let}\ z = \zeta\ \mathtt{in}\ P$ attempts evaluation of ζ: if the evaluation succeeds the result is bound to z within P, otherwise the whole process is stuck.

We shall often abbreviate $(\nu\, a)\,(\nu\, b)\, P$ as $(\nu\, a, b)\, P$. Notions of *free names* of a process P, $fn(P)$, of *bound names* of P, $bn(P)$, and of alpha-equivalence arise as expected; $n(P)$ is $fn(P) \cup bn(P)$. Often, we shall write $fn(P, Q)$ in place of $fn(P) \cup fn(Q)$ (similarly for $bn(\cdot)$ and $n(\cdot)$). Similar notations are used for formulae, expressions and messages. From now on, *we shall identify alpha-equivalent terms.*

A *substitution* σ is a finite partial map from $\mathcal{N}$ to the set of messages $\mathcal{M}$. The domain and proper co-domain of σ are written $dom(\sigma)$ and $range(\sigma)$, respectively. We let $n(\sigma) = dom(\sigma) \cup (\cup_{M \in range(\sigma)} n(M))$.

Table 1 Syntax of the calculus

$a, b \ldots, h, k, \ldots, x, y, z \ldots$			*names* $\mathcal{N}$
M, N	$::=$	$a \mid \langle M_1, M_2 \rangle \mid \{M\}_k$	*messages* $\mathcal{M}$
η, ζ	$::=$	$a \mid \{\eta\}_\zeta \mid \mathtt{dec}_\eta(\zeta)$	*expressions* $\mathcal{Z}$
		$\mid \langle \zeta_1, \zeta_2 \rangle \mid \pi_1(\zeta) \mid \pi_2(\zeta)$	
ϕ, ψ	$::=$	$tt \mid name(\zeta) \mid [\zeta = \eta]$	*formulae* Φ
		$\mid \mathtt{let}\ z = \zeta\ \mathtt{in}\ \phi \mid \phi \wedge \psi \mid \neg\phi$	
P, Q	$::=$		*processes* $\mathcal{P}$
		$\mathbf{0}$	*(null)*
	$\mid$	$\eta(x).P$	*(input prefix)*
	$\mid$	$\overline{\eta}\zeta.P$	*(output prefix)*
	$\mid$	$P + Q$	*(non − deterministic choice)*
	$\mid$	$P \mid Q$	*(parallel composition)*
	$\mid$	$(\nu a) P$	*(restriction)*
	$\mid$	$!P$	*(replication)*
	$\mid$	ϕP	*(boolean guard)*
	$\mid$	$\mathtt{let}\ z = \zeta\ \mathtt{in}\ P$	*(encryption/decryption)*

It is assumed that $\mathtt{dec}.(\cdot)$ does not occur in $name(\zeta)$, $[\zeta = \eta]$, $\eta(x).$ and $\overline{\eta}\zeta$. Operators $a(x).\cdot$, $(\nu a)\cdot$ and $\mathtt{let}\ z = \zeta\ \mathtt{in}\ \cdot$ are *binders*, with the obvious scope, for names x, a and z, respectively. In $\mathtt{let}\ z = \zeta\ \mathtt{in}\ \cdot$, it is assumed that $z \notin n(\zeta)$.

Given a tuple of distinct names $\widetilde{x} = (x_1, \ldots, x_n)$ and a tuple of messages $\widetilde{M} = (M_1, \ldots, M_n)$, the substitution mapping each x_i to M_i will be sometimes written as $[\widetilde{M}/\widetilde{x}]$ or $[M_i/x_i]_{i \in 1..n}$. When $\widetilde{x} \cap dom(\sigma) = \emptyset$ (if convenient, we regard a tuple as the set of its components), we write $\sigma[\widetilde{M}/\widetilde{x}]$ for the substitution σ' which is the union of σ and $[\widetilde{M}/\widetilde{x}]$ (in this case we say that σ' *extends* σ). For a given $V \subseteq_{\mathrm{fin}} \mathcal{N}$, we write ϵ_V for the substitution with $dom(\epsilon_V) = V$ that acts as the identity on V. For any term (name/expression/formula/process) t, $t\sigma$ denotes the term obtained by simultaneously replacing each $x \in fn(t) \cap dom(\sigma)$ with $\sigma(x)$, with renaming of bound names of t possibly involved to avoid captures.

Operational semantics

The operational semantics makes use of two evaluation functions: one for expressions, the other for formulae. The *evaluation function for expressions*, $\widehat{\cdot} : \mathcal{Z} \rightarrow \mathcal{M} \cup \{\bot\}$ (where $\bot$ is a distinct symbol), is defined by induction on ζ as follows:

- $\widehat{a} = a$

- $\widehat{\{\zeta_1\}_{\zeta_2}} = \begin{cases} \{M\}_k & \text{if } \widehat{\zeta_1} = M \text{ and } \widehat{\zeta_2} = k, \text{ for some } M \text{ and } k \in \mathcal{N} \\ \bot & \text{otherwise} \end{cases}$

- $\widehat{\mathtt{dec}_{\zeta_2}(\zeta_1)} = \begin{cases} M & \text{if } \widehat{\zeta_1} = \{M\}_k \text{ and } \widehat{\zeta_2} = k, \text{ for some } M \text{ and } k \in \mathcal{N} \\ \bot & \text{otherwise} \end{cases}$

- $\widehat{\langle \zeta_1, \zeta_2 \rangle} = \begin{cases} \langle M_1, M_2 \rangle & \text{if } \widehat{\zeta_1} = M_1 \text{ and } \widehat{\zeta_2} = M_2, \text{ for some } M_1 \text{ and } M_2 \\ \bot & \text{otherwise} \end{cases}$

- for $i = 1, 2$, $\widehat{\pi_i(\zeta)} = \begin{cases} M_i & \text{if } \widehat{\zeta} = \langle M_1, M_2 \rangle, \text{ for some } M_1 \text{ and } M_2 \\ \bot & \text{otherwise} \end{cases}$

The *evaluation function for formulae*, $[\![\cdot]\!] : \Phi \rightarrow \{tt, ff\}$, is defined by induction on ϕ. The only non-standard clause is

- $[\![\mathtt{let}\ z = \zeta\ \mathtt{in}\ \phi]\!] = \begin{cases} [\![\phi[\widehat{\zeta}/z]]\!] & \text{if } \widehat{\zeta} \neq \bot \\ ff & \text{otherwise.} \end{cases}$

The operational semantics is defined by the inference rules of Table 2. All rules in the first part of the table, but the last two, are standard from π-calculus. Rule (GUARD) says that process ϕP behaves like P provided that ϕ evaluates to true; otherwise, process ϕP is stuck. Rule (LET) attempts evaluation of expression ζ: if the evaluation succeeds, then process $\mathtt{let}\ z = \zeta\ \mathtt{in}\ P$ behaves like process $P[\widehat{\zeta}/z]$, otherwise $\mathtt{let}\ z = \zeta\ \mathtt{in}\ P$ is stuck.

The rules in the upper part of Table 2 only account for process intentions. The rules in the lower part of the table build on those in the upper part, and account for interaction with the environment. They define an environment-sensitive lts whose states are *configurations* of the form $\sigma \triangleright P$, where P is a process and substitution σ represents the environment (from now on, terms 'substitution' and 'environment' will be used interchangeably). Transitions take the form

$$\sigma \triangleright P \overset{\mu}{\underset{\delta}{\longmapsto}} \sigma' \triangleright P'$$

and represent atomic interactions between process P and environment σ, μ is the *process action* and δ is a corresponding *environment action.* Process actions, ranged over by μ, can be of three forms: τ (internal action), $a\langle M \rangle$ (input at a where message M is received), and $(\nu \widetilde{b})\, \overline{a}\langle M \rangle$ (output at a where message M containing the fresh, private names $\widetilde{b}$ is sent). We shall write $\overline{a}\langle M \rangle$ instead of $(\nu \widetilde{b})\, \overline{a}\langle M \rangle$ whenever $\widetilde{b} = \emptyset$. Environment actions, ranged over by δ, can be of three forms: $(\nu \widetilde{b})\, \overline{\eta}\langle \zeta \rangle$ (output), $\eta(x)$ (input), and $-$ (no action). Free names and bound names of δ are defined as expected, in particular $bn(\eta(x)) = \{x\}$.

Table 2 Operational semantics (symmetric of (SUM), (PAR) and (COM) omitted)

(INP) $a(x).\,P \xrightarrow{a\langle M\rangle} P[^M/_x]$ (OUT) $\overline{a}M.\,P \xrightarrow{\overline{a}\langle M\rangle} P$

$$(\text{SUM})\ \frac{P \xrightarrow{\mu} P'}{P+Q \xrightarrow{\mu} P'} \qquad (\text{IDE})\ \frac{P[\tilde{v}/\tilde{x}] \xrightarrow{\mu} P'}{A(\tilde{v}) \xrightarrow{\mu} P'}\ A(\tilde{x}) \Leftarrow P$$

$$(\text{PAR})\ \frac{P \xrightarrow{\mu} P'}{P\,|\,Q \xrightarrow{\mu} P'\,|\,Q} \qquad (\text{COM})\ \frac{P \xrightarrow{(\nu\tilde{b})\overline{a}\langle M\rangle} P' \quad Q \xrightarrow{a\langle M\rangle} Q'}{P\,|\,Q \xrightarrow{\tau} (\nu\tilde{b})(P'\,|\,Q')}$$

$$(\text{RES})\ \frac{P \xrightarrow{\mu} P'}{(\nu c)\,P \xrightarrow{\mu} (\nu c)\,P'}\ c \notin n(\mu) \qquad (\text{OPEN})\ \frac{P \xrightarrow{(\nu\tilde{b})\overline{a}\langle M\rangle} P'}{(\nu c)\,P \xrightarrow{(\nu\tilde{b}c)\overline{a}\langle M\rangle} P'}\ \begin{array}{l} c \neq a, \\ c \in n(M) - \tilde{b} \end{array}$$

$$(\text{GUARD})\ \frac{[\![\phi]\!] = tt \quad P \xrightarrow{\mu} P'}{\phi P \xrightarrow{\mu} P'} \qquad (\text{LET})\ \frac{\widehat{\zeta} \neq \bot \quad P[\widehat{\zeta}/z] \xrightarrow{\mu} P'}{\texttt{let}\ z = \zeta\ \texttt{in}\ P \xrightarrow{\mu} P'}$$

. . . .

$$(\text{E-OUT})\ \frac{P \xrightarrow{(\nu\tilde{b})\overline{a}\langle M\rangle} P' \quad \widehat{\eta\sigma} = a}{\sigma \rhd P \overset{(\nu\tilde{b})\overline{a}\langle M\rangle}{\underset{\eta(x)}{\longmapsto}} \sigma[M/x] \rhd P'} \qquad (\text{E-TAU})\ \frac{P \xrightarrow{\tau} P'}{\sigma \rhd P \overset{\tau}{\underset{-}{\longmapsto}} \sigma \rhd P'}$$

$$(\text{E-INP})\ \frac{P \xrightarrow{a\langle M\rangle} P' \quad \widehat{\eta\sigma} = a \quad M = \widehat{\zeta\sigma} \quad \tilde{b} \overset{\text{def}}{=} (n(\zeta) - dom(\sigma))}{\sigma \rhd P \overset{a\langle M\rangle}{\underset{(\nu\tilde{b})\overline{\eta}(\zeta)}{\longmapsto}} \sigma[\tilde{b}/\tilde{b}] \rhd P'}$$

Assumptions: $n(\eta) \subseteq dom(\sigma)$ and names in $\tilde{b}$ fresh for σ and P.

Let us have a closer look at the rules of the environment-sensitive lts. In rule (E-OUT), the environment receives a message M and updates its knowledge accordingly. For the transition to take place, channel a must belong to the knowledge of the environment, thus η is some expression describing how a can be obtained out of σ (this is what $\widehat{\eta\sigma} = a$ means). In rule (E-INP), the environment sends a message M to the process. Message M is not arbitrary: expression ζ describes how it is built out of σ and of the names $\tilde{b}$ ($\widehat{\zeta\sigma} = M$). The update $[\tilde{b}/\tilde{b}]$ records the creation of the new names $\tilde{b}$. Like in the previous rule, a must belong to the knowledge of σ.

Bisimilarity

Bisimilarity relies on a notion of equivalence on environments. Informally, two environments σ and σ' are equivalent whenever they are logically indistinguishable.

Definition 2.1 (equivalence on environments) *Two substitutions σ and σ' are* equivalent, *written $\sigma \sim \sigma'$, if $dom(\sigma) = dom(\sigma')$ and for each formula ϕ with $fn(\phi) \subseteq dom(\sigma)$, it holds that $[\![\phi\sigma]\!] = [\![\phi\sigma']\!]$.* ◇

This logical characterization contains a quantification on all formulae. In [4], we give an equivalent definition that avoids explicitly considering all formulae and is amenable to automatic checking. However, the above definition will suffice for the purposes of this paper.

In the following, input and output actions will be called *visible* actions and we shall use s to range over sequences of visible process actions and u to range over sequences of visible environment actions. We write $\Longmapsto$ to denote the reflexive and transitive closure of $\overset{\tau}{\underset{-}{\longmapsto}}$ and, inductively, write $\overset{s}{\underset{u}{\Longmapsto}}$ for $\Longmapsto\overset{\mu}{\underset{\delta}{\longmapsto}}\overset{s'}{\underset{u'}{\Longmapsto}}$ when $s = \mu \cdot s'$ and $u = \delta \cdot u'$. $\sigma \triangleright P \overset{s}{\underset{u}{\Longmapsto}}$ will stand for 'there are P' and σ' s.t. $\sigma \triangleright P \overset{s}{\underset{u}{\Longmapsto}} \sigma' \triangleright P'$'. Finally, $\sigma \triangleright P \overset{\widehat{\mu}}{\underset{\delta}{\Longmapsto}} \sigma' \triangleright P'$ stands for $\sigma \triangleright P \overset{\mu}{\underset{\delta}{\Longmapsto}} \sigma' \triangleright P'$ if $\mu \neq \tau$, and for $\sigma \triangleright P \Longmapsto \sigma' \triangleright P'$ if $\mu = \tau$ (the $\widehat{\cdot}$ defined here has of course nothing to do with the evaluation function on expressions previously defined).

We say that a pair of configurations $(\sigma_1 \triangleright P, \sigma_2 \triangleright Q)$ is *compatible* if $\sigma_1 \sim \sigma_2$. A relation $\mathcal{R}$ is *compatible* if it only contains compatible pairs of configurations. Given a binary relation $\mathcal{R}$, we write $(\sigma_1, \sigma_2) \vdash P \mathcal{R} Q$ if $(\sigma_1 \triangleright P\,,\, \sigma_2 \triangleright Q) \in \mathcal{R}$. The intuition of $(\sigma_1, \sigma_2) \vdash P \approx Q$, defined below, is that P and Q are indistinguishable when placed in environments σ_1 and σ_2, respectively. Thus the definition requires, at each step, checking process actions against the same environment action δ.

Definition 2.2 (weak bisimilarity) Weak bisimilarity, *written $\approx$, is the largest binary compatible relation of configurations such that whenever $(\sigma_1, \sigma_2) \vdash P \approx Q$ and $\sigma_1 \triangleright P \overset{\mu}{\underset{\delta}{\longmapsto}} \sigma_1' \triangleright P'$ then there are μ', σ_2' and Q' such that $\sigma_2 \triangleright Q \overset{\widehat{\mu'}}{\underset{\delta}{\Longmapsto}} \sigma_2' \triangleright Q'$ and $(\sigma_1', \sigma_2') \vdash P' \approx Q'$, and the converse on the transitions of Q and P.* ◇

Table 3 Basic properties of $\approx$

(REFLEXIVITY)	For any σ and P, $(\sigma, \sigma) \vdash P \approx P$.
(TRANSITIVITY)	If $(\sigma_1, \sigma_2) \vdash P \approx Q$ and $(\sigma_2, \sigma_3) \vdash Q \approx R$ then $(\sigma_1, \sigma_3) \vdash P \approx R$.
(WEAKENING)	Suppose that $(\sigma_1[{}^{M}\!/\!{}_{x}], \sigma_2[{}^{N}\!/\!{}_{x}]) \vdash P \approx Q$. Then $(\sigma_1, \sigma_2) \vdash P \approx Q$.
(CONTRACTION)	Suppose that $(\sigma_1, \sigma_2) \vdash P \approx Q$ and consider any ζ such that $n(\zeta) \subseteq dom(\sigma_1)$ and $\widehat{\zeta\sigma_1} \neq \bot$. Then $(\sigma_1[{}^{\widehat{\zeta\sigma_1}}\!/\!{}_{x}], \sigma_2[{}^{\widehat{\zeta\sigma_2}}\!/\!{}_{x}]) \vdash P \approx Q$.
(STRUCT-EQ)	Suppose that $P \equiv Q$. Then, for any σ, $(\sigma, \sigma) \vdash P \approx Q$.

In [4], we prove that weak bisimilarity is in agreement with *barbed equivalence* ($\cong$), a contextual process semantics. More precisely, weak bisimilarity is included in barbed equivalence: $(\epsilon_V, \epsilon_V) \vdash P \approx Q$ implies $P \cong Q$, for any processes P and Q and $V = fn(P, Q)$. The opposite inclusion holds only for the class of *structurally image-finite* processes (see [4]), a sufficiently broad class of processes to ensure that $\approx$ is a fairly general proof technique.

3. COMPOSITIONAL REASONING

We discuss in this section a few laws which are very are useful for compositional reasoning on processes. Table 3 contains a few basic properties of $\approx$, while Table 4 contains some congruence rules for $\approx$. In Table 3, $\equiv$ denotes *structural equivalence*, which is defined as the least equivalence relation over processes that is preserved by parallel composition and restriction, and satisfies the structural laws of [9] plus the laws:

$$(\texttt{let}\ z = \zeta\ \texttt{in}\ P) \equiv P[{}^{\widehat{\zeta}}\!/\!{}_{z}] \quad \text{if } \widehat{\zeta} \neq \bot \qquad \text{and} \qquad \phi P \equiv P \quad \text{if } [\![\phi]\!] = tt$$

The next definition formalizes the concept of *decryption closure* of an environment σ, written $dc(\sigma)$. Elements in $dc(\sigma)$ represent the basic knowledge derivable from σ, i.e. the "building blocks" the environment can use to synthesize more complex messages.

Definition 3.1 (decryption closure and knowledge) *Let W be a set of messages. The* decryption closure *of W, written $dc(W)$, is the set of messages defined inductively as follows:*

(i) $W \subseteq dc(W)$;

(ii) if $k \in dc(W)$ and $\{M\}_k \in dc(W)$ then $M \in dc(W)$;

Table 4 Congruence rules for $\approx$

(C-INP)	Suppose that for all ζ such that $\widetilde{b} \stackrel{\text{def}}{=} (n(\zeta) - dom(\sigma_1))$ are fresh and $\widehat{\zeta\sigma_1} \neq \bot$ it holds: $(\sigma_1[\widetilde{b}/\widetilde{b}], \sigma_2[\widetilde{b}/\widetilde{b}]) \vdash P[\widehat{\zeta\sigma_1}/x] \approx Q[\widehat{\zeta\sigma_2}/x]$. Suppose that $a_i = \widehat{\eta\sigma_i}$ $(i = 1,2)$ with $n(\eta) \subseteq dom(\sigma_1)$. Then $(\sigma_1, \sigma_2) \vdash a_1(x).P \approx a_2(x).Q$.
(C-OUT)	Suppose that $(\sigma_1[M_1/x], \sigma_2[M_2/x]) \vdash P \approx Q$ and that $a_i = \widehat{\eta\sigma_i}$ $(i = 1,2)$ with $n(\eta) \subseteq dom(\sigma_1)$. Then $(\sigma_1[M_1/x], \sigma_2[M_2/x]) \vdash \overline{a}_1 M_1.P \approx \overline{a}_2 M_2.Q$.
(C-PAR)	Suppose that $fn(R) \subseteq dom(\sigma_1)$ and $(\sigma_1, \sigma_2) \vdash P \approx Q$. Then $(\sigma_1, \sigma_2) \vdash P \mid R\sigma_1 \approx Q \mid R\sigma_2$.
(C-RES)	Suppose that $(\sigma_1, \sigma_2) \vdash P \approx Q$, that $\widetilde{k} \cap n(\sigma_1) = \emptyset$ and that $\widetilde{h} \cap n(\sigma_2) = \emptyset$. Then $(\sigma_1, \sigma_2) \vdash (\nu\widetilde{k})P \approx (\nu\widetilde{h})Q$.

(iii) if $\langle M_1, M_2 \rangle \in dc(W)$ then $M_1, M_2 \in dc(W)$.

The knowledge *of W, written $kn(W)$, is the set of names in $dc(W)$, i.e.: $kn(W) \stackrel{\text{def}}{=} dc(W) \cap \mathcal{N}$. Let σ be a substitution; we let $dc(\sigma) \stackrel{\text{def}}{=} dc(range(\sigma))$ and $kn(\sigma) \stackrel{\text{def}}{=} kn(range(\sigma))$.* ◇

Note that both $dc(\sigma)$ and $kn(\sigma)$ are finite, and can be effectively computed. The next proposition is useful when reasoning on the values a given expression ζ can take on when evaluated under a substitution σ.

Proposition 3.2 *Let ζ and σ be such that $n(\zeta) \subseteq dom(\sigma)$.*

- *If $\widehat{\zeta\sigma}$ is a name then $\widehat{\zeta\sigma} \in kn(\sigma)$.*
- *If $\widehat{\zeta\sigma} = \{M\}_k$ and $k \notin kn(\sigma)$ then $\{M\}_k \in dc(\sigma)$.*
- *If $\widehat{\zeta\sigma} = \langle M_1, M_2 \rangle$ then, for $i = 1,2$, there is ζ_i such that $n(\zeta_i) \subseteq dom(\sigma)$ and $\widehat{\zeta_i\sigma} = M_i$.*

Finally, we state a crucial compositionality result for $\approx$, which strengthens the congruence rule for parallel composition (C-PAR), under the assumption that the involved processes are "safe" for the appropriate environments, that is, they never increase the knowledge of these environments. Formally, we say that a process R is *σ-safe* if for each

s, whenever $\sigma \rhd R \overset{s}{\underset{u}{\Longmapsto}} \sigma' \rhd R' \overset{(\nu \tilde{b})\, \overline{a}\langle M\rangle}{\underset{\eta(x)}{\Longmapsto}}$, then $M \in dc(\sigma')$ (hence $\tilde{b} = \emptyset$).

Theorem 1 (safe composition) *Suppose that* $(\sigma_1, \sigma_2) \vdash Q_1 \approx Q_2$, *that* $(\sigma_1, \sigma_2) \vdash R_1 \approx R_2$, *and that, for* $i = 1, 2$, Q_i *and* R_i *are* σ_i*-safe. Then* $(\sigma_1, \sigma_2) \vdash Q_1 \mid R_1 \approx Q_2 \mid R_2$.

4. THE KERBEROS PROTOCOL

Specification

Consider a system where two agents A (the initiator) and B (the responder) share two long-term secret keys, k_{AS} and k_{BS} respectively, with a server S. The Kerberos protocol is designed to set up a new secret session key k_{AB} between A and B. Below, we consider a slightly simplified version of this protocol, akin to that considered in [5]. Informally, the protocol can be described as follows:

Message 1	$A \longrightarrow S:$	A, B
Message 2	$S \longrightarrow A:$	$\{n_S, k_{AB}, B, \{n_S, k_{AB}, A\}_{k_{BS}}\}_{k_{AS}}$
Message 3	$A \longrightarrow B:$	$\{n_S, k_{AB}, A\}_{k_{BS}}, \{A, n_A\}_{k_{AB}}$
Message 4	$B \longrightarrow A:$	$\{n_A\}_{k_{AB}}.$

In message 1, A starts the protocol by simply communicating to S his intention to establish a new connection with B. In message 2, S generates a fresh key k_{AB} and embodies it into an appropriate certificate, which is sent to A. The certificate uses a fresh nonce n_S, that acts as a timestamp and is meant to assure A and B about the freshness of the message (this is to prevent attacks based on replays of old messages). In message 3, A extracts B's part of the certificate ($\{\cdots\}_{k_{BS}}$) and forwards it to B, together with some challenge-information containing a new nonce n_A. The fourth message represents a synchronization between A and B: the presence of n_A assures A he is really talking to B.

Our intent here is to verify one run of the protocol, under the hypothesis that an old session key k has been compromised. We shall not consider the case of multiple runs, which requires a more complex analysis. For the sake of simplicity, we also suppose that the protocol is always started by A. In our language, the protocol translates to the process K defined in Table 5. For the sake of readability, we have used here and will be using in the sequel a few obvious notational shorthands. For example, $a(\{M\}_k).P$ stands for $a(x).\mathtt{let}\ y = \mathtt{dec}_k(x)\ \mathtt{in}\ [y = M]\ P$ and $[x \notin \{a_1, \ldots, a_n\}]$ stands for $\neg[x = a_1] \wedge \ldots \wedge \neg[x = a_n]$. An expanded version of the protocol in Table 5 is reported below:

Table 5 The Kerberos Protocol: spi-calculus specification

$$
\begin{array}{rcl}
A & \stackrel{\text{def}}{=} & \overline{c_{AS}}\langle A, B\rangle .\, c_{AS}(\{x_{n_S}, x_{k_{AB}}, B, x_{cert_B}\}_{k_{AS}}).\\
& & \overline{c_{AB}}\langle x_{cert_B}, \{A, n_A\}_{x_{k_{AB}}}\rangle .\, c_{AB}(\{n_A\}_{x_{k_{AB}}}).\mathbf{0}\\
B^V & \stackrel{\text{def}}{=} & c_{AB}(\{y_{n_S}, y_{k_{AB}}, A\}_{k_{BS}}, \{A, y_{n_A}\}_{y_{k_{AB}}}).\\
& & [y_{n_S} \notin V]\, \overline{c_{AB}}\langle \{y_{n_A}\}_{y_{k_{AB}}}\rangle .\mathbf{0}\\
S & \stackrel{\text{def}}{=} & c_{AS}(A, B).\, \overline{c_{AS}}\langle \{n_S, k_{AB}, B, \{n_S, k_{AB}, A\}_{k_{BS}}\}_{k_{AS}}\rangle .\mathbf{0}\\
K & \stackrel{\text{def}}{=} & (\nu\, k_{AS}, k_{BS})\,\big(\, \overline{lost}\langle \{n_{old}, k, A\}_{k_{BS}}, k\rangle .\mathbf{0}\\
& & \mid ((\nu\, n_A)\, A) \mid B^V \mid ((\nu\, n_S, k_{AB})\, S)\,\big)
\end{array}
$$

$$
\begin{array}{l}
A \stackrel{\text{def}}{=} \overline{c_{AS}}\langle A, B\rangle .\, c_{AS}(x_1).\\
\quad \texttt{let}\ x_{n_S} = \pi_1(\texttt{dec}_{k_{AS}}(x_1)), x_{k_{AB}} = \pi_2(\texttt{dec}_{k_{AS}}(x_1)),\\
\quad\quad x_B = \pi_3(\texttt{dec}_{k_{AS}}(x_1)), x'_{n_S} = \pi_1(\texttt{dec}_{k_{BS}}(\pi_4(\texttt{dec}_{k_{AS}}(x_1)))),\\
\quad\quad x'_{k_{AB}} = \pi_2(\texttt{dec}_{k_{BS}}(\pi_4(\texttt{dec}_{k_{AS}}(x_1)))), x_A = \pi_3(\texttt{dec}_{k_{BS}}(\pi_4(\texttt{dec}_{k_{AS}}(x_1))))\\
\quad \texttt{in}\ [x_B = B, x_{n_S} = x'_{n_S}, x_{k_{AB}} = x'_{k_{AB}}, x_A = A]\\
\quad\quad \overline{c_{AB}}\langle \{x_{n_S}, x_{k_{AB}}, A\}_{k_{BS}}, \{A, n_A\}_{x_{k_{AB}}}\rangle .\\
\quad\quad c_{AB}(x_2).\, [x_2 = \{n_A\}_{x_{k_{AB}}}].\mathbf{0}
\end{array}
$$

$$
\begin{array}{l}
B^V \stackrel{\text{def}}{=} c_{AB}(y).\\
\quad \texttt{let}\ y_{n_S} = \pi_1(\texttt{dec}_{k_{BS}}(\pi_1(x_1))), y_{k_{AB}} = \pi_2(\texttt{dec}_{k_{BS}}(\pi_1(x_1))),\\
\quad\quad y_A = \pi_3(\texttt{dec}_{k_{BS}}(\pi_1(x_1))), y'_A = \pi_1(\texttt{dec}_{y_{k_{AB}}}(\pi_2(x_1))),\\
\quad\quad y_{n_A} = \pi_1(\texttt{dec}_{y_{k_{AB}}}(\pi_2(x_1)))\\
\quad \texttt{in}\ [y_A = A, y'_A = y_A]\, [y_{n_S} \notin V]\, \overline{c_{AB}}\langle \{y_{n_A}\}_{y_{k_{AB}}}\rangle .\mathbf{0}\\
S \stackrel{\text{def}}{=} c_{AS}(z).\, \texttt{let}\ z_A = \pi_1(z), z_B = \pi_2(z)\\
\quad\quad \texttt{in}\ [z_A = A, z_B = B]\, \overline{c_{AS}}\langle \{n_S, k_{AB}, z_B, \{n_S, k_{AB}, z_A\}_{k_{BS}}\}_{k_{AS}}\rangle .\mathbf{0}\\
K \stackrel{\text{def}}{=} (\nu\, k_{AS}, k_{BS})\,\big(\, \overline{lost}\langle \{n_{old}, k, A\}_{k_{BS}}, k\rangle .\mathbf{0}\\
\quad\quad \mid ((\nu\, n_A)\, A) \mid B^V \mid ((\nu\, n_S, k_{AB})\, S)\,\big)
\end{array}
$$

As usual, all bound names in K are assumed to be distinct from one another and from the free names. Subscripts should help reminding the expected values of each variable. For instance, the expected value for x_{certB} is B's certificate, i.e. $\{n_S, k_{AB}, A\}_{k_{BS}}$. The $\overline{lost}$-output action accounts for the accidental loss of an old session key k and of the corresponding certificate for B, $\{n_{old}, k, A\}_{k_{BS}}$. Set V contains all the past nonces generated by S: thus it is assumed that $n_{old} \in V$. Note that the freshness check on n_S is performed by B in $[y_{n_S} \notin V]$ and then propagated to A, via the last synchronization on channel c_{AS}.

There several properties that one might want to check of K. Here, we will just consider two of them. We formalize each property as a process and then check barbed equivalence ($\cong$) between this process and K. In doing so, we partially follow Abadi and Gordon [2], but deviate from their presentation due to our use of an $\overline{error}\langle\rangle$-action to explicitly signal violation of a property. Indeed, the notions of authentication and secrecy considered here are in some sense less demanding than those in [2], as we are not considering implicit information flow – which does not arise in this specific example anyway.

Authentication: *"B and A only accept the key k_{AB} generated by S"*. Formally, it holds that $K \cong K_{aut}$, where K_{aut} is defined below:

$$
\begin{aligned}
A_{aut} &\stackrel{\mathrm{def}}{=} \overline{c_{AS}}\langle A, B\rangle.\, c_{AS}(\{x_{n_S}, x_{k_{AB}}, B, x_{cert_B}\}_{k_{AS}}).\\
&\qquad \overline{c_{AB}}\langle x_{cert_B}, \{A, n_A\}_{x_{k_{AB}}}\rangle.\, c_{AB}(\{n_A\}_{x_{k_{AB}}}).\\
&\qquad [x_{k_{AB}} \neq k_{AB}]\, \overline{error}\langle\rangle.\, \mathbf{0}\\
B^V_{aut} &\stackrel{\mathrm{def}}{=} c_{AB}(\{y_{n_S}, y_{k_{AB}}, A\}_{k_{BS}}, \{A, y_{n_A}\}_{y_{k_{AB}}}).\\
&\qquad [y_{n_S} \notin V]\, \overline{c_{AB}}\langle \{y_{n_A}\}_{y_{k_{AB}}}\rangle.\\
&\qquad [y_{k_{AB}} \neq k_{AB}]\, \overline{error}\langle\rangle.\, \mathbf{0}\\
S &\stackrel{\mathrm{def}}{=} c_{AS}(A, B).\, \overline{c_{AS}}\langle \{n_S, k_{AB}, B, \{n_S, k_{AB}, A\}_{k_{BS}}\}_{k_{AS}}\rangle.\, \mathbf{0}\\
K_{aut} &\stackrel{\mathrm{def}}{=} (\nu\, k_{AS}, k_{BS}, k_{AB})\, \big(\, \overline{lost}\langle \{n_{old}, k, A\}_{k_{BS}}, k\rangle.\, \mathbf{0}\\
&\qquad\qquad \mid ((\nu\, n_A)\, A_{aut}) \mid B^V_{aut} \mid ((\nu\, n_S)\, S)\,\big)\,.
\end{aligned}
$$

Output at channel *error* is used to explicitly signal that a key different from k_{AB} has been received by A or B. The fact that K and K_{aut} are equivalent means that action $\overline{error}\langle\rangle$, hence violation of authentication, never occurs.

Secrecy: *"The key k_{AB} is never revealed"*.
Formally, it holds that $K \mid E \cong K_{secr}$, where K_{secr} is defined below:

$$
\begin{aligned}
E &\stackrel{\mathrm{def}}{=} e(w).\, \mathbf{0}\\
E_{secr} &\stackrel{\mathrm{def}}{=} e(w).[w = k_{AB}]\, \overline{error}\langle\rangle.\, \mathbf{0}\\
K_{secr} &\stackrel{\mathrm{def}}{=} (\nu\, k_{AS}, k_{BS})\, \big(\, \overline{lost}\langle \{n_{old}, k, A\}_{k_{BS}}, k\rangle.\, \mathbf{0}\\
&\qquad\qquad \mid ((\nu\, n_A)\, A) \mid B^V \mid (\nu\, n_S, k_{AB})\, (S \mid E_{secr})\,\big)
\end{aligned}
$$

Here E and E_{secr} represent "eavesdroppers": each of them may at any time receive at e any message known to the environment. After doing so, E just remains silent, but E_{secr} signals an $\overline{error}\langle\rangle$ if the received message happens to be k_{AB}. The fact that $K \mid E$ and K_{secr} are equivalent means that action $\overline{error}\langle\rangle$, hence violation of secrecy, never occurs.

Verification

In this section, we will prove the above authentication and secrecy by reasoning compositionally on processes. Let $I \stackrel{\text{def}}{=} fn(K, K_{aut})$ (that is, $I = \{lost, A, B, c_{AS}, c_{BS}, c_{AB}, k, error\} \cup V$) and $I' \stackrel{\text{def}}{=} fn(K \mid E, K_{secr})$ (that is, $I' = \{e\} \cup I$). By virtue of soundness of $\approx$ w.r.t. $\cong$, to establish authentication it will be sufficient to prove that

$$(\epsilon_I, \epsilon_I) \vdash K \approx K_{aut} \tag{1}$$

while, for secrecy, it will be sufficient to establish that

$$(\epsilon_{I'}, \epsilon_{I'}) \vdash K \mid E \approx K_{secr}. \tag{2}$$

Let M_0 be the message issued in the $\overline{lost}$-action, and M_i, $i = 1, \ldots, 4$ be the four messages exchanged in the protocol, that is, let us define:

$$\begin{array}{lcl} M_0 & = & \langle \{n_{old}, k, A\}_{BS}, k \rangle \\ M_1 & = & \langle A, B \rangle \\ M_2 & = & \{n_S, k_{AB}, B, \{n_S, k_{AB}, A\}_{k_{BS}}\}_{k_{AS}} \\ M_3 & = & \langle \{n_S, k_{AB}, A\}_{k_{BS}}, \{A, n_A\}_{k_{AB}} \rangle \\ M_4 & = & \{n_A\}_{k_{AB}}. \end{array}$$

We also define $\sigma \stackrel{\text{def}}{=} \epsilon_I[M_i/w_i]_{i \in \{0,\ldots,4\}}$ and $\sigma' \stackrel{\text{def}}{=} \epsilon_{I'}[M_i/w_i]_{i \in \{0,\ldots,4\}}$, where the w_i's are fresh names.

Authentication

Let us start proving (1). Suppose we can prove that:

$$(\sigma, \sigma) \vdash A \approx A_{aut} \tag{3}$$

and

$$(\sigma, \sigma) \vdash B^V \approx B^V_{aut} \tag{4}$$

and that each of A, A_{aut}, B^V, B^V_{aut} and S is σ-safe. Then, since $(\sigma, \sigma) \vdash S \approx S$ and $(\sigma, \sigma) \vdash \overline{lost}\langle M_0 \rangle.\mathbf{0} \approx \overline{lost}\langle M_0 \rangle.\mathbf{0}$ trivially hold (REFLEXIVITY), using three times Theorem 1, we get that:

$$(\sigma, \sigma) \vdash \overline{lost}\langle M_0 \rangle.\mathbf{0} \mid A \mid B^V \mid S \approx \overline{lost}\langle M_0 \rangle.\mathbf{0} \mid A_{aut} \mid B^V_{aut} \mid S.$$

We can now use (WEAKENING) so as to discard the w_i-entries of σ, and then use (C-RES), with $(\nu\, k_{AS}, k_{BS}, k_{AB}, n_A, n_S)$, so as to get:

$$\begin{array}{ll} (\epsilon_I, \epsilon_I) \vdash & (\nu\, k_{AS}, k_{BS}, k_{AB}, n_A, n_S)\left(\overline{lost}\langle M_0 \rangle.\mathbf{0} \mid A \mid B^V \mid S\right) \approx \\ & (\nu\, k_{AS}, k_{BS}, k_{AB}, n_A, n_S)\left(\overline{lost}\langle M_0 \rangle.\mathbf{0} \mid A_{aut} \mid B^V_{aut} \mid S\right). \end{array}$$

Applying repeatedly the structural equivalence axiom $(\nu a)(P \mid Q) \equiv P \mid ((\nu a) Q)$ if $a \notin fn(P)$, we can push each restriction onto the appropriate position, both in the RHS and LHS. By (STRUCT-EQ) we get thus the wanted equality (1).

So the proof reduces to establishing equations (3,4) above, and to proving that each of A, A_{aut}, B^V, B^V_{aut}, S and $\overline{lost}\langle M_0\rangle.\mathbf{0}$ is σ-safe. This we do in detail below.

In the sequel, we shall write $R(w)$ to indicate that name w may occur free in R, and, for any message M, we write $R(M)$ in place of $R[M/w]$.

Equation (3). Let $A'(x_1)$ and $A'_{aut}(x_1)$ be the parts of A and A_{aut}, respectively, underneath the input prefix $c_{AS}(x_1)$., that is let $A = \overline{c_{AS}}\langle A, B\rangle.\, c_{AS}(x_1).A'(x_1)$ and $A_{aut} = \overline{c_{AS}}\langle A, B\rangle.\, c_{AS}(x_1).A'_{aut}(x_1)$. By virtue of (C-OUT), it suffices to show that

$$(\sigma, \sigma) \vdash c_{AS}(x_1).A'(x_1) \approx c_{AS}(x_1).A'_{aut}(x_1)\,. \tag{5}$$

We prove (5) using (C-INP). Consider any $M \overset{\text{def}}{=} \widehat{\zeta\sigma}$ s.t. $\widetilde{b} \overset{\text{def}}{=} n(\zeta) - dom(\sigma)$ are fresh names; we have to check that

$$(\sigma[\widetilde{b}/\widetilde{b}], \sigma[\widetilde{b}/\widetilde{b}]) \vdash A'(M) \approx A'_{aut}(M)\,. \tag{6}$$

By definition, if $A'(M)$ is stuck so is $A'_{aut}(M)$ – and vice-versa, thus in these cases (6) follows trivially. Suppose now that $A'(M)$ and $A'_{aut}(M)$ are not stuck. By definition, this implies that $M = \{M'\}_{\widehat{k_{AS}}}$, for some M' (this also implies that $\widetilde{b} = \emptyset$). Now, since $M = \zeta\sigma[\widetilde{b}/\widetilde{b}]$ and $k_{AS} \notin dc(\sigma[\widetilde{b}/\widetilde{b}])$, Proposition 3.2 implies that $M \in dc(\sigma[\widetilde{b}/\widetilde{b}])$: the only possibility is therefore that $M = M_2$. Now, using the structural equivalence rules for `let` and matching, $A'(M)$ and $A'_{aut}(M)$ get simplified as follows:

$$\begin{aligned} A'(M) &\equiv \overline{c_{AB}}\langle M_3\rangle.\, c_{AB}(x_2).[x_2 = \{n_A\}_{k_{AB}}]\mathbf{0} \\ A'_{aut}(M) &\equiv \overline{c_{AB}}\langle M_3\rangle.\, c_{AB}(x_2).[x_2 = \{n_A\}_{k_{AB}}][k_{AB} \neq k_{AB}]\overline{error}\langle\rangle.\,\mathbf{0}. \end{aligned} \tag{7}$$

The RHS of these two equalities are clearly bisimilar under $\sigma[\widetilde{b}/\widetilde{b}]$, because the mismatch-guarded output in the second expression ($[k_{AB} \neq k_{AB}]\overline{error}\langle\rangle.\,\mathbf{0}$) is actually stuck. By (STRUCT-EQ), we therefore get (6) for this case.

***A* and A_{aut} are σ-safe.** Consider A (the proof for A_{aut} is the same). The first output action clearly does not violate σ-safety. Consider now the second output, that is consider the case when

$$(\sigma \rhd A) \xrightarrow[\eta(t)]{\overline{c_{AS}}\langle M_1\rangle} \xrightarrow[(\nu\widetilde{b})\,\overline{\eta'}\langle\zeta\rangle]{c_{AS}\,(M)} (\sigma_1 \rhd A'(M)) \xrightarrow[\eta''(t')]{\overline{c_{AB}}\langle M'\rangle} .$$

(for some η, η', η'') where $M = \widehat{\zeta\sigma}$. Since $A'(M)$ is not stuck, we know from the above analysis that (7) holds, hence it must be that $M' = M_3 \in dc(\sigma) \subseteq dc(\sigma_1)$. Thus A is σ-safe.

Equation (4). This is similar to the proof of (3). Let us call $B'(y)$ and $B'_{aut}(y)$ the parts of B^V and B^V_{aut} underneath the input prefix, that is assume that $B^V = c_{AB}(y).B'(y)$ and that $B^V_{aut} = c_{AB}(y).B'_{aut}(y)$. Using (C-INP), it suffices to check that for any $M \stackrel{\text{def}}{=} \widehat{\zeta\sigma}$ s.t. $\widetilde{b} \stackrel{\text{def}}{=} n(\zeta) - dom(\sigma)$ are fresh names, it holds

$$(\sigma[^{\widetilde{b}}/_{\widetilde{b}}], \sigma[^{\widetilde{b}}/_{\widetilde{b}}]) \vdash B'(M) \approx B'_{aut}(M). \tag{8}$$

For $B'(M)$ and $B'_{aut}(M)$ not being stuck, M must be a pair, say $M = \langle N_1, N_2\rangle$, and N_1 must be an encryption under k_{BS}. Since $M = \widehat{\zeta\sigma[^{\widetilde{b}}/_{\widetilde{b}}]}$ and $k_{BS} \notin dc(\sigma[^{\widetilde{b}}/_{\widetilde{b}}])$, Proposition 3.2 implies that $N_1 \in dc(\sigma[^{\widetilde{b}}/_{\widetilde{b}}])$: there are two possibilities, either $N_1 = \{n_{old}, k, A\}_{BS}$ or $N_1 = \{n_S, k_{AB}, A\}_{BS}$. In the first case, both $B'(M)$ and $B'_{aut}(M)$ are stuck, hence $\approx$-equivalent, because the mismatch sequence $[y_{n_S} \notin V]$ gets instantiated to $[n_{old} \notin V]$, which evaluates to false. In the second case, for the two processes not being stuck, it must be that N_2 is a pair encrypted under k_{AB}. Again Proposition 3.2 tells us that the only possibility is $N_2 = \{A, n_A\}_{k_{AB}}$, thus $M = M_3$. Using the structural rules for `let` and matching, it is easily seen that $B'(M)$ and $B'_{aut}(M)$ are $\approx$-equivalent under $\sigma[^{\widetilde{b}}/_{\widetilde{b}}]$.

B^V and B^V_{aut} are σ-safe. Let us call M the message received at c_{AB}. The previous analysis shows that the only case in which $B'(M)$ and $B'_{aut}(M)$ are not stuck is when $M = M_3$, which implies that the message sent along $\overline{c_{AB}}$ is M_4. This easily implies that B^V and B^V_{aut} are σ-safe.

S and $\overline{lost}\langle M_0\rangle.\mathbf{0}$ are σ-safe. Trivial.

Secrecy

The crucial step is showing that:

$$(\sigma', \sigma') \vdash E \approx E_{secr} \tag{9}$$

and that E and E_{secr} are σ'-safe. Otherwise, the proof proceeds like in the case of authentication. To prove (9), it will be sufficient to use (C-INP), and check that for any $M \stackrel{\text{def}}{=} \widehat{\zeta\sigma'}$ s.t. $\widetilde{b} \stackrel{\text{def}}{=} n(\zeta) - dom(\sigma')$ are fresh names, it holds that:

$$(\sigma'[^{\widetilde{b}}/_{\widetilde{b}}], \sigma'[^{\widetilde{b}}/_{\widetilde{b}}]) \vdash \mathbf{0} \approx [M = k_{AB}]\,\overline{error}\langle\rangle.\mathbf{0}.$$

Since $M = \widetilde{\zeta\sigma'[\tilde{b}/\tilde{b}]}$ and $k_{AB} \notin dc(\sigma'[\tilde{b}/\tilde{b}])$, Proposition 3.2 implies that $M \neq k_{AB}$. Hence, $[M = k_{AB}]\, \overline{error}\langle\rangle.\mathbf{0}$ is stuck and equivalent to $\mathbf{0}$. This analysis also shows that E_{secr} is σ'-safe. Finally, E is trivially σ'-safe.

5. TRACE ANALYSIS

An alternative approach to the analysis of crypto-protocols is based on analyzing the sequences of actions (*traces*) executable by a given spi-calculus specification. We briefly touch upon some ongoing work by Martin Abadi, Marcelo Fiore and one of us (M.B.) in this direction.

A sensible way of formalizing secrecy of the session key in our version of Kerberos, is requiring that action $\overline{error}\langle\rangle$ is never executed by the specification K_{secr}: formally, for no s, u, σ, P and δ, it is the case that $\epsilon_I \rhd K_{secr} \overset{s}{\underset{u}{\longmapsto}} \sigma \rhd P \overset{\overline{error}\langle\rangle}{\underset{\delta}{\longmapsto}}$. Trace analysis is less amenable to compositional reasoning than equivalence-based analysis, but is well-suited for mechanical checking. We discuss this below.

The presence of input actions causes any nontrivial process to be infinite-state, as each input variable can be instantiated to any of infinitely many values (see rules (INP) and (E-INP) of Table 2). It is possible to avoid this kind of state-explosion by equipping spi-calculus with a *symbolic* operational semantics. Given a generic input-prefixed process $a(x).R$, the idea is to incrementally constrain the set of possible values x can take on, rather than to immediately instantiate x. Constraints are expressed as unifiers. To see how this can be done, let us consider simplified versions of both configurations and transitions: in fact, the 'environment action' δ is not necessary anymore and the first component of any configuration is assumed to be simply a trace of input/output actions. As an example, consider $P \stackrel{\text{def}}{=} a(x).R$, where $R \stackrel{\text{def}}{=} \texttt{let}\ x' = \texttt{dec}_k(x)\ \texttt{in}\ \overline{a}\langle x'\rangle.\,R'$. The two initial symbolic transitions of $\sigma_0 \rhd P$ are reported below. Here σ_0 is any trace of ouput actions that export names in $fn(P)$, thus encoding the initial knowledge of the environment:

$$\sigma_0 \rhd P \xrightarrow{a\,\langle x\rangle} (\sigma_0 \cdot a\,\langle x\rangle) \rhd R \xrightarrow{\overline{a}\langle x'\rangle} (\sigma_0 \cdot a\,\langle\{x'\}_k\rangle \cdot \overline{a}\langle x'\rangle) \rhd R'\,.$$

In the second step, decryption of x with key k is resolved as unification of x with a term $\{x'\}_k$, for a fresh x'. The resulting substitution $([\{x'\}_k/x])$ is propagated through the whole configuration, and causes the first input action of the trace, $a\,\langle x\rangle$, to be 'refined' into $a\,\langle\{x'\}_k\rangle$. A generic matching $[M_1 = M_2]$ would be similarly resolved as unification between M_1 and M_2.

Let us consider again the process K_{secr} that specifies secrecy of the session key in Kerberos. Suppose $\sigma_0 \triangleright K_{secr} \xrightarrow{\sigma} (\sigma_0 \cdot \sigma) \triangleright P$, where again σ_0 represents the initial knowledge of the environment (i.e. $fn(K_{secr})$) and:

$$
\begin{aligned}
\sigma \quad = \quad & \overline{lost}\langle M_0\rangle \cdot \\
& \overline{c_{AS}}\langle A, B\rangle \cdot c_{AB}\,\langle A, B\rangle \cdot \\
& \overline{c_{AS}}\langle \{n_S, k_{AB}, B, \{n_S, k_{AB}, A\}_{k_{BS}}\}_{k_{AS}}\rangle \cdot c_{AS}\,\langle \{x_{n_S}, x_{k_{AB}}, B, x_{cert_B}\}_{k_{AS}}\rangle \cdot \\
& \overline{c_{AB}}\langle x_{cert_B}, \{A, n_A\}_{x_{k_{AB}}}\rangle \cdot c_{AB}\,\langle \{y_{n_S}, y_{k_{AB}}, A\}_{k_{BS}}, \{A, y_{n_A}\}_{y_{k_{AB}}}\rangle \cdot \\
& \overline{c_{AB}}\langle \{y_{n_A}\}_{y_{k_{AB}}}\rangle \cdot c_{AB}\,\langle \{n_A\}_{x_{k_{AB}}}\rangle \cdot \\
& e\,\langle k_{AB}\rangle \cdot \overline{error}\langle\rangle
\end{aligned}
$$

It can be seen that the final trace $\sigma_0 \cdot \sigma$ is not *consistent*: this means that it cannot be instantiated to any *concrete* trace, i.e. a trace generable by the 'concrete' operational semantics of Table 2. In fact, it can be proven that no symbolic trace of K_{secr} leading to $\overline{error}\langle\rangle$ is consistent in this sense. Thus, K_{secr} will never execute $\overline{error}\langle\rangle$. In general, it can be shown that consistency of symbolic traces can be effectively decided.

6. CONCLUSIONS

We have outlined some recent approaches to the analysis of crypto-protocols, centered around concepts derived from process algebras, such as labelled transition systems and observational semantics.

When compared to more traditional methods – like CSP-based model checking [7, 13] and theorem proving [12] – major benefits of the equivalence-based approach seem to be a fully satisfactory formalization of many important properties (including implicit information flow), and compositional proof rules. Symbolic trace analysis appears to be similar in spirit to model checking, but it does not suffer from the limitation of model checking, which requires the model to be 'cut' to a suitable finite size, even in the presence of only a limited number of participants. On the other hand, both model checking and theorem proving have proven very effective in practice in finding bugs of crypto-protocols. Real-life case studies are necessary to determine whether process algebraic approaches may represent a valid alternative to established techniques.

Acknowledgments

This work has been partially supported by MURST projects SALADIN and TOSCA.

References

[1] M. Abadi, A.D. Gordon. Reasoning about cryptographic protocols in the spi calculus. *CONCUR'97, Proceedings* (A. Mazurkiewicz, J. Winkowsky, Eds.), *LNCS* 1243, pp.59-73, Springer-Verlag, 1997.

[2] M. Abadi, A.D. Gordon. A calculus for cryptographic protocols: The spi calculus. *Information and Computation*, 148(1):1-70, Academic Press, 1999.

[3] M. Boreale, R. De Nicola. Testing equivalence for mobile processes. *Information and Computation*, 120:279-303, Academic Press, 1995.

[4] M. Boreale, R. De Nicola, R. Pugliese. Proof Techniques for Cryptographic Processes. In *Proc. of the 14th IEEE Symposium Logic In Computer Science (LICS'99)*, IEEE Computer Society Press, pp.157-166, 1999. (Full paper available at: `http://music.dsi.unifi.it`)

[5] M. Burrows, M. Abadi, R. Needham. A Logic of Authentication. *ACM Transactions on Computer Systems*, 8(1):18-36, 1990.

[6] R. De Nicola, M.C.B. Hennessy. Testing Equivalence for Processes. *Theoretical Computers Science*, 34:83-133, Elsevier, 1984.

[7] G. Lowe. Breaking and Fixing the Needham-Schroeder Public-Key Protocol Using FDR. *TACAS'96, Proceedings* (T. Margaria, B. Steffen, Eds.), *LNCS* 1055, pp. 147-166, Springer-Verlag, 1996.

[8] S.P. Miller, C. Neumann, J.I. Schiller, J.H. Saltzer. Kerberos authentication and authorization system. In Project Athena Technical Plan, Section E.2.1, MIT, 1987.

[9] R. Milner. The Polyadic π-calculus: a Tutorial. In *Logic and Algebra of Specification* (F.L. Hamer, W. Brauer, H. Schwichtenberg, Eds.), Springer-Verlag, 1993.

[10] R. Milner, J. Parrow, D. Walker. A calculus of mobile processes, (Part I and II). *Information and Computation*, 100:1-77, Academic Press, 1992.

[11] R. Milner, D. Sangiorgi. Barbed Bisimulation. *ICALP'92, Proceedings* (W. Kuich, Ed.), *LNCS* 623, pp.685-695, Springer-Verlag, 1992.

[12] L.C. Paulson. Provin Security Protocols Correct. In *Proc. of the 14th IEEE Symposium Logic In Computer Science (LICS'99)*, IEEE Computer Society Press, pp.370-381, 1999.

[13] S. Schneider. Verifying Authentication Protocols in CSP. *IEEE Transactions on Software Engineering*, 24(8):743-758, IEEE Computer Society Press, 1998.

A LOGIC OF BELIEF AND A MODEL CHECKING ALGORITHM FOR SECURITY PROTOCOLS

M. Benerecetti
DISA - Università degli Studi di Trento, Via Inama 5, 38100 Trento, Italy
bene@na.infn.it

F. Giunchiglia
DISA - Università degli Studi di Trento, Via Inama 5, 38100 Trento, Italy
fausto@cs.unitn.it

M. Panti
Istituto di Informatica, Università di Ancona, via Brecce Bianche,60131 Ancona, Italy
panti@inform.unian.it

L. Spalazzi
Istituto di Informatica, Università di Ancona, via Brecce Bianche,60131 Ancona, Italy
spalazzi@inform.unian.it

Abstract Model checking is a very successful technique which has been applied in the design and verification of finite state concurrent reactive processes. In this paper we show how this technique can be used for the verification of security protocols using a logic of belief. The underlying idea is to treat separately the temporal evolution and the belief aspects of principals. In practice, things work as follows: when we consider the temporal evolution of a principal we treat belief atoms (namely, atomic formulae expressing belief) as atomic propositions. When we deal with the beliefs of a principal A, we model its beliefs about another principal B as the fact that A has access to a representation of B as a process. Then, any time it needs to verify the truth value of some belief atom about B, e.g., $B_B\phi$, A simply tests whether, e.g., ϕ holds in its (appropriate) representation of B. Beliefs are essentially used to control the "jumping" among processes. Our approach allows us to reuse the technology and tools developed in model checking.

Tommaso Bolognesi and Diego Latella (Eds.), *Formal Methods for Distributed System Development.*

1. INTRODUCTION

In this paper we show how model checking (see, e.g. [9]) can be used for the verification of security protocols using a logic of belief and time [2]. Our approach allows us to reuse with almost no variations all the technology and tools developed in model checking and it is on the line of the work by Fagin et al. [7]. We model *principals* participating to a protocol session as (concurrent reactive finite state) processes able to have beliefs. The specification of a principal has therefore two orthogonal aspects: a temporal aspect and a belief aspect. The key idea underlying our approach is to keep these two aspects separated. In practice things work as follows:

- When we consider the temporal evolution of a principal we treat belief atoms (namely, atomic formulae expressing belief) as atomic propositions. The fact that these formulae talk about beliefs is not taken into consideration.

- When we deal with the beliefs of a principal A, we model its beliefs about another principal B as the fact that A has access to a representation of B as a process. Then, any time it needs to verify the truth value of some belief atom about B, e.g., $B_B\phi$, a simply tests whether, e.g., ϕ holds in its (appropriate) representation of B. Beliefs are essentially used to control the "jumping" among processes. This operation is iterated in the obvious way in case of nested beliefs.

The application of model checking to security protocol verification is not new (see. e.g., [8, 10]). However, in the previous work, security protocols are verified by introducing the notion of intruder and, then, by verifying whether the intruder can attack a given protocol. This approach makes possible to directly find a trace of a possible attack but it may not be clear what the protocol flaw is. This work usually employs temporal logics or process algebras. Our approach is different, we do not use any model of intruders and our formalism is able to represent both time and belief. This work builds on the work described in [3], where model checking is applied to BDI attitudes (namely, Belief, Desire, and Intention) of agents. Various papers can be found in literature, where belief logics have been applied to testing security in protocol verification (see for instance [4]). However, in all the previous work (see, e.g., [4, 1]) verification is performed proof-theoretically.

The paper is structured as follows. In Section 2 we describe the logic of belief and time we employ (called MATL, MultiAgent Temporal Logic). We also briefly introduce a well-known protocol, the CCITT X.509 protocol, as a running example. Section 3 introduces finite presentations of MATL (called MAFSM, MultiAgent Finite State Machines). Section 4 describes the model checking procedure we propose. Finally, some conclusions are drawn in Section 5.

2. A LOGIC OF BELIEF AND TIME

We model principals engaged in an authentication session as finite state processes. We build the notion of principal incrementally over the notion of process. Suppose we have a set I of principals. Each principal is seen as a process having beliefs about (itself and) other principals. We adopt the usual syntax for beliefs: $B_i\phi$ means that principal i believes ϕ, and ϕ is a belief of i. B_i is the belief operator for i.

The idea is to associate to each (level of) nesting of belief operators a process evolving over time. Let $B = \{B_1, ..., B_n\}$, where each index $1, ..., n \in I$ corresponds to a principal. Let B^* denote the set of finite strings of the form $B_1, ..., B_n$ with $B_i \in B$. We call any $\alpha \in B^*$, a *view*. Figure 1 depicts the general structure of the views. Each view in B^* corresponds to a possible nesting of belief operators. We also allow for the empty string, ϵ. The intuition is that ϵ represents the view of an external observer (e.g., the designer) which, from the outside, "sees" the behavior of the overall protocol. The beliefs of principal A correspond to the view B_A and are modeled by a process playing A's role in a protocol. The beliefs of principal B (the view B_B) are modeled similarly. The beliefs that A has about (the behavior of) principal B correspond to the view B_AB_B and are modeled by a process playing B's role in the protocol. Things work in the same way for any arbitrary nesting of belief operators.

We associate a language $\mathcal{L}_\alpha$ to each view $\alpha \in B^*$. Intuitively, each $\mathcal{L}_\alpha$ is the language used to express what is true (and false) of the process of view α. We employ the logic CTL [6], a well known propositional branching-time temporal logic widely used in formal verification. For each α, let P_α be a set of propositional atoms. Each P_α allows for the definition of a different language, called a MultiAgent Temporal Logic (MATL) language (on P_α). A MATL language $\mathcal{L}_\alpha$ on P_α is the smallest CTL language containing the set of propositional atoms P_α and the belief atoms $B_i\phi$, for any formula ϕ of $\mathcal{L}_{\alpha B_i}$. In particular, $\mathcal{L}_\epsilon$ is used to speak about the whole protocol. The language $\mathcal{L}_{B_i}$ ($\mathcal{L}_{B_j}$) is the language adopted to represent i's (j's) beliefs. i's beliefs about j's beliefs are specified by the language of the view B_iB_j. Given a family $\{P_\alpha\}$ of sets of propositional atoms, the family of MATL languages on $\{P_\alpha\}$ is the family of CTL languages $\{\mathcal{L}_\alpha\}$. We write $\alpha : \phi$ (called labeled wff) to mean that ϕ is a formula of $\mathcal{L}_\alpha$.

Example 2.1 MATL is expressive enough to capture most security logics. For instance, we can easily express in MATL the analysis of the CCITT X.509 protocol, as described in [4]. N_i denotes a nonce (a fresh message) newly created by principal i for the current session; T_i is a timestamp which represents the time at which the message was generated; X_i and Y_i are user data. In public key cryptography, each principal i has a *public key*, denoted K_i, which any other principal can obtain from a key server; it also has a *private secret key*,

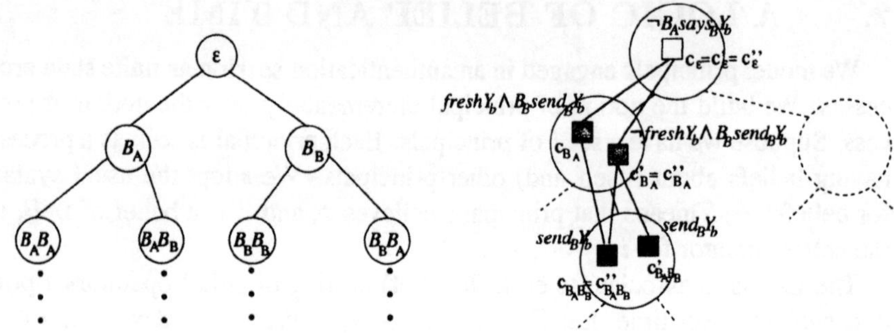

Figure 1 The set of views for the CCITT protocol. *Figure 2* Some chains of the CCITT protocol.

K_i^{-1}, which is the inverse of K_i. Any message encrypted with the public key, $\{M\}_{K_i}$, can be decrypted only by i with its own private key; vice versa, any message encrypted (signed) by i with its private key, $\{M\}_{K_i^{-1}}$, can be decrypted by any principal with the public key K_i. The CCITT X.509 protocol can be formulated as follows:

$$\begin{array}{ll} 1\ A \rightarrow B: & \{T_a, N_a, X_a, \{Y_a\}_{K_b}\}_{K_a^{-1}} \\ 2\ B \rightarrow A: & \{T_b, N_b, N_a, X_b, \{Y_b\}_{K_a}\}_{K_b^{-1}} \\ 3\ A \rightarrow B: & \{N_b\}_{K_a^{-1}} \end{array}$$

Intuitively, with Message 1, A sends B the data X_a and Y_a. Y_a must be secret and thus it is encrypted with K_b. The whole message is signed with K_a^{-1} to assure that the sender is A. Similarly, B sends A the data X_b and Y_b with Message 2. We then set the atoms P_α for the view B_A as follows:

$$P_{B_A} = \left\{ \begin{array}{l} send_B\, \{T_a, N_a, X_a, \{Y_a\}_{K_b}\}_{K_a^{-1}}, \\ rec\, \{T_b, N_b, N_a, X_b, \{Y_b\}_{K_a}\}_{K_b^{-1}}, \\ fresh\, \{T_b, N_b, N_a, X_b, \{Y_b\}_{K_a}\}_{K_b^{-1}}, \\ fresh\, N_a, \\ pubk_B\, K_b, \\ \ldots \end{array} \right\}$$

where the variables of the form *rec* M or $send_B\, M$ (where M is a (sub)message of the CCITT protocol) are called *message variables* and represent the act of receiving M, and sending M to B, respectively. For instance, the atom $send_B\, \{T_a, N_a, X_a, \{Y_a\}_{K_b}\}_{K_a^{-1}}$ in view B_A represent A sending Message 1 to B. Variables of the form *fresh* M (e.g., *fresh* N_a in view B_A) mean that M (e.g., N_a) is a fresh (sub)message. Variables of the form $pubk_i\, K$ (e.g., $pubk_B\, K_b$in view B_A) mean that K is the public key of i. The atomic propositions of the other views can be defined similarly. Since each view αB_i (with $i = A, B$) models the (beliefs about the) behavior of principal i, the set of atomic proposition will be that of view B_i (see [2]).

In MATL we can formalize how messages sent and received during a protocol session by a party may affect its beliefs about the other parties. For instance, one might want to express the following property: at the end of a protocol session, A believes that B recently "said" X_b and Y_b. Formally:

$$\epsilon : \mathsf{AG}\, send_{A,B}\, \{N_b\}_{K_a^{-1}} \rightarrow B_A says_B X_b \wedge B_A says_B Y_b \quad (1)$$

where, following [1], for each message M, in the view modeling the beliefs of principal j, the proposition $says_i\, M$ is defined as $fresh M \wedge B_i send_j M$.

Semantics. To understand the semantics of the family of languages $\{\mathcal{L}_\alpha\}_{\alpha \in B^*}$ (hereafter we drop the subscript), we need to understand two key facts. On the one hand the semantics of formulae depend on the view. For instance, the formula p in the view B_i expresses the fact that i believes that p is true. The same formula in the view B_j expresses the fact that j believes that p is true. As a consequence, the semantics associates *locally* to each view α a set of pairs $\langle m, s \rangle$, where: $m = \langle S, J, R, L \rangle$ is a CTL structure, with S a set of states, $J \subseteq S$ the set of *initial states*, R the transition relation, and $L : S \rightarrow \mathcal{P}(P)$ the *labeling function*; and s is a *reachable state* of m (a state s of a CTL structure is said to be reachable if there is a path leading from an initial state of the CTL structure to state s). On the other hand there are formulae in different views which have the same intended meaning. For instance $B_j p$ in view B_i, and p in view $B_i B_j$ both mean that i believes that j believes that p is true. This implies that only certain interpretations of different views are *compatible* with each other, and these are those which agree on the truth values of the formulae with the same intended meaning. To capture this notion of compatibility we introduce the notion of chain.

Let α be any view, a α*−chain* c is a finite sequence $\langle c_\epsilon, ..., c_\beta, ..., c_\alpha \rangle$, where $c_\beta = \langle m, s \rangle$ is an interpretation for $\mathcal{L}_\beta$ and β is a prefix of α (i.e., $\alpha = \beta\gamma$ for some index $\gamma \in B^*$). A *compatibility relation* C *on* $\{\mathcal{L}_\alpha\}$ is a set of α−chains, for every α. Intuitively, C will contain all those c's whose elements c_α, c_β (where α, β are two views in B^*) assign the same truth values to the formulae with the same intended meaning.

Example 2.2 Figure 2 (see previous page) shows some possible chains of the MATL structure for the CCITT protocol. The boxes in each view represent interpretations of the language associated with the corresponding view. The figure shows an interpretation for view ϵ, two interpretations for view B_A and two interpretations for view $B_A B_B$. Links connecting boxes in different views represent $B_A B_B$−chains. Figure 2 shows three $B_A B_B$−chains, $c = \langle c_\epsilon, c_{B_A}, c_{B_A B_B} \rangle$, $c' = \langle c'_\epsilon, c'_{B_A}, c'_{B_A B_B} \rangle$ and $c'' = \langle c''_\epsilon, c''_{B_A}, c''_{B_A B_B} \rangle$, where $c_\epsilon = c'_\epsilon = c''_\epsilon$, $c'_{B_A} = c''_{B_A}$ and $c_{B_A B_B} = c''_{B_A B_B}$. Let us assume that the each interpretation satisfies the formula written close to it in figure. Therefore, the interpretation labeled c'_{B_A} ($= c''_{B_A}$) satisfies the formula $B_B send_A Y_b$. The intended meaning of this formula in view B_A is that A believes that B believes it has sent message Y_b to A. The formula $send_A Y_b$ in view $B_A B_B$ has the same intended meaning, as $B_A B_B$ models the beliefs of B seen from (the beliefs of) A. Therefore any $B_A B_B$–chain passing through the interpretation c'_{B_A} must reach an interpretation in $B_A B_B$ which satisfies the argument $send_A Y_b$, as shown in the figure.

Let us now define the notion of satisfiability. We start with satisfiability local to views (first step) and suppose that for each view α there is a satisfiability relation between CTL structures and formulae of $\mathcal{L}_\alpha$. With an abuse of notation, we denote all these satisfiability relations with the same symbol $\models$. The context always makes clear which relation we mean. The second step is to define (global) satisfiability taking into account chains. To do this we need some further notation. Let $\models$ denote satisfiability also on chains. For any α–chain c and for any formula in $\mathcal{L}_\beta$, satisfiability relation $\models$ is defined only when either α is a prefix of β or β is a prefix of α. (i.e., when either $\alpha = \beta\gamma$ or $\beta = \alpha\gamma$). If $\alpha = \beta\gamma$ then $c_\beta \models \phi$ iff ϕ is true in c_β. If $\beta = \alpha\gamma$ then $c_\beta \models \phi$ for any ϕ. In other words, if a chain stops at a given level (e.g. α), then it satisfies every formula of the views (e.g., $\alpha\gamma$) which are below that level in the tree. Let us extend the satisfiability relation to sets of formulae: $x \models Y$ if and only if for any $y \in Y$, $x \models y$.

We are now ready to define the notion of model for MATL (called MATL structure), and then that of satisfiability between MATL structures and formulae of a view.

Definition 2.1 (MATL structure) *A nonempty compatibility relation C for a family of MATL languages on $\{P_\alpha\}$ is a* MATL structure *on $\{P_\alpha\}$ if for any $\alpha\beta$–chain $c \in C$,*

1. *$c_\alpha \models B_i\phi$ iff for every $\alpha\gamma$–chain $c' \in C$, $c'_\alpha = c_\alpha$ implies $c'_{\alpha B_i} \models \phi$;*
2. *if $c_\alpha = \langle m, s\rangle$, then for any state s' of m, there is a $\alpha\beta$–chain $c' \in C$ such that $c'_\alpha = \langle m, s'\rangle$.*

Briefly: the nonemptyness condition for C guarantees that the external observer has a consistent view of the world. The *only if* part in condition 1 guarantees that each view has correct beliefs, i.e., any time $B_i\phi$ holds at a view then ϕ holds in the view one level down in the chain. The *if* part is the dual property and ensures the completeness of each view. Condition 2 can be understood on the basis of two crucial observations, concerning the mutual nesting of CTL operators and belief operators. The first, concerning the nesting of CTL operators inside belief operators is that $c_\alpha \models \phi$ is computed using the notion of satisfiability in a CTL structure. Therefore, a chain links the fact that a belief atom holds in a state of a CTL structure in one view with the fact that its argument holds in a state of a CTL structure in the view below. The second observation concerns the nesting of belief operators inside temporal operators (temporal operators which involve no belief atoms are treated as in CTL structures, that is, without jumping among views). Consider for instance the formula $\mathsf{EX}\, B_i p$. To assess the truth of $\mathsf{EX}\, B_i p$ we need to be able to assess the truth of $B_i p$ in some future (reachable) state s' of the CTL structure we are considering, e.g., $\langle\langle S, J, R, L\rangle, s\rangle$. The only way to establish this is to request that in s' we have a chain c' which gives access to a CTL structure in the view below. Given the fact that chains connect CTL structures only for what holds in their (reachable) states, the only solution is to request that s' is the state component of the structure $c'_\epsilon = \langle\langle S, J, R, L\rangle, s'\rangle$ with $c' \in C$. Given the fact that temporal operators allow us to state facts about all the states in a CTL structure, this operation must be repeated for each state $s \in S$. But this is exactly what Item 2 of Definition 2.1 says.

Example 2.3 Consider again Figure 2. The formula $\neg B_A says_B Y_b$ is satisfied by the interpretation in view ϵ. The formula $B_A says_B Y_b$ is defined as $B_A(\mathit{fresh} Y_b \wedge B_B send_A Y_b)$. By Item 1 of Definition 2.1 (only if direction), there must be a chain starting from the interpretation in view ϵ whose component for view B_A does not satisfy the argument of the belief, namely $\mathit{fresh} Y_b \wedge B_B send_A Y_b$. This is indeed the case for both c' and c'', as $c'_{B_A} = c''_{B_A}$ and both of them do not satisfy $\mathit{fresh} Y_b$. Assume now that c' and c'' are the only two chains passing through interpretation c'_{B_A}. The components for view $B_A B_B$ of both chains passing through that interpretation, namely $c'_{B_A B_B}$ and $c''_{B_A B_B}$, satisfy the argument $send_B Y_b$. Thus, by Item 1 of Definition 2.1 (if direction), c'_{B_A} must satisfy $B_B send_B Y_b$, as shown in Figure 2.

Given a MATL structure C, a formula ϕ and a view α, $C \models \alpha : \phi$ is read as ϕ is true in C (or equivalently, ϕ holds in C, or ϕ is satisfied by C) at view α, and it is defined as follows:

$$C \models \alpha : \phi \text{ iff for all } c \in C \text{ s.t. } c_\alpha = \langle\langle S, J, R, L\rangle, s\rangle \text{ and } s \in J,\ c_\alpha \models \phi \tag{2}$$

The intuition is that in order to check the satisfiability of ϕ at the view α we need to check all the interpretations of $\mathcal{L}_\alpha$ allowed by the compatibility imposed by the chains we are considering.

3. MULTIAGENT FINITE STATE MACHINES

We are interested in extending CTL model checking to the model checking of belief formulae. In model checking, finite state processes are modeled as finite state machines. A *finite state machine* (FSM) is a finite CTL structure $f = \langle S, J, R, L\rangle$. Our solution is to extend the notion of FSM to that of *Multi-*

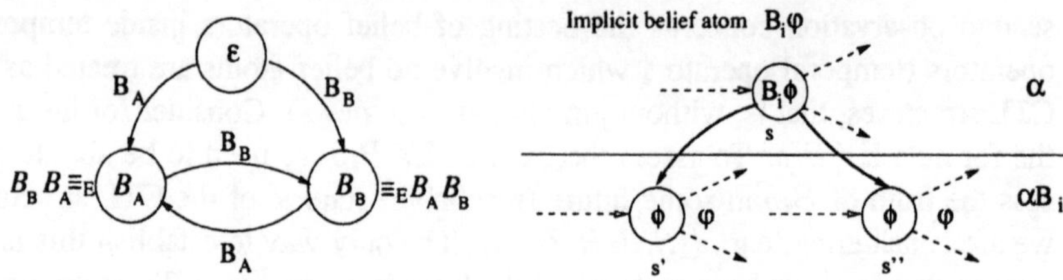

Figure 3 The set of views for the CCITT protocol.

Figure 4 Explicit belief atoms and satisfiability.

Agent Finite State Machine (MAFSM), where, roughly speaking, a MAFSM is a finite set of FSMs.

A first step in this direction is to restrict ourselves to a finite number of views α. Let B^n denote a finite subset of B^* obtained by taking the views in any finite subtree of B^* rooted at ϵ. This restriction is not enough, as a finite set of views still allows for an infinite number of belief atoms. Even if we had a finite number of processes we would not be able to model them as FSMs. This problem can be solved introducing the notion of *explicit belief atoms* as a finite subset of the set of belief atoms. Explicit belief atoms are the only belief atoms which are explicitly represented in a FSM, allowing for a finite set of states. As we explain below, the truth value of all the other belief atoms not explicitly represented in a FSM is computed starting from explicit beliefs. Formally, if $\mathcal{L}_\alpha$ is a MATL language, then for each belief operator B_i, the set of *explicit belief atoms* of B_i for α (in symbols $Expl(\textsc{b}_i, \alpha)$) is a (possibly empty) *finite* subset of the belief atoms of $\mathcal{L}_\alpha$.

In this way, though, one can only model systems whose specifications must be expressed with an "a priori" bound on the nesting of belief operators. As a consequence, it would not be possible to deal with, e.g., (infinite) positive/negative introspection or an arbitrary nesting of belief operators. Moreover, each view must be specified separately even when different views of principals happen to behave in the same way. One way to overcome these limitations is to extend the finite tree of views to a graph of views. Cycles in the graph allow us to represent arbitrary nesting of beliefs and to model equivalent views using the same specification.

Example 3.1 Consider for instance the CCITT protocol as represented in Figure 3. The view $\textsc{b}_B$ is modeled by a process whose (possible) behaviors are those of B as specified by the protocol. View $\textsc{b}_A\textsc{b}_B$ is modeled by a process whose (possible) behaviors are those of B as seen from A's point of view. Since protocols are usually supposed to be publicly known, and each honest principal's behavior is assumed to comply to the protocol, those two views in the CCITT protocol should exhibit the same set of behaviors. In other words, the two views are behaviorally equivalent. Therefore, both views can be modeled by the same process. Similar argument can be given for B_A and for any level of nesting of belief operators. The resulting graph is represented in Figure 3.

Let $Front(B^n)$ be the set of views of B^* not in B^n which are immediately below a view in B^n. Each view in $Front(B^n)$ will be the first view, in a path of the tree of views, that do not belong to B^n. Formally $Front(B^n) = \{\alpha B_i \mid \alpha \in B^n \text{ and } \alpha B_i \notin B^n\}$.

Let E be any functional relation included in $Front(B^n) \times B^n$. The intuitive interpretation of the relation E between two views α and β (i.e., $E(\alpha, \beta)$) is that view α is "behaviorally equivalent" to view β. In other words α can be modeled by the same process as β. Intuitively, every time we need to verify satisfiability of a formula of (the process modelling) view α, we need to jump to view β to assess the truth of that formula. We write $\equiv_E$ to mean the smallest equivalence relation on B^* including E. Let $[\ ]_E$ be the equivalence class induced by $\equiv_E$ on B^*. By means of the equivalence relation induced by E, any view below α can be considered behaviorally equivalent to the appropriate view below β. That is, suppose we have to verify a formula in a given view α which doesn't belong to B^n but is equivalent (via $\equiv_E$) to a view $\beta \in B^n$ (that is, there is a view $\beta \in B^n$ with $[\beta]_E = [\alpha]_E$). In this case, we would jump to β and verify the given formula in that view.

Example 3.2 See again Figure 3. The protocol considered in the example involves two principals, A and B. Therefore, we have $I = \{A, B\}$. We can consider only three views in B^n, namely ϵ, B_A and B_B. ϵ (the external observer) is modeled as a process which "sees" all the messages sent and received by the principals. B_A models the behavior of principal A, while B_B models the behavior of principal B. In this case, we set $Front(B^n) = \{B_A B_A, B_A B_B, B_B B_A, B_B B_B\}$. The beliefs of B (A, respectively) about A (about B, respectively) are then modeled by the view B_A (B_B, respectively). Therefore, both $E(B_A, B_B B_A)$ and $E(B_B, B_A B_B)$ holds. Figure 3 illustrates this intuition by representing the relation E as arrows connecting the views B_A with B_B. Each arrow connects two views and is labeled with a belief operator. The arrow with label B_A from view ϵ to view B_A tells us that the beliefs the external observer has about principal A are represented by view B_A. The arrow from view B_A to view B_B labeled B_B means that the beliefs that A has about B (corresponding to view $B_A B_B$ in MATL) are modeled by view B_B, as expressed by the relation $E(B_A, B_B B_A)$. Similarly for the arrow linking views B_B and B_A in the other direction. As a consequence, the equivalence class $[B_B]_E$ contains, for instance, the view $B_A B_B$, while $[B_A]_E$ contains view $B_B B_A$. Whenever we need to verify a formula in the language of $B_A B_B$ ($B_B B_A$, respectively) we need to jump to view B_B (B_A, respectively) and verify the formula in that view.

Thus, we have the following:

Definition 3.1 (MAFSM) *Let $B^n \subset B^*$ be a finite subtree of views rooted at ϵ. A* **MAFSM** *is a pair $MF = \langle E, F \rangle$ where: E is any functional relation included in $Front(B^n) \times B^n$ and $F = \{F_\alpha\}$ is total recursive function such that:*

1 $F_\epsilon \neq \emptyset$;

2 for all views $\alpha \in B^n$, it associates a finite set F_α of FSMs on the MATL language on the following atoms: P_α and, for all $i \in I$, $Expl(B_i, \alpha)$;

3 for all the views $\alpha \in B^* \setminus B^n$, $F_\alpha = \emptyset$.

The first condition on F ensures that the protocol specification is not empty; the second allows us to deal, in each view, with finite sets of FSMs; and the third allows us to deal with a finite number of views. In general, there may be more than one FSM associated to each view. This allows for situations in which a view can be only partially specified, and consequently there can be more than one process modeling that view. If it is completely specified, a view contains only one FSM.

Let us extend the set $Expl(B_i, \alpha)$ and the function F to the equivalence classes induced by E over B^* in the following way:

$$Expl(B_i, [\alpha]_E) = \begin{cases} Expl(B_i, \beta) & \text{if } \beta \in [\alpha]_E \cap B^n \\ \emptyset & \text{otherwise} \end{cases} \qquad F_{[\alpha]_E} = \begin{cases} F_\beta & \text{if } \beta \in [\alpha]_E \cap B^n \\ \emptyset & \text{otherwise} \end{cases}$$

Intuitively, $Expl(B_i, [\alpha]_E)$ ($F_{[\alpha]_E}$, respectively) is the set of explicit belief atoms (set of FSMs, respectively) of a view β belonging to B^n in the equivalent class of α, if it exists, the empty set, otherwise. Notice that, if it exists, β must be unique. This follows from the fact that $\equiv_E$ is smallest equivalence relation including E and from the fact that E is a functional relation. Therefore there can be at most one view belonging to B^n equivalent to a view in $B^* \setminus B^n$.

In the rest of the paper we assume that each set of explicit belief atoms only contains belief atoms of depth one, i.e., atoms of the form $B_i\phi$, where ϕ does not contain other belief atoms. This is clearly a restriction in general, but it greatly simplifies the framework when cycles are allowed, while being sufficient for modeling most security protocols (see, e.g., the informal discussion in [2]).

Example 3.3 We need to choose the explicit belief atoms of each view. In general, the choice of the appropriate set of explicit belief atoms of (the views in) a MAFSM depends on what kind of aspects of the protocol one wants to analyze, and on what kind of properties need to be verified. In the case of security protocols, principals can only gain information carried by the messages they receive. We choose, therefore, the beliefs about other principals having sent or received a given message as explicit beliefs atoms. In our case, we have:

$$Expl(B_B, B_A) = \left\{ \begin{array}{l} B_B send_A \{T_b, N_b, N_a, X_b, \{Y_b\}_{K_a}\}_{K_b^{-1}}, \\ \ldots \end{array} \right\}$$

where, for instance, $B_B send_A \{T_b, N_b, N_a, X_b, \{Y_b\}_{K_a}\}_{K_b^{-1}}$ in B_A intuitively means that A believes that B believes that it has sent Message 2 of the CCITT protocol to A. Similar explicit belief atoms are contained in view B_A and in view ϵ, the letter concerning beliefs of each principal about messages it sends and receives.

Given the notion of MAFSM, the next step is to give a notion of satisfiability in a MAFSM. We start from the notion of satisfiability of CTL formulae in an FSM at a state (defined as in CTL structures). Since FSMs are built on the propositional and explicit belief atoms of a view, to assess satisfiability of the propositional and explicit belief atoms (and the CTL formulae build out of them) we do not need to use the machinery associated to belief operators. However, this machinery is needed in order to deal with the (infinite) number of belief atoms which are not memorized anywhere in MAFSM. Let the set of *implicit belief atoms* of a view α, written $Impl(B_i, \alpha)$, be defined as the (infinite) subset of all belief atoms of $\mathcal{L}_\alpha$ which are not explicit belief atoms. Formally:

$$Impl(B_i, \alpha) = \{B_i\phi \in \mathcal{L}_\alpha \setminus Expl(B_i, \alpha)\}$$

The idea is to use the information explicitly contained in the labeling function of each state s of a FSM f of a view α to assess the truth value of the implicit belief atoms at a state s. Figure 4 illustrates the underlying intuition. Intuitively, the principal modeled by FSM f (in view α), when in state s, ascribes to principal i the explicit belief atoms of the form $B_i\phi$ true at s. This means that the FMSs of view αB_i, which model the beliefs of i, must be in any of the states (s' and s'' in Figure 4) in which the formulae ϕ, occurring as arguments of the explicit belief atoms, are true. This motivates the following definitions. Let $ArgExpl(B_i, \alpha, s)$ be defined as follows:

$$ArgExpl(B_i, \alpha, s) = \{\phi \in \mathcal{L}_{\alpha B_i} \mid B_i\phi \in L(s) \cap Expl(B_i, \alpha)\}$$

Thus, $ArgExpl(B_i, \alpha, s)$ consists of all the formulae $\phi \in \mathcal{L}_{\alpha B_i}$ such that the explicit belief atom $B_i\phi$ is true at state s (i.e., it belongs to the labeling function of s). $ArgExpl(B_i, \alpha, s)$ contains the formulae which identify the states in which the FSMs in view αB_i can be, whenever the process in view α is in state s.

We are now ready to define the notion of satisfiability of implicit belief atoms. Let $B_i\psi$ be an implicit belief atom of a view α. For each state s of a FSM of α, we can compute $ArgExpl(B_i, \alpha, s)$. As shown in Figure 4, we just need to check whether all the reachable states of the FSMs of view αB_i, which satisfy $ArgExpl(B_i, \alpha, s)$ (namely, the set $\{\phi\}$ in Figure 4), also satisfy the argument ψ of the implicit belief atom. If this is the case, then s satisfies $B_i\psi$.

Definition 3.2 (Satisfiability in a MAFSM) *Let MF be a MAFSM, α a view, $f = \langle S, J, R, L\rangle \in F_{[\alpha]}$ an FSM, and $s \in S$ a state. Then, for any formula ϕ of $\mathcal{L}_\alpha$, the satisfiability relation $MF, \alpha, f, s \models \phi$ is defined as follows:*

1. *$MF, \alpha, f, s \models p$, where p is a propositional atom or an explicit belief atom: the same as FSM satisfiability;*
2. *satisfiability of propositional connectives and CTL operators: the same as FSM satisfiability;*
3. *$MF, \alpha, f, s \models B_i\phi$ if and only if for all $f' \in F_{[\alpha B_i]}$ and s' state of f', $MF, \alpha B_i, f', s' \models \bigwedge ArgExpl(B_i, \alpha, s) \rightarrow \phi$, where $B_i\phi$ is an implicit belief atom,*

In the definition of satisfiability above, Item 3 is the crucial step. In the definition $\bigwedge ArgExpl(B_i, \alpha, s)$ denotes the conjunction of all the elements of the set $ArgExpl(B_i, \alpha, s)$. Notice that, under the assumption that the set of explicit belief atoms contains only belief atoms of depth one, the definition of satisfiability of implicit belief atoms above is well defined.

4. THE MODEL CHECKING ALGORITHM

The basic operation of a standard CTL model checking algorithm is to extend the labeling function of an FSM (which considers only propositional atoms) to all the (atomic and not atomic) subformulae of the formula being model checked. Let us call Extended FSM (or, simply, FSM when the context makes clear what we mean) the result of this operation. The generation of an extended FSM relies on the fact that the labeling function explicitly defines the truth value of all atoms. The problem is that in the FSMs of a MAFSM the labeling function is not defined on implicit belief atoms, whose truth value is therefore left undefined; and that we need to know the truth values of the implicit belief atoms occurring in the formula to be model checked. The definition of satisfiability in a MAFSM (Item 3 in Definition 3.2) tells us how to solve this problem.

The crucial observation is that $ArgExpl(B_i, \alpha, s)$, in Item 3 of Definition 3.2, is generated from the formulae in $Expl(B_i, \alpha)$ and the labeling functions of the FSMs in α; $ArgExpl(B_i, \alpha, s)$ is a finite set; and it only depends on the MAFSM specified (and thus independent of the formula to be model checked). For each belief operator B_i, C_{B_i} is called the *(MAFSM) compatibility relation* of B_i, and it is a relation defined as follows. Let $ex \subseteq Expl(B_i, \alpha)$ be a subset of the explicit belief atoms of a view $\alpha \in B^n$ and $MF = \langle E, F\rangle$ a MAFSM. The notion of compatibility relation just defined for a MAFSM is the dual notion of the compatibility relation defined in MATL. Then:

$$C_{B_i}(\alpha, ex) = \left\{ \begin{array}{c} \langle f', s'\rangle \mid f' \in F_{\alpha B_i}, s' \text{ a reachable state of } f' \text{ and} \\ MF, \alpha B_i, f', s' \models \{\phi \mid B_i\phi \in ex \end{array} \right\}$$

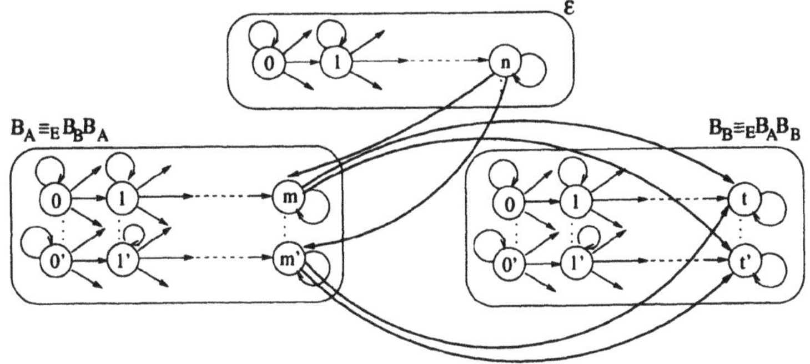

Figure 5 The MAFSM for the CCITT protocol.

Starting from a view α and a subset of explicit belief atoms ex of α, $C_{B_i}(\alpha, ex)$ collects all the FSMs f' and reachable states s' of f' (in the view α_{B_i}) which satisfy the arguments of the explicit belief atoms ex (formally: $\{\phi \mid B_i\phi \in ex\}$). Intuitively, $C_{B_i}(\alpha, ex)$ contains all the pairs $\langle f', s' \rangle$ of view α_{B_i} where the process associated to view α_{B_i} can be, whenever the process modeling view α is in a state that satisfies all the explicit belief atoms in ex. It can be easily seen that, for f' a FSM of view α_{B_i}, s' any reachable state of f', and s a state of a FSM of view α:

$$\langle f', s' \rangle \in C_{B_i}(\alpha, L(s) \cap Expl(B_i, \alpha)) \text{ iff } s' \text{ is a reachable state of } f' \text{ and } MF, \alpha_{B_i}, f', s' \models \bigwedge ArgExpl(B_i, \alpha, s)$$

where $L(s) \cap Expl(B_i, \alpha)$ is the set of explicit belief atoms true at state s in the view $\alpha \in B^n$. Hence, the following holds:

$$MF, \alpha, f, s \models B_i\phi \text{ iff for all } \langle f', s' \rangle \in C_{B_i}([\alpha]_E, L(s) \cap Expl(B_i, [\alpha]_E)), MF, \alpha_{B_i}, f', s' \models \phi \quad (3)$$

where $\alpha \in B^*$ and $B_i\phi$ is any implicit belief atom of view α. For any set Γ of formulae of $\mathcal{L}_\alpha$, $C_{B_i}([\alpha]_E, \Gamma)$ is the extension of the compatibility relation C_{B_i} to the equivalence classes induced by $\equiv_E$:

$$C_{B_i}([\alpha]_E, \Gamma) = \begin{cases} C_{B_i}(\beta, \Gamma) & \text{if } \beta \in [\alpha]_E \cap B^n \\ \emptyset & \text{otherwise} \end{cases}$$

Example 4.1 Figure 5 represents the underlying intuition on the MAFSM for the CCITT protocol. The relation $C_{B_i}(\alpha, L(s) \cap Expl(B_i, \alpha))$ is depicted as arrows connecting states of different views. Let us consider view ϵ. When the process associated to ϵ is in state n, the external observer believes that principal A, modeled by view B_A, can be in any one of the states m and m' of the FSM of that view, which are compatible with state n. The states of view B_A are completely identified by the explicit beliefs of ϵ true at state n. Therefore, ϵ

```
Algorithm MAMC-View(α, Γ)
  Sub := ⋃{sub(φ) | φ ∈ Γ}                                                    }Phase A
  for each B_i do                                                              }
      ArgImpl(B_i, α, Sub) := {φ | B_iφ ∈ Sub \ Expl(B_i, [α]_E)}              }
      if ArgImpl(B_i, α, Sub) ≠ ∅ then                                         }Phase B
          MAMC-View(αB_i, ArgImpl(B_i, α, Sub))                                }
      endif                                                                    }
  end                                                                          }
  for each f ∈ F_[α]_E do                          /* f = ⟨S, J, R, L⟩ */       }
      if ⟨Sub contains implicit belief atoms⟩ then                             }
          for each s ∈ S do                                                    }
              for each B_i do                                                  }
                  ArgImpl(B_i, α, Sub) := {φ | B_iφ ∈ Sub \                    }
                  Expl(B_i, [α]_E)}                                            }
                  for each ⟨f', s'⟩ ∈ C_B_i([α]_E, L(s) ∩ Expl(B_i, [α]_E)) do }Phase C1
                  /* f' = ⟨S', J', R', L'⟩ */                                  }
                      ArgImpl(B_i, α, Sub) := ArgImpl(B_i, α, Sub) ∩           }
                      L'(s')                                                   }
                  end                                                          }
                  L(s) := L(s) ∪ B_iArgImpl(B_i, α, Sub)                       }
              end                                                              }
          end                                                                  }
      endif                                                                    }
      CTLMC(f, Γ)                                                              }Phase C2
  end
```

Figure 6 The model checking algorithm.

believes $B_i\phi$ in state n if and only if each state of B_A, in which ϵ believes A to be (i.e., states m and m' in Figure 5), satisfies ϕ. Given a state s' of a FSM f' of view αB_i, we say that s' is *compatible* with a state s of a FSM of view α if the pair $\langle f', s'\rangle$ belongs to $C_{B_i}(\alpha, L(s) \cap Expl(B_i, \alpha))$.

The model checking algorithm MAMC(α, ϕ) takes two arguments, namely a view α and the MATL formula $\phi \in \mathcal{L}_\alpha$ that we want to model check. It calls the algorithm MAMC-View (shown in Figure 6) on view α and the set of formulae $\{\phi\}$. As a result, after this step, MAMC can return the appropriate truth value simply by testing whether ϕ is contained in the label set of the initial states J of all the FSMs $f \in F_{[\alpha]_E}$.

Notationally, let $sub(\phi)$ denote the set of subformulae of ϕ (remember that $B_i\phi$ is atomic and that, therefore, it is the only subformula of itself). The algorithm MAMC-View(α, Γ) takes two arguments: a view α, and a set of MATL formulae $\Gamma \subset \mathcal{L}_\alpha$. MAMC-View($\alpha, \Gamma$) performs the following phases: **Phase A.** This phase, corresponding to the first line of the algorithm, collects in Sub all the subformulae of the formulae in Γ.

Phase B. MAMC-View(α, Γ) considers in turn the belief operators B_i. For each of them, $ArgImpl(B_i, \alpha, \Gamma)$, the set of all the formulae ϕ which are arguments of the implicit belief atoms $B_i\phi$ occurring in Γ, is computed. MAMC-View(α, Γ) calls itself recursively on the view below (e.g., αB_i) and on the set $ArgImpl(B_i, \alpha, \Gamma)$. In this process, the algorithm recursively descends the tree structure which needs to be model checked. The leaves of this tree are the views for which there is no need to model check implicit belief atoms, as there are no more implicit belief atoms occuring in Γ.
Phase C. This phase is a loop over all the FSMs f of the equivalence class of current view α to extend the labeling functions of the visited FSMs. This loop iteratively performs the following two phases:
Phase C.1. In this phase, all the states of f of the (equivalence class of the) current view α, where the algorithm is, are labeled with the implicit belief atoms. This phase is executed only if there occur implicit belief atoms in the input formulae. The labeling of states of f is computed according to definition of satisfiability of implicit belief atoms in a MAFSM. Therefore, for each reachable state s of f, the set $L(s) \cap Expl(B_i, \alpha)$ is computed. Then, the implicit belief atoms occurring in Γ are added to the labeling function of s. Therefore, for each implicit belief atom $B_i\phi$, $B_i\phi$ is added to $L(s)$ if ϕ is satisfied by all the pairs $\langle f', s' \rangle$ belonging to the compatibility relation $C_{B_i}([\alpha]_E, L(s) \cap Expl(B_i, [\alpha]_E))$.
Phase C.2. This phase simply calls a standard CTL model algorithm on the FSM f of the current view. Indeed, at this point every state s in the current FSM f is labeled (by phase C.1) with all the atoms (i.e, propositional atoms, explicit and implicit belief atoms) occurring in the input formulae. Notice that in this phase we can employ any model checker (in our case NuSMV [5]) as a black box.

The following result states that MAMC-View actually solves the model checking problem for MATL.

Theorem 1 (Correctness of MAMC-View) *Let* $f = \langle S, J, R, L \rangle \in F_{[\alpha]_E}$, *with* α *any view, and* s *a state in* S. $MF, \alpha, f, s \models \phi$ *iff MAMC-View*$(\alpha, \{\phi\})$ *applied to* MF *constructs a new* MF *such that* $\phi \in L(s)$, *with* s *state of* f.

This result extends a similar result proved in the case of trees of views and reported in [3]. The termination of MAMC–View is guaranteed by the fact that input formulae always have beliefs of finite depth.

5. CONCLUSIONS

In this paper we have described a model-checking based verification procedure for security protocols employing a logic of belief. Our approach allows us to reuse the technology and tools developed in model checking. To model beliefs in security protocols, we have defined MATL – MultiAgent Temporal Logic – a logic where the temporal aspect and the belief aspect are kept separated. This semantics is the basis for introducing the notion of MultiAgent Finite State Machine (MAFSM) as an extension of the usual notion of Finite State Machine. Then, we have described a model checking algorithm (MAMC) which allows us to verify formulae containing belief (sub)formulae in a MAFSM.

A semantics for MATL was originally proposed in [3] for MultiAgent systems. The version proposed in this paper extends the work in [3] and makes it applicable to security protocols. [2] describes in some detail how a specific security protocol (the Andrew protocol) can be modeled and model checked using the logic defined in [3]. The analysis reported in [2] has some limitations due to the limited expressibility of the logic defined in [3].

References

[1] M. Abadi and M. Tuttle. A semantics for a logic of authentication. In *Proceedings of the 10th Annual ACM Symposium on Principles of Distributed Computing*, pages 201–216, 1991.

[2] M. Benerecetti and F. Giunchiglia. Model checking security protocols using a logic of belief. In *Proceedings of the Sixth International Conference on Tools and Algorithms for the Construction and Analysis of Systems (TACAS 2000). March 27th - April 1st, 2000, Berlin, Germany*. To appear.

[3] M. Benerecetti, F. Giunchiglia, and L. Serafini. Model Checking Multiagent Systems. *Journal of Logic and Computation, Special Issue on Computational & Logical Aspects of Multi-Agent Systems*, 8(3):401–423, 1998.

[4] M. Burrows, M. Abadi, and R. M. Needham. A logic of authentication. *ACM Transactions on Computer Systems*, 8(1):18–36, 1990.

[5] A. Cimatti, E. Clarke, F. Giunchiglia, and M. Roveri. Nusmv: a new symbolic model verifier. In *Proceedings of the International Conference on Computer-Aided Verification (CAV'99) Trento, Italy. July 1999.*, 1999.

[6] E.A. Emerson. Temporal and Modal Logic. In J. van Leeuwen, editor, *Handbook of Theoretical Computer Science*, volume B, pages 995–1072. Elsevier Science Publisher B.V., 1990.

[7] R. Fagin, J.Y. Halpern, Y. Moses, and M. Y. Vardi. *Reasoning about knowledge*. MIT Press, 1995.

[8] W. Marrero, E.M. Clarke, and S. Jha. Model checking for security protocols. In *DIMACS Workshop on Design and Formal Verification of Security Protocols*, 1997.

[9] K.L. McMillan. *Symbolic Model Checking*. Kluwer Academic Publ., 1993.

[10] J.C. Mitchell, M Mitchell, and U. Stern. Automated Analysis of Cryptographic Protocols Using Murphi. In *IEEE Symp. Security and Privacy*, pages 141–153, 1997.

GPSR Compliance
The European Union's (EU) General Product Safety Regulation (GPSR) is a set of rules that requires consumer products to be safe and our obligations to ensure this.

If you have any concerns about our products, you can contact us on

ProductSafety@springernature.com

In case Publisher is established outside the EU, the EU authorized representative is:

Springer Nature Customer Service Center GmbH
Europaplatz 3
69115 Heidelberg, Germany

www.ingramcontent.com/pod-product-compliance
Ingram Content Group UK Ltd.
Pitfield, Milton Keynes, MK11 3LW, UK
UKHW020153250726
13967UKWH00003B/1030
* 9 7 8 1 4 7 5 7 5 2 6 3 2 *